T0100475

Nanoscale Photonics and Optoelectronics

Lecture Notes in Nanoscale Science and Technology

Volume 9

Previous Volumes in This Series:

Volume 1: Self-Assembled Quantum Dots, Wang, Z.M., 2007

Volume 2: Nanoscale Phenomena: Basic Science to Device Applications, Tang, Z., and Sheng, P., 2007

Volume 3: One-Dimensional Nanostructures, Wang, Z.M., 2008

Volume 4: Nanoparticles and Nanodevices in Biological Applications: The INFN Lectures – Vol I, Bellucci, S., 2009

Volume 5: Toward Functional Nanomaterials Wang, Z.M., 2009

Volume 6: B-C-N Nanotubes and Related Nanostructures, Yap, Y.K., 2009

Volume 7: Crystallization and Growth of Colloidal Nanocrystals. Leite, E.R., 2009

Volume 8: Epitaxial Semiconductor Nanostructures, Wang, Z.M., and Salamo, G., 2009

Volume 9: Nanoscale Photonics and Optoelectronics: Science and Technology, Wang, Z.M., and Neogi, A., 2010

Zhiming M. Wang · Arup Neogi
Editors

Nanoscale Photonics and Optoelectronics

Editors
Zhiming M. Wang
Arkansas Institute for Nanoscale Materials
Science and Engineering
Fayetteville, AR 72701, USA
zmwang@uark.edu

Arup Neogi
Department of Physics
University of North Texas
Denton, TX 76203, USA
arup@unt.edu

ISBN 978-1-4419-7233-0 e-ISBN 978-1-4419-7587-4
DOI 10.1007/978-1-4419-7587-4
Springer New York Dordrecht Heidelberg London

Library of Congress Control Number: 2010938644

Printed on acid-free paper

Springer is part of Springer Science+Business Media (www.springer.com)

Preface

Photonics, the science of light generation, manipulation, and detection, has long been the basis for a wide range of key technological applications, including light-emitting diodes (LEDs), lasers, optical fibers, amplifiers, and photodetectors. Exponentially growing interests in photonics has been driven in recent years mostly by novel properties and processes of devices and structures with at least one geometrical dimension between 1 and 100 nm. This book, stimulated by a joint United States and Japan workshop on Nanophotonics in 2008, aims to capture a current snapshot of the burgeoning frontier of nanoscale photonics.

Surface plasmon (SP) is one of the nanophotonic fields experiencing tremendous renewed interest. In "Spontaneous Emission Control in a Plasmonic Structure," Iwase et al. discuss the quantum electrodynamics of SP polaritons coupling of excitons near metal-layer surfaces, and an exciton embedded in a metal microcavity [??]. Their research should fuel new applications in photoelectric devices and bio-sensing technology. In "Surface Plasmon Enhanced Solid-State Light-Emitting Devices," Okamoto demonstrates that the SP enhancement of photoluminescence intensities of light emitters is a very promising method for developing high-efficiency LEDs.

In "Polariton Devices Based on Wide-Bandgap Semiconductor Microcavities," Shimada et al. review the recent progress on cavity polaritons and their applications based on GaN or ZnO wide-bandgap materials. Their research now focuses on the realization of room temperature polariton devices based on these wide-bandgap semiconductor microcavities.

In "Search for Negative Refraction in the Visible Region of Light [??] by Fluorescent Microscopy of Quantum Dots Infiltrated into Regular and Inverse Synthetic Opals," Moussa et al. reveal the possibility of using infiltrated quantum dots as internal light sources inside the porous photonic crystal for the study of negative-index material effects.

The last four chapters emphasize the diversity and control of photonic properties of different nanostructures from the materials perspective.

In "Self-Assembled Guanosine-Based Nanoscale Molecular Photonic Devices," Li et al. describe novel hybrid photonic crystals employing a unique material system consisting of a DNA base encapsulated within highly polar GaN nanoscale confined structures with the aim of developing integrated nanophotonic and biomolecular devices.

The chapter "Carbon Nanotubes for Optical Power Limiting Applications" covers the origin and mechanism of optical power limiting (OPL). Mirza et al. then explore the potential of carbon nanotubes and their composites in OPL applications.

In "Field Emission Properties of ZnO, ZnS, and GaN Nanostructures," Mo et al. review the growth and field emission properties of ZnO, ZnS, and GaN nanostructures, with an emphasis on ZnO nanorods.

In "Growth, Optical, and Transport Properties of Self-Assembled InAs/InP Nanostructures," Bierwagen et al. provide an in-depth review of the formation and properties of InAs/InP epitaxial nanostructures, a material system very important for fiber-optic communication operated at 1.55 μm.

Finally, we would like to thank the chapter authors for their efforts and patience which was essential in making this book possible.

Fayetteville, Arkansas — Zhiming M. Wang
Denton, Texas — Arup Neogi

Contents

Contributors

Oliver Bierwagen Department of Physics, Humboldt-Universitätzu Berlin, Newtonstrasse 15, 12489 Berlin, Germany, bierwage@physik.hu-berlin.de

Q.Y. Chen Departments of Physics and Materials and Optoelectronic Sciences, and Center for Nanoscience and Nanotechnology, National Sun Yat-Sen University, Kaohsiung 80424, Taiwan; Department of Physics and Texas Center for Superconductivity, University of Houston, Houston, TX 77004, USA, qchen@uh.edu

P.A. Ecton Department of Physics, University of North Texas, Denton, TX 76203, USA, pae0031@unt.edu

Dirk Englund Ginzton Laboratory, Stanford University, Stanford, CA 94305, USA, englund@columbia.edu

Y. Fujita Interdisciplinary Faculty of Science and Engineering, Shimane University, Matsue, Shimane Prefecture 690-8504, Japan, fujita@ecs.shimane-u.ac.jp

Yiyang Gong Ginzton Laboratory, Stanford University, Stanford, CA 94305, USA, yiyangg@stanford.edu

N.J. Ho Departments of Physics and Materials and Optoelectronic Sciences, and Center for Nanoscience and Nanotechnology, National Sun Yat-Sen University, Kaohsiung 80424, Taiwan, njho@mail.nsysu.edu.tw

Hideo Iwase Production Engineering Research Laboratory, Canon Inc., Saiwai-ku, Kawasaki-shi, Kanagawa 212-8602, Japan, iwase.hideo@canon.co.jp

A. Kuznetsov UTD-Nanotech Institute, University of Texas at Dallas, Richardson, TX 75083, USA, aak036000@utdallas.edu

Jianyou Li Department of Physics, University of North Texas, Denton, TX, USA, lijianyou@hotmail.com

M.H. Lynch Department of Physics, Angelo State University, San Angelo, TX 76909, USA, m.lynch@yahoo.com

Yuriy I. Mazur Department of Physics, University of Arkansas, 226 Physics Building, Fayetteville, AR 72701, USA, ymazur@uark.edu

Shamim Mirza Luminit, LLC, 1850 W. 250th Street, Torrance, CA 90501, USA, smirza@luminitco.com

Y. Mo Department of Physics, University of North Texas, Denton, TX 76203, USA, mott168@hotmail.com

Hadis Morkoç Department of Electrical and Computer Engineering, Virginia Commonwealth University, Richmond, VA 23284, USA, hmorkoc@vcu.edu

R. Moussa UTD-Nanotech Institute, University of Texas at Dallas, Richardson, TX 75083, USA, rabia.moussa@utdallas.edu

E. Neiser UTD-Nanotech Institute, University of Texas at Dallas, Richardson, TX 75083, USA, erica.neiser@student.utdallas.edu

Arup Neogi Department of Physics, University of North Texas, Denton, TX, USA, arup@unt.edu

Koichi Okamoto Department of Electronic Science and Engineering, Kyoto University, Katsura, Nishikyo-ku, Kyoto 615-8510, Japan; PRESTO, Japan Science and Technology Agency, 4-1-8 Honcho Kawaguchi, Saitama 332-0012, Japan, k.okamoto@hy4.ecs.kyoto-u.ac.jp

Ümit Özgür Department of Electrical and Computer Engineering, Virginia Commonwealth University, Richmond, VA 23284, USA, uozgur@vcu.edu

J.M. Perez Department of Physics, University of North Texas, Denton, TX 76203, USA, jperez@unt.edu

Salma Rahman Michigan Molecular Institute, Midland, MI 48640, USA, rahman@mmi.org

George Rayfield Department of Physics, University of Oregon, Eugene, OR 97403, USA, rayfield@uoregon.edu

Gregory J. Salamo Department of Physics, University of Arkansas, 226 Physics Building, Fayetteville, AR 72701, USA, salamo@uark.edu

Abhijit Sarkar Michigan Molecular Institute, Midland, MI 48640, USA, sarkar@mmi.org

J.J. Schwartz Department of Physics, University of Texas at Dallas, Richardson, TX 75080, USA, schwartz@physics.ucla.edu

H.W. Seo Department of Physics, University of Arkansas, Little Rock, AR 72204, USA, fireflyshw@yahoo.com

Ryoko Shimada Department of Mathematical and Physical Sciences, Japan Women's University, 2-8-1 Mejirodai, Bunkyo-ku, Tokyo, 112-8681, Japan, shimadar@fc.jwu.ac.jp

Georgiy G. Tarasov Institute of Semiconductor Physics, National Academy of Sciences, pr. Nauki 45, Kiev-03028, Ukraine, tarasov@isp.kiev.ua

W. Ted Masselink Department of Physics, Humboldt-Universitätzu Berlin, Newtonstrasse 15, 12489 Berlin, Germany, massel@physik.hu-berlin.de

L.W. Tu Departments of Physics and Materials and Optoelectronic Sciences, and Center for Nanoscience and Nanotechnology, National Sun Yat-Sen University, Kaohsiung 80424, Taiwan, lwtu@faculty.nsysu.edu.tw

Jelena Vučković Ginzton Laboratory, Stanford University, Stanford, CA 94305, USA, jela@stanford.edu

A.A. Zakhidov UTD-Nanotech Institute, University of Texas at Dallas, Richardson, TX 75083, USA, zakhidov@utdallas.edu

Chapter 1
Spontaneous Emission Control in a Plasmonic Structure

Hideo Iwase, Yiyang Gong, Dirk Englund, and Jelena Vučković

Abstract Surface plasmon polaritons (SPPs) are electromagnetic waves at optical frequencies that propagate at the surface of a conductor [1]. SPPs can trap optical photons far below their diffraction limit. The field confinement of SPP provides the environment for controlling the interaction between light and matter. In this chapter, we discuss the quantum electrodynamics (QED) of SPP coupling of excitons near a metal-layer surface, and an exciton embedded in a metal microcavity. We analyze the enhanced spontaneous emission (SE) rate of the exciton coupled to a large number of SPP modes near a uniform or periodically patterned metal layer traveling with extremely slow group velocities. Combining the effects of quality factor (Q) and ohmic losses for each SPP mode, we explain how various loss mechanisms affect the SE rate of excitons in such structures. Similarly, we consider the Q-factor and mode volume of a cavity mode formed by a defect in a grating structure and again investigate the enhancement of SE for excitons lying in a metal cavity. Because defect cavity designs confine modes in all three dimensions, we observe that such a structure of extremely small mode volume could reach various regimes of cavity quantum electro-dynamics (cavity QED). Controlling the SE properties of emitters through the exciton–SPP coupling is great promise for new types of opto-electronic devices overcoming the diffraction limit.

Keywords Purcell effect · Plasmonics · QED

1 Introduction

There has been tremendous renewed interest in SPPs in recent years, fueled by greatly improved nanofabrication, characterization, and computer-assisted design of devices. Today, SPPs have found a range of promising applications in

H. Iwase (✉)
Production Engineering Research Laboratory, Canon Inc., 70-1, Yanagi-cho, Saiwai-ku, Kawasaki-shi, Kanagawa 212-8602, Japan
e-mail: iwase.hideo@canon.co.jp

Z.M. Wang, A. Neogi (eds.), *Nanoscale Photonics and Optoelectronics*, Lecture Notes in Nanoscale Science and Technology 9,
DOI 10.1007/978-1-4419-7587-4_1,

subdiffraction optics, including magneto-optic data storage, microscopy, solar cells [2, 3], sensors for biological and chemical substances [4, 5], light sources such as light-emitting diodes [6] and quantum cascade lasers [7], and nonlinear processes including harmonic generation [8].

SPPs can provide tighter confinement than corresponding modes in dielectric, as the effective index of the plasmonic mode is given by

$$n_{\mathrm{eff}} = \frac{k}{k_0} = \sqrt{\frac{\varepsilon_{\mathrm{d}}\varepsilon_{\mathrm{m}}}{\varepsilon_{\mathrm{d}} + \varepsilon_{\mathrm{m}}}} > \sqrt{\varepsilon_{\mathrm{d}}}. \tag{1.1}$$

Because of such a strong field confinement in a metallic structure (plasmonic structure), the spontaneous emission (SE) rate of emitters near the structure is enhanced through the Purcell effect [9]. First, we will develop a theoretical description for this SE rate enhancement of an emitter near a plasmonic structure. The description applies to periodically patterned metallic surfaces, which will be central in the remainder of this chapter. The formulation begins with the Purcell modifications of coupling rates to traveling SPP modes, derived from quantum electrodynamics (QED). The full SE rate modification (Purcell enhancement) is then expressed as a summation over all traveling modes, each of which has a quality factor (Q) to describe its radiative and resistive losses. This summation of coupling to a set of modes gives easy physical insight into the resulting Purcell enhancement between the emitter and the plasmonic structure. The enhancement we discuss in this chapter numerically agrees with the description for Purcell enhancement based on electric currents [10–12].

In Section 2, we experimentally study the Purcell enhancement with InP-based quantum well (QW). We consider a novel hexagonal plasmonic crystal coupled to QWs emitting near 1.3 μm via an antisymmetric mode with low resistive losses in metals [13]. By direct time domain measurements of emitter decay, we deduce Purcell SE rate enhancements (Purcell enhancement factor) up to 3.7 for QW-confined excitons coupled to SPP states of the plasmonic crystal [14]. This rate enhancement is one of the highest reported in the infrared so far, similar to the measured rate enhancement of single CdSe quantum dots coupled to metal nanowires [15]. The plasmonic crystal serves the dual purpose of enhancing spontaneous emission and also directing this emission off the chip. Such plasmonic crystals are interesting for improving efficiency and brightness of LEDs and for designing entirely new devices such as a plasmon laser [16, 17].

If the period of the periodically patterned media approaches a half-wavelength of the plasmon, in-plane standing waves are formed, and a stop-band for the propagation of plasmonic waves opens [18], akin to distributed Bragg reflection (DBR) in dielectric structures. This in-plane DBR effect can be used to create tight lateral confinement for surface plasmon waves, while in the vertical direction, a fraction of the SPP's energy can be radiated (scattered) off the chip. In an unpatterned metal layer, the SPP would be confined to the metal surface (i.e., would not radiate off-chip) and would eventually loose all its energy to material absorption. For reasonable off-chip collection efficiency of SPP modes, the vertical scattering rate through the DBR should exceed the material loss rate in the metal structure. In Section 3, we describe a proposal for a metal grating structure that uses the DBR confinement to confine light to a mode volume of only $V_{\text{mode}} = (50\text{ nm})^3 = 0.001 \times (\lambda/2)^3$ with a quality factor $Q\sim$ 1,000. The extremely small mode volume opens the possibility of reaching the strong coupling regime between emitters and cavities, even with modest Q values [19, 20]. Strong coupling occurs when the coupling rate g between an emitter and the plasmon mode is greater than the decoherence rates of the system, i.e., the dipole decay rate γ and the cavity field decay rate $\kappa = \omega/2Q$. In the case of many solid-state emitters such as InAs quantum dots, γ is much smaller than κ. In this case, strong coupling requires $g \geq \kappa/2$ [21, 22]. In dielectric cavities with host refractive index n, whose mode volume must be beyond $(\lambda/2n)^3$, large Q values greater than 10^4 are required to reach strong coupling with InAs QDs. The much smaller mode volume in plasmonic structures enables the strong coupling regime even with moderate Q values on the order of 10^2.

2 Purcell Enhancement Effect in a Uniform and Periodically Patterned Metal Surface

In this section, we study on exciton coupled to SPP traveling on a uniform and periodically patterned metal–dielectric interface.

2.1 *Quantization of Plasmonic Field*

The electromagnetic field of a mode with a particular **k**-vector can be described as follows

$$\mathbf{E}(\mathbf{r},t) = \frac{1}{2}\left[D\mathbf{f_k}(\mathbf{r})\eta(t) + cc\right], \tag{1.2}$$

$$\mathbf{H}(\mathbf{r},t) = \frac{1}{2}\frac{1}{\mu_0}\left[\frac{D}{\omega_\mathbf{k}}\nabla \times \mathbf{f_k}(\mathbf{r})\chi(t) + cc\right], \tag{1.3}$$

with $\eta(t) = q(t) + ip(t)$, and $\chi(t) = p(t) - iq(t)$. Here, $q(t) \equiv \cos(\omega_k t)$ and $p(t) \equiv -\sin(\omega_k t)$ describe time-varying (oscillating) parts, μ_0 is the magnetic permeability of free space, and D is a constant. $\mathbf{f_k}(\mathbf{r})$ is a normalized time-independent part of electric field: $\mathbf{f_k}(\mathbf{r}) = \mathbf{u_k}(z)e^{i\mathbf{k}\cdot\mathbf{x}}$ with $\iiint \partial(\varepsilon\omega)/\partial\omega|\mathbf{u_k}|^2\, d\mathbf{r} = 1$. The total energy can then be expressed as

$$W = \frac{1}{2}\iiint \left[\partial(\varepsilon\omega)/\partial\omega \mathbf{E}(\mathbf{r},t)^2 + \mu_0 \mathbf{H}(\mathbf{r},t)^2\right] d\mathbf{r} = \frac{1}{2} D^2 \frac{1+\Theta_\mathrm{k}}{2}\left(p^2+q^2\right) \tag{1.4}$$

where $1/(1+\Theta_\mathbf{k})$ describes the ratio of the electric field energy to the total field energy ($\Theta_\mathbf{k} \equiv \int\int\int (1/\mu_0)|\nabla \times (\mathbf{f}_\mathbf{k}/i\omega_\mathbf{k})|^2 d\mathbf{r}$). The choice of $D = -i\sqrt{2\omega_\mathbf{k}/(1+\Theta_\mathbf{k})}$ satisfies Hamilton's equations: $-\partial W/\partial q = -\omega_\mathbf{k} q = \dot{p}$, and $-\partial W/\partial p = \omega_\mathbf{k} p = \dot{q}$. Then the mode can be represented as a harmonic oscillator, the Hamiltonian quantized , and the electric field operator expressed as

$$\mathbf{E}(\mathbf{r},t) = i\sqrt{\frac{\hbar\omega_\mathbf{k}}{1+\Theta_\mathbf{k}}}\mathbf{f}_\mathbf{k}(\mathbf{r})a + H.C. \tag{1.5}$$

where H.C. is the hermitian conjugate. In a dielectric structure $1/(1+\Theta_\mathbf{k})$ equals 1/2. However, in a plasmonic structure (a structure with metal), the introduction of the $1/(1+\Theta_\mathbf{k})$ term is critical, as the energy is not equally distributed between electric and magnetic fields [23]. At the metal-dielectric boundaries, the oscillation of the magnetic field induces surface charges because of the discontinuity of the conductivity on metal surface. The distribution of these charges produces an electric field $\mathbf{E}_\mathbf{k}^{//}$ parallel to the metal surface by Coulomb's low. At a frequency close to the SPPs' resonant frequency υ_{sp}, high density of the surface charges induces a large $\mathbf{E}_\mathbf{k}^{//}$ around the metal-dielectric boundary. This field-charge interaction increases $1/(1+\Theta_\mathbf{k})$ from 1/2, storing more energy in the electric field than in the magnetic field.

2.2 *Quantum Electrodynamics of the Exciton Lying in the Vicinity of a Uniform Metal Surface*

The SE rate Γ for the exciton lying in the vicinity of a metal surface is enhanced from the SE rate $n\Gamma_0$ in a dielectric bulk with host index n due to the exciton–SPP coupling (Purcell enhancement effect) [24–26]. Here, Γ_0 is the SE rate for the exciton lying in vacuum. To analyze the SE rate enhancement for the exciton with a radiative frequency υ coupled to traveling SPPs, the Purcell enhancement factor $F(\upsilon)$, defined as the Γ normalized by $n\Gamma_0$, has been studied (in this definition we ignore the nonradiative decay of the exciton). As opposed to Purcell factor for the exciton coupled to a cavity mode, $F(\upsilon)$ describes the electrodynamics of the exciton coupled to a number of traveling SPP modes. In a uniform metal surface, the SPP frequency ω is almost independent of its propagation constant k at the resonant frequency $\upsilon_\mathrm{sp} \equiv \omega_\mathrm{p}/\sqrt{1+n^2}$, where ω_p is a plasma frequency of the metal [27]. This k-independence of ω increases the density of SPP states $D(\omega)$ and supports a large number of SPPs coupled to excitons at υ_sp [24]. While this increase of $D(\omega)$ accounts for part of the Γ enhancement, the field confinement of the SPP at the metal–dielectric interface also contributes to the Purcell effect.

The field confinement of SPP increases the photon energy density (PED), enhancing the coupling between excitons near the metal surface and SPP modes relative to the coupling between the same excitons and non-SPP modes. The dissipation of the photon energy from each SPP mode has an additional effect on Γ. It gives the SE into each SPP mode a spectral width, analogous to Purcell effect in a cavity. In an unpatterned metal–dielectric layer structure, the dispersion diagram of SPP lies below the light line, so that such a dissipation of the photon energy dominantly comes from ohmic loss in metal. For investigation of $F(\upsilon)$ in this chapter, we consider the mechanisms above according to quasi-quantum electrodynamics (QED) analysis.

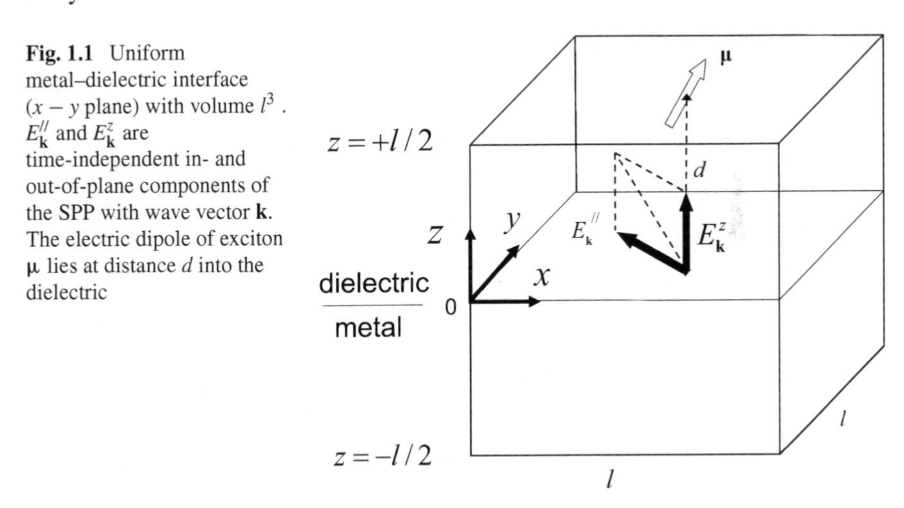

Fig. 1.1 Uniform metal–dielectric interface ($x - y$ plane) with volume l^3. $E_{\mathbf{k}}^{//}$ and $E_{\mathbf{k}}^{z}$ are time-independent in- and out-of-plane components of the SPP with wave vector **k**. The electric dipole of exciton μ lies at distance d into the dielectric

Here, we derive the theoretical formulation of $F(\upsilon)$ for the exciton lying in the vicinity of a uniform metal–dielectric interface shown in Fig. 1.1 [23]. Including the effect of the ohmic loss in the metal, we assume a dissipative SPP mode with the following time evolution of its E-field:

$$E_{\mathbf{k}}(\mathbf{r},t) = E_{\mathbf{k}}(\mathbf{r}) \times \exp\{-\mathrm{i}(\omega_{\mathbf{k}} - \mathrm{i}\omega_{\mathbf{k}}/Q_{\mathbf{k}})t\}, \tag{1.6}$$

where $\omega_{\mathbf{k}}$ and $Q_{\mathbf{k}}$ are the frequency and the quality factor of **k**-mode, respectively [6, 28]. For simplicity, we only consider the SPP-related part $F^{\mathrm{sp}}(\upsilon)$ defined by $F^{\mathrm{sp}}(\upsilon) \equiv \Gamma^{\mathrm{sp}}/n\Gamma_0$, where Γ^{sp} is the SE rate into SPP modes. $F(\upsilon)$ is obtained by summing $F^{\mathrm{sp}}(\upsilon)$ and the non-SPP-related part $F^{\text{non-sp}}(\upsilon)$: $F(\upsilon) = F^{\mathrm{sp}}(\upsilon) + F^{\text{non-sp}}(\upsilon)$ [25]. By solving the Jaynes–Cummings Hamiltonian with the multimode field expressed by Eq. (1.6), $F^{\mathrm{sp}}(\upsilon)$ then reads

$$F^{\mathrm{sp}}(\upsilon) = \sum\nolimits_{\mathbf{k}} \frac{2\pi}{n\Gamma_0(\upsilon)} \left|g_{\mathbf{k}}(d)\right|^2 D_{\mathbf{k}}(\upsilon), \tag{1.7}$$

$$D_{\mathbf{k}}(\upsilon) = \frac{1}{\pi} \frac{\omega_{\mathbf{k}}/2Q_{\mathbf{k}}}{(\omega_{\mathbf{k}} - \upsilon)^2 + (\omega_{\mathbf{k}}/2Q_{\mathbf{k}})^2}, \tag{1.8}$$

where $g_{\mathbf{k}}(d)$ is a coupling strength between **k**-mode and the exciton located at $z = d$ defined as $g_{\mathbf{k}} = g_0 \psi(z) \cos(\xi)$, where $g_0 = \frac{\mu}{\hbar}\sqrt{\frac{\hbar\upsilon}{(1+\Theta_{\mathbf{k}})\varepsilon V_{\mathbf{k}}}} \frac{\mu}{\hbar}\sqrt{\frac{\hbar\upsilon}{2\varepsilon V_{\mathbf{k}}}}$, $\psi(z) = \frac{|\mathbf{E}_{\mathbf{k}}(z)|}{\max|\mathbf{E}_{\mathbf{k}}(z)|}$, and $\cos(\xi) = \mathbf{e}_{\mathbf{k}} \cdot \mathbf{e}_{\mu}$ [29]. Here, $V_{\mathbf{k}}$ is a mode volume of each SPP mode, and $\mathbf{e}_{\mathbf{k}}$ and $\mathbf{e}_{\mu}$ are unit vectors for the SPP electric field and the electric dipole $\boldsymbol{\mu}$, respectively [23]. $D_{\mathbf{k}}(\upsilon)$ shown in Eq. (1.8) gives the Lorentzian spectrum for the SE into **k**-mode. The density of the SPP states does not show up in Eq. (1.7). However, it is noted that the $\sum_{\mathbf{k}} D_{\mathbf{k}}(\upsilon)$ becomes the density of the SPP states in case $D_{\mathbf{k}}(\upsilon) \to \delta(\omega_{\mathbf{k}} - \upsilon)$ with $Q_{\mathbf{k}} \to \infty$. To investigate the enhanced SE rate into each **k**-mode, we introduce the distributed Purcell factor defined by the following equation:

$$F^{\text{sp-dis}}(\upsilon, \mathbf{k}) \equiv \frac{1}{\Delta k^x \Delta k^y} \frac{2\pi}{n\Gamma_0(\upsilon)} |g_{\mathbf{k}}(d)|^2 D_{\mathbf{k}}(\upsilon), \tag{1.9}$$

where $\Delta k^x \Delta k^y \equiv (2\pi)^2 / l^2$ is the reciprocal space area taken by an SPP mode. If the real-space area l^2 is large enough that **k** can be considered continuous, Eq. (1.9) is rewritten as $F^{\text{sp-dis}}(\upsilon, \mathbf{k}) = \partial^2 F^{\text{sp}}(\upsilon) / \partial k^x \partial k^y$. In that case, the SE rate into a small reciprocal area $\Delta_{\mathbf{k}}$, normalized by $n\Gamma_0$, is expressed as $F^{\text{sp-dis}}(\upsilon, \mathbf{k})\Delta_{\mathbf{k}}$. The mode volume for each traveling SPP mode is expressed as $V_{\mathbf{k}} = l^2 L_{\mathbf{k}}$, where $L_{\mathbf{k}}$ is defined as the 1D integral of the PED [25]:

$$L_{\mathbf{k}} = \frac{\int \{\partial(\omega\varepsilon)\partial\omega\}_{\omega=\upsilon} |\mathbf{E}_{\mathbf{k}}(z)|^2 \, \mathrm{d}z}{\max\left[\{\partial(\omega\varepsilon)/\partial\omega\}_{\omega=\upsilon} |\mathbf{E}_{\mathbf{k}}(z)|^2\right]}. \tag{1.10}$$

By combining Eqs. (1.9) and (1.10), $F^{\text{sp-dis}}(\upsilon, \mathbf{k})$ is rewritten as follows:

$$F^{\text{sp-dis}}(\upsilon, \mathbf{k}) = \frac{3}{2} \frac{1}{n^3} \frac{c^3}{\upsilon^2} \frac{H_{\mathbf{k}}(d)}{L_{\mathbf{k}}} \frac{1}{1 + \Theta_{\mathbf{k}}} (\mathbf{e}_{\mathbf{k}} \cdot \mathbf{e}_{\mu})^2 D_{\mathbf{k}}(\upsilon), \tag{1.11}$$

where

$$H_{\mathbf{k}}(z) = \frac{\varepsilon_{\omega=\upsilon} \left|\mathbf{E}_{\mathbf{k}}(z)\right|^2}{\max\left[\{\partial(\omega\varepsilon)/\partial\omega\}_{\omega=\upsilon} \left|\mathbf{E}_{\mathbf{k}}(z)\right|^2\right]}. \tag{1.12}$$

$H_{\mathbf{k}}(d) \cdot (\mathbf{e}_{\mu} \cdot \mathbf{e}_{\mathbf{k}})^2$ determines the position dependence of the coupling of the dipole to a **k**-mode, and $L_{\mathbf{k}} \Big/ \left\{H_{\mathbf{k}}(d) \cdot (\mathbf{e}_{\mu} \cdot \mathbf{e}_{\mathbf{k}})^2\right\}$ corresponds to an effective mode volume for 1D-field confinement (effective mode length) [4, 30].

At a uniform metal–dielectric interface, $\omega_{\mathbf{k}}$, $Q_{\mathbf{k}}$, and $L_{\mathbf{k}}$ are independent of the direction of **k**. By rewriting Eq. (1.7) as the integral in the polar coordinates $\mathbf{k} = (k, \phi)$, $F^{\text{sp}}(\upsilon)$ can be expressed as a function of the propagation constant $k = |\mathbf{k}|$ below the light line:

$$F^{\text{sp}}(\upsilon) = \int_{k_0(\upsilon)}^{\infty} F^{\text{sp-dis}}_{\text{uniform}}(\upsilon, k) dk, \tag{1.13}$$

where $k_0(\upsilon)$ is k on the light line. $F^{\text{sp-dis}}_{\text{uniform}}(\upsilon, k)$ is the integral of $F^{\text{sp-dis}}(\upsilon, \mathbf{k})$ over ϕ:

$$F^{\text{sp-dis}}_{\text{uniform}}(\upsilon, k) \equiv \int_0^{2\pi} kF^{\text{sp-dis}}(\upsilon, \mathbf{k})d\phi = \frac{3\pi}{n^3}\frac{c^3}{\upsilon^2}\frac{1}{1+\Theta_{\mathbf{k}}} \cdot \frac{h_k(d)}{L_k} \cdot kD_k(\upsilon)^{\cdot} \tag{1.14}$$

Here, $h_k(z)$ is defined as an integral of $1/2\pi \cdot H_{\mathbf{k}}(d) \cdot (\mathbf{e}_\mu \cdot \mathbf{e}_{\mathbf{k}})^2$ over ϕ:

$$h_k(z) = \alpha \frac{\varepsilon_{\omega=\upsilon}\left|\mathbf{E}^i_k(z)\right|^2}{\max\left[\left\{\partial(\omega\varepsilon)/\partial\omega\right\}_{\omega=\upsilon}\left|\mathbf{E}_k(z)\right|^2\right]}, \tag{1.15}$$

where $\alpha = 1/2$ and $\mathbf{E}^i_k(z) = \mathbf{E}^{//}_k(z)$ when the dipole moment of the emitter μ is in the plane of the metal surface, and $\alpha = 1$ and $\mathbf{E}^i_k(z) = \mathbf{E}^z_k(z)$ when μ is out of the plane.

In Fig. 1.2, we show $F^{\text{sp}}(\upsilon)$, Q_k, and L_k/h_k for the exciton lying 10 nm from the metal surface with various plasma lifetimes τ and in Fig. 1.3 for the exciton lying at different distances d from the metal surface. The dipole moment of the emitter, μ is set normal to the metal surface, to maximize coupling to SPP modes. Dispersion diagrams of SPP were obtained by solving Maxwell's equations under the condition of $\varepsilon''_{\text{m}} << |\varepsilon'_{\text{m}}|$, where ε'_{m} and ε''_{m} are real and imaginary parts of the complex permittivity of metal [27]. SPP field patterns and an imaginary part of the propagation constant k'' are analytically obtained under the same condition. The value of Q_k is estimated with k'' and a group velocity v^{g}_k : $Q_k = \omega_k/(2k''v^{\text{g}}_k)$ [23]. As seen in Figs. 1.2b and 1.3b, the value, of Q_k and L_k/h_k are dependent on τ and d,. respectively. The different values of τ and d mostly affect $F^{\text{sp}}(\upsilon)$ at υ_{sp} through the values of Q_k and L_k/h_k with a large number of k's.

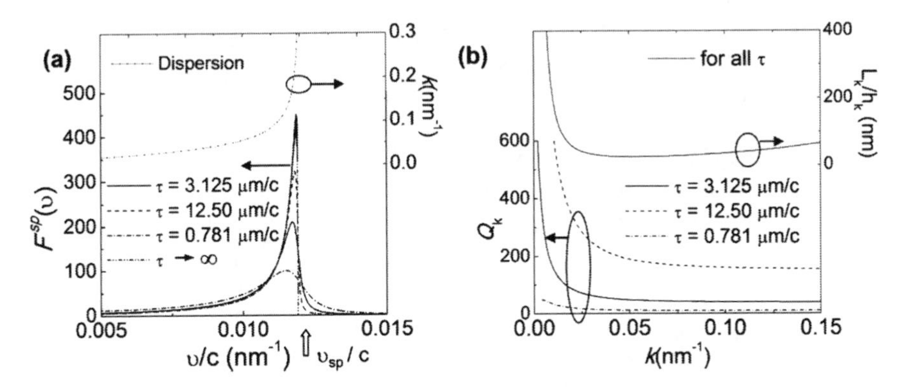

Fig. 1.2 (**a**) $F^{\text{sp}}(\upsilon)$ and the dispersion of SPP and (**b**) k-dependence of Q_k and L_k/h_k for the exciton lying 10 nm apart from a uniform metal surface with various plasma lifetimes $\tau = 3.125$ (for Au at room temperature [RT]), 12.50, and 0.781 μm/c. The electric dipole μ is oriented normal to the metal surface. L_k/h_k takes the same values independent of the values of τ. The exciton's host index n and the plasma frequency ω_{p} are set as $n = 3.2$, and $2\pi c/\omega_{\text{p}} = 156$ nm, respectively

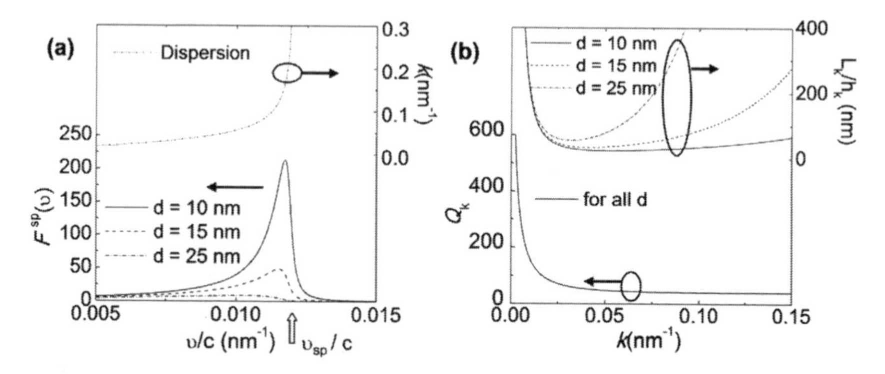

Fig. 1.3 (**a**) $F^{\mathrm{sp}}(\upsilon)$ and the dispersion of SPP, and (**b**) k-dependence of Q_k and L_k/h_k for the exciton lying at various distances $d = 10$, 15, and 25 nm from a uniform metal surface. The electric dipole μ is oriented normal to the metal surface. Q_k takes the same value independent of the values of d. The exciton's host index n, the plasma frequency ω_{p}, and plasma lifetime τ are set as $n = 3.2$, $2\pi c/\omega_{\mathrm{p}} = 156\,\mathrm{nm}$, and $\tau = 3.125\,\mu\mathrm{m/c}$, respectively

2.3 Purcell Effect for the Exciton in a Periodically Patterned Structure and Its Relation to Purcell Effect in a Cavity

Periodical patterning in a metal or dielectric waveguide redistributes the electrical field and periodically localizes the PED. The excitons lying at the position of the enhanced PED in a patterned structure gain larger Purcell effect than in an unpatterned structure [33]. This Purcell enhancement effect is similar to Purcell effect in a cavity: in a cavity, the field confinement with a small volume grants an exciton coupled to a cavity mode a large Purcell factor $F^{\mathrm{cav}}(\upsilon)$ defined as the SE rate into a cavity mode, normalized by $n\Gamma_0$ [34]. We can relate such Purcell enhancement effects in a periodically patterned structure to $F^{\mathrm{cav}}(\upsilon)$ by considering the localized PED in a unit cell of a periodical structure [23]. To study it, we investigate the Purcell enhancement factor $F_{\mathrm{peri}}(\upsilon)$ for a periodically patterned metal surface (plasmonic structure). We take the same approach as in the previous section, noting that the periodic structure folds the dispersion diagram into the first Brillouin zone, and the higher order bands above the light line correspond to scattering the SPPs to non-SPP far fields. The electric fields of SPP modes $\mathbf{E}_{\mathbf{k}',j}(\mathbf{r})$ in such a 2D-plasmonic structure are represented as Bloch states [35]:

$$\begin{aligned}\mathbf{E}_{\mathbf{k}',j}(\mathbf{r}) &= \mathbf{u}_{\mathbf{k}',j}(\mathbf{r})\exp\left(\mathrm{i}\mathbf{k}'\cdot\mathbf{x}\right)\\ \mathbf{u}_{\mathbf{k}',j}(\mathbf{r}) &= \textstyle\sum_{\mathbf{G}}\mathbf{\Phi}_{\mathbf{k}',\mathbf{G}}(\mathbf{r})\exp\left(\mathrm{i}\mathbf{G}\cdot\mathbf{x}\right)\end{aligned}, \tag{1.16}$$

where $\mathbf{k}'$, $\mathbf{G}$, and $\mathbf{x}$ denote an in-plane wave vector in the first Brillouin zone, a reciprocal lattice vector, and an in-plane position vector, respectively. Subscript j labels a dispersion branch, which is identified by the mode profiles $\left\{\mathbf{\Phi}_{\mathbf{k}',\mathbf{G}}(\mathbf{r})\right\}$. Note that the number of unit cells in space is l^2/S_{cell}, where S_{cell} is the area of one unit cell in the $x-y$ plane, and the mode volume $V_{\mathbf{k}',j}$ for Bloch state $(\mathbf{k}',j)$ is $V_{\mathbf{k}',j}=V^{\text{cell}}_{\mathbf{k}',j}\frac{l^2}{S_{\text{cell}}}$. Here, $V^{\text{cell}}_{\mathbf{k}',j}$ is the mode volume defined by the integral of PED in a unit cell:

$$V^{\text{cell}}_{\mathbf{k}',j} \equiv \frac{\iiint_{\text{unit cell}} p(\omega,\ \mathbf{r})\mathrm{d}^3\mathbf{r}}{\max p(\omega,\ \mathbf{r})}, \tag{1.17}$$

where $p(\omega,\mathbf{r})$ is the PED function, defined as $p(\omega,\ \mathbf{r}) = \{\partial(\omega\varepsilon)/\partial\omega\}|\mathbf{E}_{\mathrm{k}'},\hat{j}(\mathbf{r})|^2$. The mode's quality factor $Q_{\mathbf{k}',j}$ is defined by the ratio of the energy dissipation rate and stored energy of the mode: $Q_{\mathbf{k}',j}=\omega_{\mathbf{k}',j}\Omega_{\mathbf{k}',j}/\left(\Delta^{\text{ab}}_{\mathbf{k}',j}+\Delta^{\perp}_{\mathbf{k}',j}\right)$, where $\Omega_{\mathbf{k}',j}$, $\Delta^{\text{ab}}_{\mathbf{k}',j}$, $\Delta^{\perp}_{\mathbf{k}',j}$, and $\omega_{\mathbf{k}',j}$ are the stored energy, energy absorption rate in metal, out-of-plane leakage rate, and frequency of the mode, respectively. Replacing $V_{\mathbf{k}}$, $Q_{\mathbf{k}}$, and $\sum_{\mathbf{k}}$ in Eqs. (1.7) and (1.8) with $V_{\mathbf{k}',j}$, $Q_{\mathbf{k}',j}$, and $\sum_j\sum_{\mathbf{k}'}$, we obtain the enhancement factor $F_{\text{peri}}(\upsilon)$ and distributed enhancement factor $F^{\text{dis}}_{\text{peri}}(\upsilon,\ \mathbf{k}',j)$ for the exciton at position $\mathbf{r}_A$ near a periodically patterned metal surface:

$$F_{\text{peri}}(\upsilon) = \Delta k^x\Delta k^y\sum_j\sum_{\mathbf{k}'}F^{\text{dis}}_{\text{peri}}(\upsilon,\ \mathbf{k}',j)+\Delta k^x\Delta k^y\sum_j\sum_{\mathbf{k}'}O_{\mathbf{k}',j}(\upsilon), \tag{1.18}$$

$$F^{\text{dis}}_{\text{peri}}(\upsilon,\ \mathbf{k}',j) = \frac{3}{2}\frac{1}{n^3}\frac{c^3}{\upsilon^2}\frac{1}{1+\Theta_{\mathbf{k}',j}}\frac{S_{\text{cell}}}{V^{\text{cell}}_{\mathbf{k}',j}}H_{\mathbf{k}',j}(\mathbf{r}_{\text{A}})\cdot\left(\mathbf{e}_\mu\cdot\mathbf{e}_{\mathbf{k}',j}\right)^2D_{\mathbf{k}',j}(\upsilon), \tag{1.19}$$

where $H_{\mathbf{k}',j}(\mathbf{r}) = \frac{\varepsilon\left|\mathbf{E}_{\mathbf{k}',j}\right|^2}{\max p(\upsilon,\mathbf{r})}$. The spectrum $D_{\mathbf{k}',j}(\upsilon)$ is defined by Eq. (1.8) with $Q_{\mathbf{k}}\to Q_{\mathbf{k}',j}$. The second term in Eq. (1.18), shown by the summation of $O_{\mathbf{k}',j}(\upsilon)$, represents the contribution from the modes which are not sufficiently bound to the metal–dielectric interface, so that their $V^{\text{cell}}_{\mathbf{k}',j}$ depend on the real-space length l. The contributions from non-SPP modes can be included in the second term. We only use the properties of 2D-Bloch functions to derive Eqs. (1.18) and (1.19). Therefore, these expressions are valid for any modes described as 2D-Bloch states, no matter if the structure is a plasmonic or photonic crystal. For photonic crystals with negligible material absorption, $Q_{\mathbf{k}',j}\to Q^{\perp}_{\mathbf{k}',j}\equiv\omega_{\mathbf{k}',j}\ \Omega_{\mathbf{k}',j}/\Delta^{\perp}_{\mathbf{k}',j}$. In this case, $Q^{\perp}_{\mathbf{k}',j}$ for the modes below the light line approaches infinite, and $D_{\mathbf{k}',j}(\upsilon)$ becomes a δ-function [36–38].

To clarify the relation between Eq. (1.18) and $F^{\text{cav}}(\upsilon)$, we analyze the Purcell effect for coupled cavity modes comprising periodically arranged metal or dielectric cavities, shown in Fig. 1.4 [23]. In Fig. 1.4, we only consider a dispersion branch m whose fields are tightly confined around the coupled cavities. The typical field shape function $\mathbf{u}_{\mathbf{k}',m}(\mathbf{r})$ and periodicity dependence of the dispersion diagram are also drawn in Fig. 1.4. As shown in Fig. 1.4b, if the coupled cavities are far enough from each other, $Q_{\mathbf{k}',m}$, $V^{\text{cell}}_{\mathbf{k}',m}$, and $\mathbf{u}_{\mathbf{k}',m}(\mathbf{r})$ become independent of $\mathbf{k}'$ (those values are represented by Q_m, V^{cell}_m, and $\mathbf{u}_m(\mathbf{r})$, respectively) and their dispersion relations become flat, because the interactions among the coupled cavities are so weak that

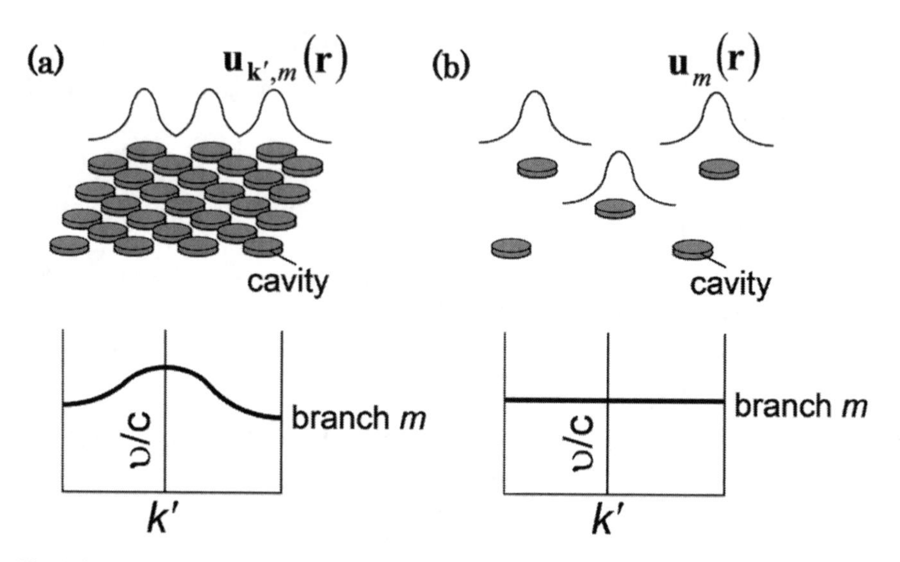

Fig. 1.4 Coupled cavity arrays (*top*) and their typical dispersion diagrams (*bottom*). The dispersion flattens as the separation increases from (**a**) to (**b**). In the bottom drawings, we only show dispersion diagrams for a branch m whose fields are tightly confined around the cavities; the dispersion is almost independent of $\mathbf{k}'$ when the cavities are far enough apart from each other, shown in (**b**). $\mathbf{u}_{\mathbf{k}',m}(\mathbf{r})$ shown in (**a**) and $\mathbf{u}_m(\mathbf{r})$ shown in (**b**) are $\mathbf{k}'$-dependent and $\mathbf{k}'$-independent field shape functions of Bloch states, depending on their periodicities

the optical properties approach those of a cavity mode [23, 39]. Assuming that the exciton is located at the field maximum of $\mathbf{u}_m(\mathbf{r})$ and $\upsilon = \omega_m$ (a resonant frequency of the cavity) with $p(\omega, \mathbf{r}) = \left\{\partial(\omega\varepsilon)/\partial\omega\right\} \left|\mathbf{u}_{\mathbf{k}',j}(\mathbf{r})\right|^2$ and $1/(1+\Theta_{\mathbf{k}',j}) = 1/2$ [31], the summation of $F_{\text{peri}}^{\text{dis}}\left(\upsilon, \ \mathbf{k}', j\right)$ for branch m in the first Brillouin zone leads to

$$\begin{aligned} F_m(\upsilon) &\equiv \Delta k^x \Delta k^y \textstyle\sum_{\mathbf{k}'} F_{\text{peri}}^{\text{dis}}\left(\upsilon, \mathbf{k}', m\right) \\ &= \frac{3}{4\pi^2}\left(\frac{2\pi c}{n\upsilon}\right)^3 \frac{Q_m}{V_m^{\text{cell}}} \end{aligned}, \tag{1.20}$$

where $F_m(\upsilon)$ is the normalized SE rate into the modes in branch m. Equation (1.20) is identical to the expression for $F^{\text{cav}}(\upsilon)$, regarding V_m^{cell} as a mode volume of a cavity mode [23, 34, 40]. This agreement indicates that $\mathbf{u}_m(\mathbf{r})$ and V_m^{cell} can be regarded as a cavity field and its mode volume, in case the interaction among coupled cavities is negligible. In other words, the field $\mathbf{u}_m(\mathbf{r})$ created by the composition of the scattered SPP waves approaches a cavity field with the scattered SPP tightly confined in each unit cell.

2.4 Experimental Study of Purcell Enhancement Effect of the Exciton Lying in a Metal–Insulator–Metal Heterostructure

We experimentally investigated the Purcell enhancement effect for the exciton in the InP–TiO–Au–TiO–Si heterostructure [14]. We replace the top layer of a symmetric InP–Au–InP structure with a slightly higher index silicon layer, which confines the antisymmetric-like SPP mode close to the Au layer in the infrared. The resulting underlying InP–TiO–Au–TiO–Si structure is shown as the inset of Fig. 1.5a, where TiO layers are added as protection against interdiffusion. The corresponding dispersion relations of SPP modes are shown in Fig. 1.5a. The solid lines show the dispersion as derived analytically from Maxwell's equations, ignoring metal absorption; we verified these by finite difference time domain (FDTD) simulations, shown in the dotted line. The antisymmetric-like SPP modes appear above the light line for the top layer (Si) and below the light line for the bottom layer (InP) because of their index mismatch ($n_{\mathrm{Si}} = 3.4$ and $n_{\mathrm{InP}} = 3.2$), but are confined by the waveguide consisting of a Si layer. We refer to these modes as the waveguide-confined (WC) antisymmetric modes. The symmetric branch is shifted toward lower frequencies in the structure with Si because of SPP overlap with a Si layer [32]. Figure 1.5b shows the distributions of PED for symmetric and WC-antisymmetric modes in the unpatterned InP–TiO–Au–TiO–Si structure. The PED for the WC-antisymmetric mode is asymmetric relative to the middle of the Au layer, and thus has a minimum inside of the Au layer, leading to low absorption losses. The mode confinement to an ultra-small region around metal–dielectric interface, well below the optical diffraction

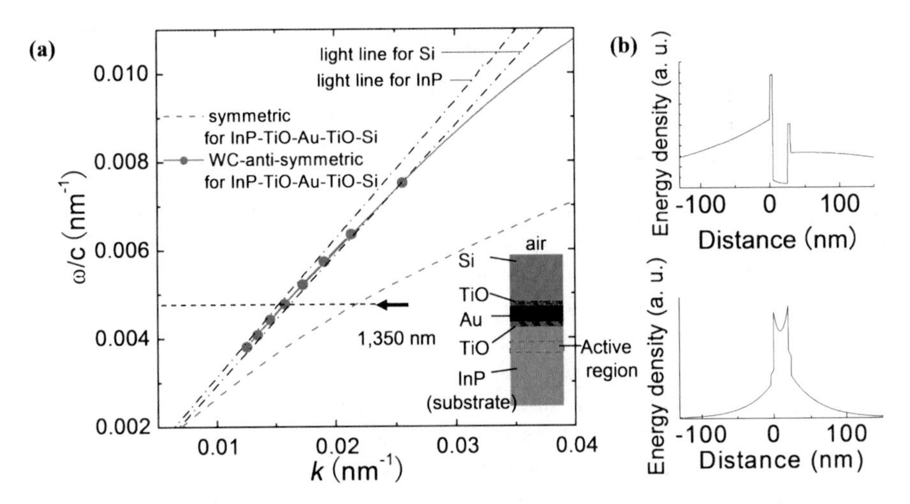

Fig. 1.5 Dispersion diagrams, field patterns, and distributions of PED in the proposed metal–dielectric heterostructure. (**a**) Dispersion diagrams for the unpatterned InP–TiO–Au–TiO–Si structure (inset), and for an InP–TiO–Au–TiO structure, obtained by analytically solving Maxwell equations (*lines*) and verified by FDTD simulation (*dots*). The thicknesses of layers in the inset are 4, 20, 4, and 400 nm, respectively, from the bottom. (**b**) Distributions of PED for symmetric (*bottom*) and WC-antisymmetric modes (*top*) for InP–TiO–Au–TiO–Si at 1,350 nm

limit, provides large Purcell enhancement for excitons in a nearby quantum well stack.

To theoretically estimate the SE enhancement, we assume the InP-based QW stack with light hole (LH) transition at 1,350 nm lies 30 nm from the Au surface. Since the excitons lying beneath the Au layer are coupled to both SPP and non-SPP modes, the total Purcell enhancement factor is represented by summing all of these contributions:$F = F_{\text{non - SPP}} + F_{\text{WC - antisym - SPP}} + F_{\text{sym - SPP}}$. From FDTD simulations, the coupling of excitons to the parallel component of the SPP field is negligible because $E_{//}^2/E_{\perp}^2 < 0.1$ at 1,350 nm and at the MQW position. To simplify the analysis, we ignore the ohmic loss ($Q_{\mathbf{k}} \to \infty$). In the infrared, this approximation gives $F^{\text{sp}}(\upsilon)$ almost the same values as with the ohmic loss at room temperature (see Fig. 1.2). From Eqs. (1.8), (1.10), (1.12), (1.13), and (1.14), we obtain Purcell enhancement rate for SPP modes:

$$F_{\text{SPP}} = \frac{2}{3} \times \frac{3\pi c^3 k E_{\perp}^2(a)}{2n\omega^2 \int_{-\infty}^{\infty} \left[\partial(\omega\varepsilon)/\partial\omega\right] E^2(z)dz} \frac{dk}{d\omega}, \tag{1.21}$$

where we approximate $1/(1+\Theta_{\mathbf{k}}) = 1/2$ because the frequency of the emitted light is far from υ_{sp} [23]. The factor 2/3 in Eq. (1.21) comes from the ratio of the coupling strengths of $\boldsymbol{\mu}$ parallel to $E_{\perp}$ and $E_{\|}$ for LH transition [6, 41]. $dk/d\omega$ in Eq. (1.21) comes from $\int kD_{\mathbf{k}}(\upsilon)dk \to \int k\delta(\omega_k - \upsilon)dk = \int k\,(dk/d\omega)\,\delta(\omega - \upsilon)d\omega = k\,(dk/d\omega)|_{\omega=\upsilon}$ with $Q_{\mathbf{k}} \to \infty$. By using the dispersion relations shown in Fig. 1.5a and the spatial distribution of PED shown in Fig. 1.5b, $F_{\text{WC - antisym - SPP}}$ and $F_{\text{sym - SPP}}$ were theoretically estimated to be 0.8 and 2.8, respectively. A smaller group velocity and strong field confinement (which can be seen in Fig. 1.5b) for symmetric modes lead to $F_{\text{sym - SPP}}$ larger than $F_{\text{WC - antisym - SPP}}$. With $F_{\text{non - SPP}} \approx 1$, an overall Purcell enhancement factor is $F \approx 4.6$.

2.4.1 Sample Preparation

In our experiment, the InP–TiO–Au–TiO–Si structure is patterned with a hexagonal array to form a plasmonic crystal. Inset of Fig. 1.6a shows a scanning electron microscope (SEM) image of the fabricated hexagonal plasmonic crystal. The introduction of plasmonic crystals folds the branches of the dispersion diagram shown in Fig. 1.5a back into the first Brillouin zone and opens mini-bandgaps at high symmetry points [6, 42]. This approach redirects the emission coupled to the SPP modes into the vertical direction (Γ-point), which is collected in our experiment.

The active region consists of five 0.9% tensile-strained InP-based multiquantum wells (MQWs) located 20 nm apart from the TiO layer: MQWs are located 24–102 nm away from the Au surface. The Au, Si, and 2 nm thick Ti layers were deposited by evaporation. Ti layers were subsequently oxidized in oxygen plasma to produce 4 nm thick TiO layers that prevent Au migration into Si and InP during the fabrication and the measurement. We confirmed that this oxidation process does not affect the spectrum of MQWs relative to its intensity and shape before fabrication of TiO layers. The hexagonal pattern in the Au layer (the inset of Fig. 1.6a)

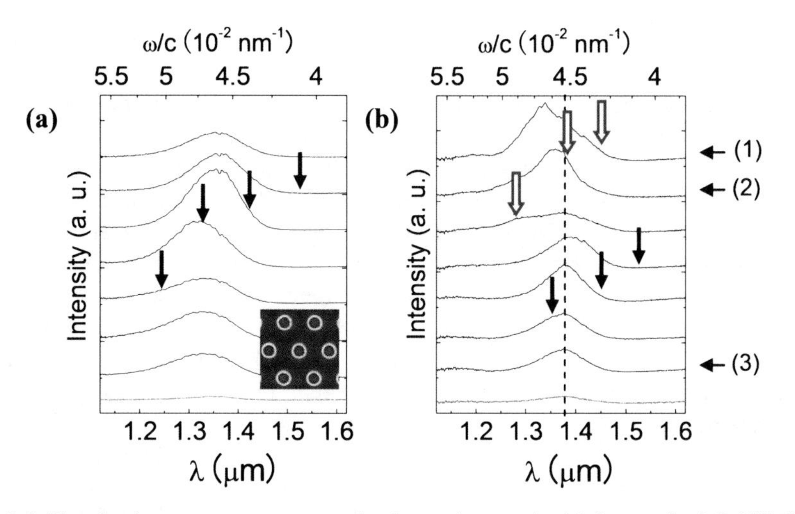

Fig. 1.6 Photoluminescence measurements in plasmonic crystals. (**a**) Spectra for InP–TiO–Au–TiO hexagonal plasmonic crystals. (**b**) Spectra for InP–TiO–Au–TiO–Si hexagonal plasmonic crystals. Lattice constants (a) of the crystals are 540, 500, 460, 420, 380, 340, and 300 nm, respectively, from the top. The bottom plots show the spectra for the samples without hexagonal patterns. *Vertical dashed line* indicates the wavelength of 1,350 nm for which the time-resolved PL measurements in Fig. 1.7 are taken. *Arrows* indicate the theoretically estimated frequencies of symmetric (*solid*) and WC-antisymmetric (*blank*) modes with $k = 4\pi/\sqrt{3}a$ in the structure without hexagonal pattern. The inset in (**a**) is a SEM image of a hexagonally patterned Au layer

was fabricated by electron beam lithography with ZEP-520 resist and subsequent Ar-ion milling in an area of 27 × 27 μm^2. We prepared samples with different periodicities of the hexagonal arrays to study the effect on vertical photoluminescence (PL) extraction efficiency. The hole radius in hexagonal arrays is chosen to keep the filling factor at 11%. The filling factor was chosen as small as possible to extract enough PL for decay-time measurement. Subsequent deposition of Si on top of the patterned Au layer leads to Si-filled holes in the Au plasmonic crystal. For comparison, we also prepared InP–TiO–Au–TiO structures and samples without the hexagonal arrays.

2.4.2 Photoluminescence Measurement

We first consider the coupling between the MQWs and the plasmonic crystal by measuring PL intensity of crystals with varying periodicities. The excitation wavelength was 980 nm. The PL from the sample was collected with the same objective lens and detected with a spectrum analyzer or a streak camera. We observed transverse magnetic (TM)-polarized PL due to the LH transition near 1,350 nm for the bare MQW wafer; the PL due to the heavy-hole (HH) transition was completely suppressed at room temperature. Figures 1.6a, b show the spectra for InP–TiO–Au–TiO and InP–TiO–Au–TiO–Si structures with and without hexagonal arrays of various lattice constants (a). The solid and blank arrows in Figs. 1.6a, b indicate the theoretically estimated frequencies of the symmetric and WC-antisymmetric modes with $k = 4\pi/\sqrt{3}a$ in the structure without hexagonal pattern. Those points are of particular interest because they are folded back to the Γ-point in plasmonic crystal, corresponding to vertical emission [6]. In the PL spectra for InP–TiO–Au–TiO shown in Fig. 1.6a, we observe an enhanced PL signal which shifts toward lower frequencies with an increase in a. These enhanced PL spectra have a maximum when the frequencies indicated by the solid arrows cross the spectrum of the MQW gain around 1,350 nm, which proves that these PL signals correspond to the symmetric SPP modes (see Fig. 1.5) vertically scattered by the hexagonal array with the same reciprocal vector as the wave vector of SPPs (Γ-point). The background signals independent of a, which were extremely small in the PL spectrum for the structure without hexagonal pattern (bottom plots), arise from uncoupled excitons located beneath the holes in plasmonic crystal. In the PL spectra for InP–TiO–Au–TiO–Si, we expect to see two kinds of enhanced PL signals corresponding to the two SPP modes: symmetric and WC-antisymmetric (see Fig. 1.5). These are visible in the spectra with $a = 380$–500 nm in Fig. 1.6b. Comparing Fig. 1.6a, b, we note that the PL signals corresponding to symmetric modes are shifted toward lower frequencies because of approximate index matching with the extra Si layer, as expected from Fig. 1.5a. The intensities of the signals corresponding to the symmetric modes were smaller than those corresponding to WC-antisymmetric modes because of their larger metal absorption loss. The PL peak observed in the spectrum for $a = 540$ nm (top plot in Fig. 1.6b) coincides with the light line for TiO layers ($n = 2.5$), and we expect that the structural periodicity of the Si waveguide induced by the hexagonal arrays probably scatters the light traveling through the TiO layers.

2.4.3 Time-Resolved (Decay-Time) Measurement

The coupling efficiency of PL into plasmon modes can be increased substantially in our structure by enhancing the spontaneous emission rate via the Purcell effect. In Fig. 1.7a, b, we directly measure this Purcell enhancement with time-resolved PL measurements. Figure 1.7a shows the initial PL decay rates defined as normalized time derivatives of the initial PL decay $(\delta PL/\delta t)_{t=0}/PL_{t=0}$ measured at the wavelength of 1,350 nm for spectra (1), (2), and (3) shown in Fig. 1.6b and for a bare MQW wafer, with different pumping powers I. Figure 1.7b shows the PL decay curves for the spectrum (2) (top plot) and for a bare MQW wafer (bottom plot), with the pumping power of $I = 6$ mW. To experimentally evaluate the Purcell enhancement factor from the time-resolved PL measurements shown in Fig. 1.7, we considered the Boltzmann approximation for decay rate of the carrier density N. In this approximation, the time derivative of N in undoped semiconductors is expressed as a polynomial of N: $dN/dt = -AN - BN^2 - CN^3$ where the coefficients A, B, and C correspond to surface- or defect-related recombination, bimolecular recombination, and an Auger process, respectively [41]. Among the three recombination processes, only the second term contributes to PL radiation: $PL \propto BN^2$, and thus $(dPL/dt)_{t=0}/PL_{t=0}$ is expressed by the following equation:

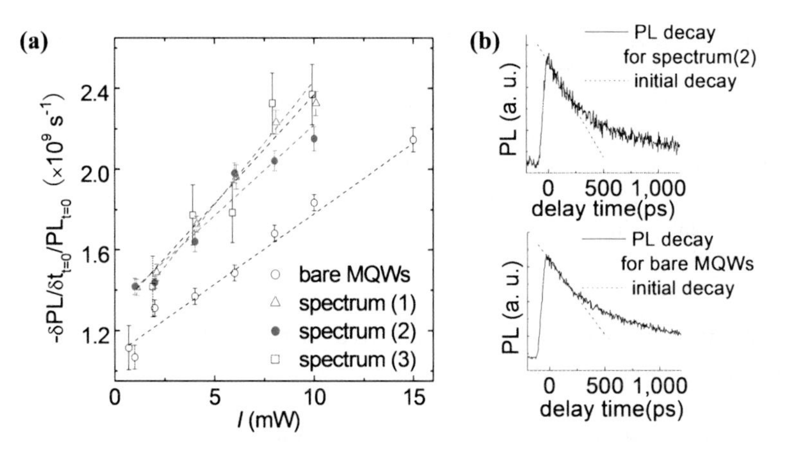

Fig. 1.7 Time-resolved photoluminescence measurements. (**a**) Initial slope of the PL decay $(\delta PL/\delta t)_{t=0}/PL_{t=0}$ plotted as a function of the pump power I, for a bare MQW wafer and for spectra (1), (2), and (3) in Fig. 1.6b at the denoted wavelength of 1,350 nm. *Dashed lines* are fits to the experimental data. (**b**) PL decay curves for a bare MQW wafer (*bottom plot*) and for spectrum (2) (*top plot*) with $I = 6$ mW. The excitation wavelength was 750 nm

$$(dPL/dt)_{t=0}\Big/PL_{t=0} = -2A - 2BN_0 - 2CN_0^2. \tag{1.22}$$

The values of I for the measurement were much lower than the saturation power for excitation of MQWs, and thus the initial carrier density N_0 could be considered proportional to I ($N_0 \cong \eta I$, where η is the excitation efficiency). Hence, Eq. (1.22) can be also rewritten as a polynomial of I. Since the observed values of $(\delta PL/\delta t)_{t=0}\big/PL_{t=0}$ change linearly with I in Fig. 1.7a, we could neglect the Auger process represented by the third term in Eq. (1.22) under our measurement condition. In this case, from the slope and offset of results in Fig. 1.7a, we estimate $A^b = 0.54 \times 10^9\ s^{-1}$, $A^m = (0.63 \pm 0.03) \times 10^9\ s^{-1}$, and $B^m\eta^m\big/B^b\eta^b = 1.5 \pm 0.2$, where superscripts b and m indicate the values for a bare MQW wafer and the InP–TiO–Au–TiO–Si structures, respectively. The large values of A could be due to the deformation caused by the large tensile strain applied in MQWs. The slopes of the fitting lines for spectra (1), (2), and (3) are almost the same within plotting errors. We can expect the position dependence of the radiative decay of the excitons within a unit cell of our plasmonic crystal [33]. However, the thermal diffusion length of carriers in GaInAsP during the typical decay time of 100 ps is estimated on the order of 10 μm, much longer than the radii of the holes in the hexagonal pattern [43]. Hence the position dependence of the radiative decay of the excitons within a unit cell of our plasmonic crystal is not observable in this measurement. We extract the Purcell enhancement factor from $B^m\eta^m\big/B^b\eta^b$. $\eta^m\big/\eta^b$ can be replaced by the ratio of the transmissions of the pump beams through plasmonic crystal and bare MQW wafer, which is estimated to be 0.45 by spatial averaging. Therefore, the Purcell enhancement factor with nonradiative decay ignored, which should be identical to the ratio of $B^m\big/B^b$, is estimated to be 2.9–3.7. Our experimentally estimated value of F falls within 75–80% of the theoretical estimate. One contribution to the reduction in F is InP oxide formed on the InP surface during fabrication, which lowers the refractive index and shifts the mode toward these layers and away from the MQW position (such a mode shift is also visible for TiO layers, as shown in Fig. 1.5b). We speculate that another contribution is the presence of uncoupled excitons located under dielectric holes in the hexagonal array due to large leakage losses in the dielectric band, which redirect part of the PL emission into the non-SPP modes (background PL).

3 Spontaneous Emission Modification in a Surface Plasmon Defect Cavity

In this section, we discuss cavity designs in metallic grating structure to achieve spatial control over the exciton–SPP coupling.

3.1 Spontaneous Emission Control in an SPP Defect Cavity

Continuing with the theme of SE rate enhancement, a cavity design offers more variety and control of quantum emitters. In photonic crystal designs, a cavity mode lies within a photonic bandgap; thus the emission into the cavity mode is enhanced while the emission into free-space modes is suppressed. In the same manner as electromagnetic modes in dielectrics, metal–dielectric structures can control the enhancement of emitters by directing emission to surface plasmon cavity modes. As with photonic crystal designs, we wish to investigate the cavity characteristics of the surface plasmon modes, namely, quality factor and mode volume. We would expect that absorption losses in the metallic portion of the cavity would generally be higher than absorption in a dielectric, and thus would decrease quality factor. On the other hand, we may try to exploit this small mode volume in cavity design by creating defect modes on a patterned surface.

As discussed in the previous sections, the modification of SE is related to the quality factor and mode volume of electromagnetic modes that interact with an active medium. Much like the reduced mode volumes of the propagating modes on a patterned metallic surface, a defect mode supported by a metallic structure could also be used to enhance the emission of emitters. Such metallic cavity structures would sacrifice ultra-high quality factors for low mode volumes, and maintain the ease of fabrication. While there have been reports of enhancement of photoluminescence by coupling emitters to regions of high SPP density of states (DOS) [24] and reports of coupling emitters to metallic nanowires and nanotips to enhance Purcell factors, a plasmonic crystal cavity would allow greater control of emission properties with the same fabrication processes [44].

In order to create a strongly confined mode, we would like to construct mirrors that are highly reflective over the bandwidth of a cavity. Several authors have demonstrated decreased transmission by using periodic structures to manipulate SPPs [45, 46]. These experiments confirm the existence of backscattering and a plasmonic bandgap in metallic gratings. In addition, other groups have demonstrated that SPP interfere as normal waves and set up standing waves under certain conditions [47]. Given such properties, it is easy to conceive a cavity that is the marriage of the previous two devices, a cavity that contains the electromagnetic field of the plasmon mode with metallic distributed Bragg reflector (DBR) gratings on either side of the cavity. While some plasmonic DBR cavities have been proposed in previous works [47, 48], the designs are often impractical to fabricate.

In the section below, we propose such a metallic grating cavity. The structure is shown in Fig. 1.8a and is composed of gratings with thin slices of metals on either side of an uninterrupted surface, which forms the cavity. Such a grating will open a plasmonic bandgap at a frequency to be determined by the grating periodicity (a). The periodicity of the grating that opens a plasmonic bandgap at frequency ω may be determined from the dispersion relationship of SPPs at a metal–dielectric interface [27]:

$$k_{\mathrm{sp}} = \frac{\pi}{a} = \frac{\omega}{c}\sqrt{\frac{\varepsilon_{\mathrm{d}}\varepsilon_{\mathrm{m}}}{\varepsilon_{\mathrm{d}}+\varepsilon_{\mathrm{m}}}}. \tag{1.23}$$

In this work, we assume that the dielectric is GaAs, with permittivity $\varepsilon_{\mathrm{d}} = \varepsilon_{\mathrm{GaAs}} = 12.96$ and the metal is silver, with permittivity estimated from the Drude model as $\varepsilon_{\mathrm{m}} = \varepsilon_{\infty} - (\omega_p/\omega)^2$ with $\varepsilon_{\infty} = 1$ and the plasmon energy of silver as $\hbar\omega_p = 8.8\mathrm{eV}$ ($\lambda_p = 140\mathrm{nm}$) [6]. Setting an operation energy of $\hbar\omega = 1.2\mathrm{eV}$, we determined the grating periodicity to be $a = 116$ nm. Although the metal is only 30 nm thick, coupled modes between the air–metal interface and GaAs–metal interfaces have a negligible impact on the dispersion relation.

2D finite difference time domain (FDTD) simulations with discretization of 1 unit cell per 2 nm were conducted with 5 periods of the DBR gratings on either side of a cavity and using the Drude model for the metal [6]. The depth of grooves in GaAs (filled with metal) and the metal slab layer thickness were both set at 30 nm

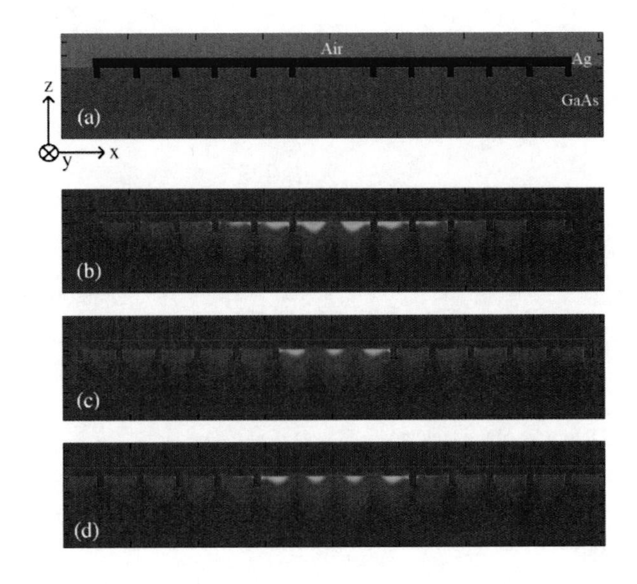

Fig. 1.8 (**a**) The proposed structure. (**b**)–(**d**) Mode profiles ($|E|^2$) with cavity lengths 216, 328, and 440 nm, respectively. These correspond to 2, 3, and 4 peaks of the electric field intensity inside the cavity

while the groove width was set at 20 nm. In order to analyze the radiative characteristics of the cavity mode, we first reduce the absorption losses of the metal to 2,000 times less than the losses at room temperature (RT). Such a reduction is noted from here on as the loss factor ξ and corresponds to a damping energy of $\hbar\eta = \hbar\eta_{\mathrm{RT}}/\xi = 2.5 \cdot 10^{-5}$eV. The cavity length was then varied over multiple grating periods to determine its effect on the modes. Three prospective modes with their electric field profiles are shown in Fig. 1.8 and the influence of cavity length on the mode quality factor (Q) and frequency is shown in Fig. 1.9. The Q factors were calculated using $Q_{r,nr} = \omega_0 U / P_{r,nr}$, where ω_0 is the frequency of the mode, U is the electromagnetic energy of the mode, and $P_{r,nr}$ is the power lost from the mode via radiative or nonradiative channels. First, we see that indeed the modes display standing wave patterns inside the cavities. Moreover, the peak quality factors of the modes are separated by grating periods, again supporting the idea that a standing wave is formed by the reflectors on either side of the cavity. The peak quality factor is approximately 1,000, and losses are dominated by radiation through the dielectric. The peaks of the quality factor all occur around $\omega = 0.153\omega_p$ suggesting a bandgap around that frequency. It is also noteworthy that the radiation parallel to the metal–dielectric interface is not the dominant pathway for losses, suggesting that increasing the number of DBRs would not enhance the overall Q of the modes. Finally, we see that the plasmonic modes of the DBR cavity exhibit a donor tendency, as the modes decrease in frequency as the cavity length increases.

The calculated Purcell enhancement of a quantum emitter (such as an InAs/GaAs quantum dot – QD) per cavity width Y (in μm) in the y-direction normal to the plane of the 2D simulation is shown in Fig. 1.10 (see Fig. 1.8). The Purcell enhancement, assuming negligible spectral detuning and nonradiative emitter decay (Γ_{nr}) is

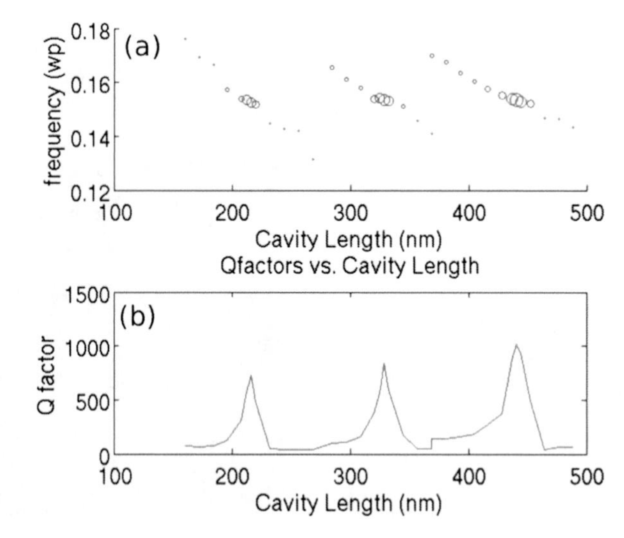

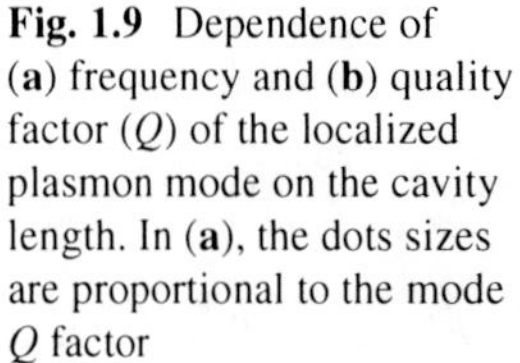

Fig. 1.9 Dependence of (**a**) frequency and (**b**) quality factor (Q) of the localized plasmon mode on the cavity length. In (**a**), the dots sizes are proportional to the mode Q factor

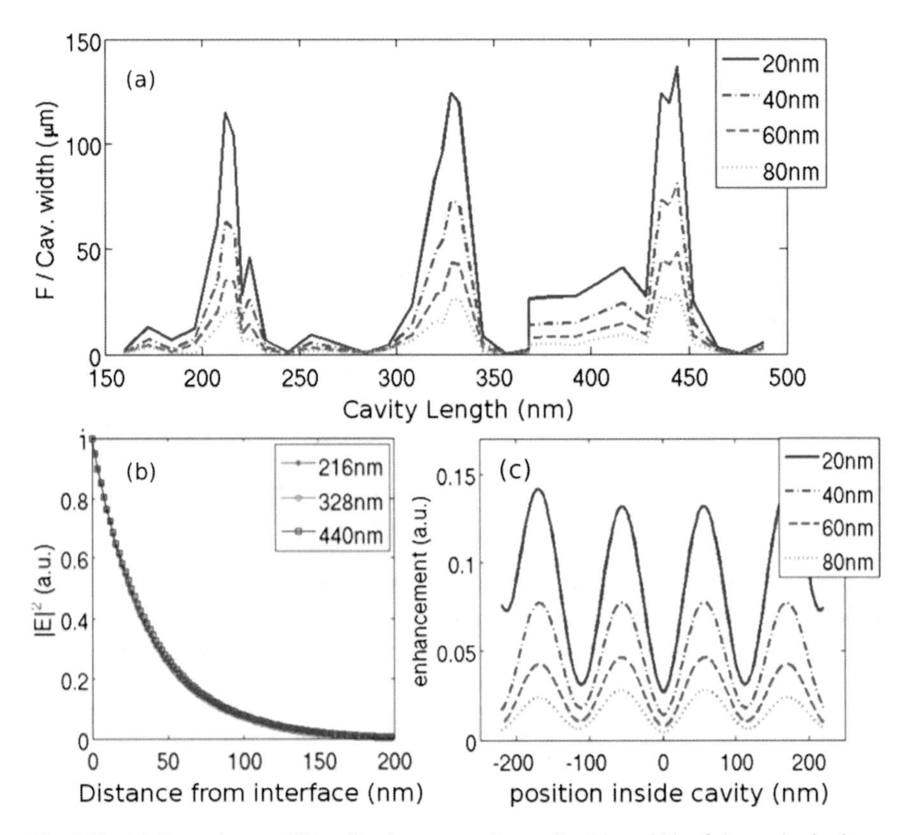

Fig. 1.10 (**a**) Dependence of Purcell enhancement (normalized by width of the cavity in the y-direction of Fig. 1.8 on cavity length for various emitter positions relative to the metal–dielectric interface (z-direction). (**b**) Exponential decay of the electric field in the dielectric ($|E|^2$) away from the metal–dielectric interface, plotted for three different cavity lengths corresponding to maximum Purcell enhancements. The decay constant of 36 nm is consistent with the plasmon modes in the bandgap at the $k_{\mathrm{sp}} = \pi/a$ point. (**c**) Normalized Purcell enhancement as a function of emitter position (in the x-direction) inside the 440 nm cavity for four different emitter distances from the metal–dielectric interface

$$F = \frac{\Gamma_b + \Gamma_{\mathrm{nr}} + \Gamma_{\mathrm{pl}}}{\Gamma_b + \Gamma_{\mathrm{nr}}} \approx 1 + \frac{\Gamma_{\mathrm{pl}}}{\Gamma_b} = 1 + \frac{3}{4\pi}\left(\frac{\lambda}{n}\right)^3 \frac{Q}{V}\left|\frac{E}{E_{\max}}\right|^2 \tag{1.24}$$

where Γ_{b} is the emitter spontaneous emission rate in bulk, Γ_{pl} is the emission rate of the QD coupled to the plasmon cavity mode, Q is the quality factor, and V is the mode volume defined for 2D simulations as

$$V_{\mathrm{m}} = \frac{\iiint \varepsilon_E(x,z)\left|E(x,z)^2\right| \mathrm{d}x\mathrm{d}z}{\max[\varepsilon_E(x,z)\,|E(x,z)|^2]} Y \tag{1.25}$$

Here, Y is the width of the cavity and the effective dielectric constant is as above [31]:

$$\varepsilon_{\mathrm{E}}(x,z) = \frac{d(\omega\varepsilon(x,z))}{d\omega} = \begin{cases} \varepsilon_{\infty} + (\omega_p/\omega)^2 & metal \\ \varepsilon_{\mathrm{d}} & non\text{-}metal \end{cases} \tag{1.26}$$

The curves pictured in Fig. 1.10 are simulated for various depths of a quantum emitter relative to the metal–dielectric interface. The larger of the cavities seem to have large tolerances of cavity lengths that could lead to the high Purcell factors. As shown in Fig. 1.10b, the electric field amplitude decays exponentially away from the metal–dielectric interface, as expected for SPP modes. Moreover, the decay profiles of all three modes are very similar, and the decay constants of $|E|^2$of 36 nm is consistent with the plasmon mode at the $k = \pi/a$ point. This again shows that a plasmon frequency is selectively contained by the bandgap created from the gratings. The high confinement of the plasmonic mode corresponds to a mode area of 50 nm^2 for 2D simulations, much smaller than the $(\lambda/n)^2$ area achieved for photonic crystal cavities. If we were able to contain the field in the y-direction to 50 nm as well, we could in theory achieve the strong coupling regime. Namely, for an InAs/GaAs quantum dot with a dipole moment of $\mu = 10^{-28}$ cm [49] positioned 20 nm from the metal–dielectric interface in the tail of E-field antinode and resonant with the field, the emitter–cavity field coupling is

$$g = \mu\sqrt{\frac{\omega}{2\varepsilon\hbar V_{\mathrm{m}}}}\left|\frac{E}{E_{\max}}\right| = 2\pi \cdot 170\ \mathrm{GHz} \tag{1.27}$$

Note that because $E_{\max}$ is not located in the middle of the cavity, but at corners of the cavities, the coupling factor is degraded from maximum values. For comparison, γ (the emitter decay rate without a cavity, dominated by radiative decay, Γ_{b}) and $\kappa = \omega/(2Q)$ (the cavity field decay rate) are $2\pi \cdot 1$ and $2\pi \cdot 160$ GHz, respectively. For such a set of parameters, $g > \gamma, \kappa$, and the onset of the strong coupling regime is reached. While degradation of Q may result from fabrication imperfections, even a tenfold drop in Q may still result in Purcell enhancements of hundreds, enabling such a device to be used in quantum information applications [50].

Another interesting property of the cavity is the effective range for high Purcell effect. As shown in Fig. 1.10c, the electrical field amplitude follows a standing wave profile of an even mode in the x-direction and exponentially decays in the z-direction. While the E_z field dominates for this mode, the contribution of the E_x field increases as we approach the surface, making the electric field amplitude near the surface significant throughout the cavity. This effect may allow for easier coupling to quantum dots embedded throughout the substrate.

In plasmonics, one great concern is the losses of the metal at optical frequencies. However, after locating the modes with peak Purcell factors, we can slowly increase losses from the previous structures and observe effects on the Purcell enhancement. First, we note that the mode volume does not change by much for most loss factors (by 1% at $\xi = 25$). The modification of Purcell enhancement then derives mostly from the change in quality factor. As shown in Fig. 1.11, there are two regimes of

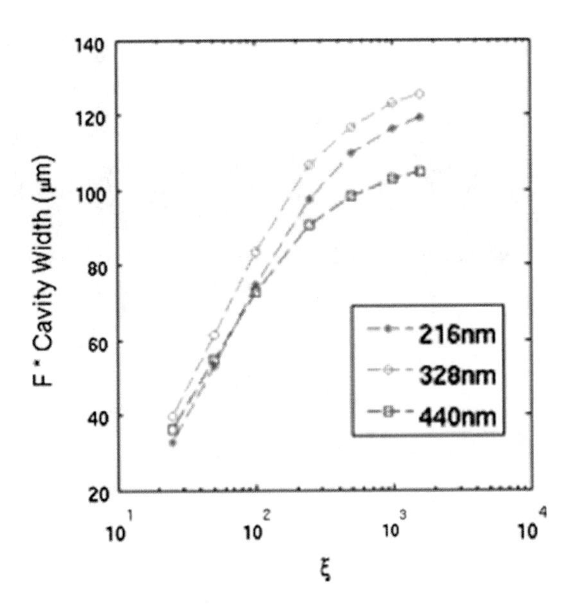

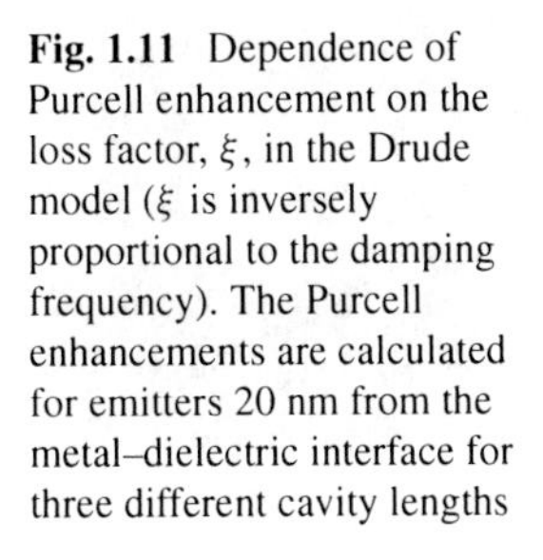
Fig. 1.11 Dependence of Purcell enhancement on the loss factor, ξ, in the Drude model (ξ is inversely proportional to the damping frequency). The Purcell enhancements are calculated for emitters 20 nm from the metal–dielectric interface for three different cavity lengths

behavior explained by describing the total Q as the parallel combination of Q_{rad} (radiative quality factor) and Q_{abs} (absorptive quality factor):

$$\frac{1}{Q_{\text{tot}}} = \frac{1}{Q_{\text{rad}}} + \frac{1}{Q_{\text{abs}}} \tag{1.28}$$

For $\xi > 100$, the quality factor of the absorption mechanism is much higher than the radiative quality factor, and the near constant Q_{rad} dominates. For $\xi < 100$, the absorption quality factor dominates and results in the linear decrease in the total Q_{tot} factor. For loss factors less than 25, the mode is changed, resulting in a diminished Q and negligible Purcell enhancement. It has been noted recently that high loss factors may be difficult to obtain, but we could still potentially achieve high Purcell factors as we decrease the width of the cavity in the lateral directions.

3.2 Surface Plasmon Grating as a Cavity

In addition to the previously described defect cavity, we could also consider a small (truncated) metallic grating as a nanocavity and compute its cavity characteristics. We modify the design from above to have the grating in air. Such a design serves two purposes. First, the emission of such a structure is directed away from the metal and toward the collection optics in air. Second, because the confinement of the SPP mode is inversely dependent on the index of the dielectric material, using a lower index material helps to increase the size of the structure. This eases the restriction on fabrication, as devices operating around 900 nm now have characteristic lengths of hundreds of nanometers.

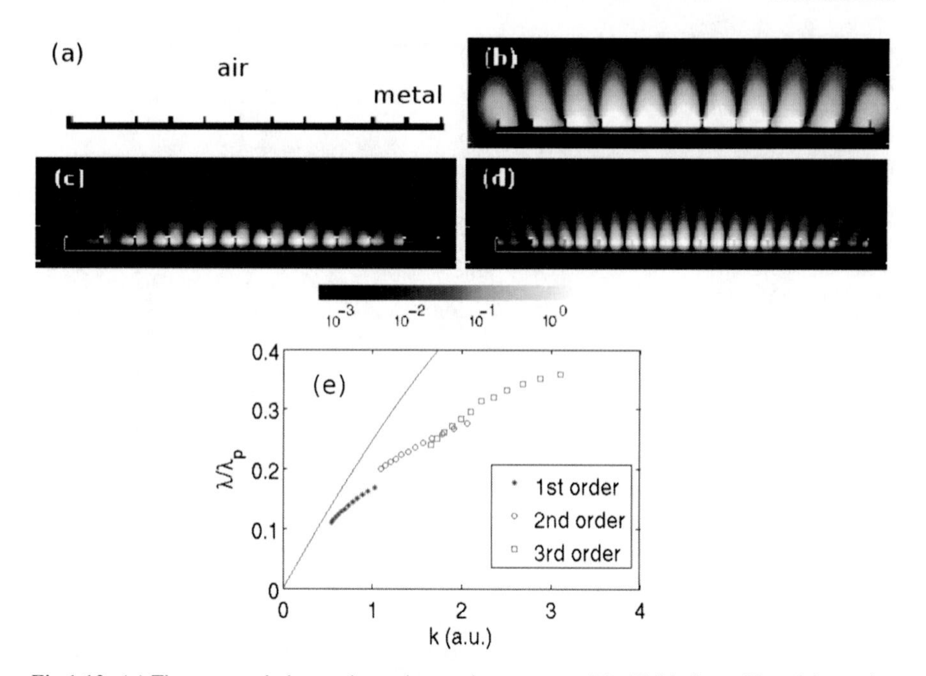

Fig 1.12 (**a**) The proposed plasmonic grating-cavity structure. (**b**)–(**d**) Mode profiles of the cavity (magnetic field intensity $|H|^2$). The fundamental, second-, and third-order modes that match the grating periodicity are shown. These modes correspond to 1, 2, and 3 peaks of magnetic field intensity per grating period. (**e**) Band edge points as the **k** vector of the grating is varied. The *star*, *circle*, and *square* markers represent the first-, second-, and third-order modes, respectively. The normalized frequencies of the modes are plotted against the one, two, or three times the **k** vector of the grating, depending on the order of the mode. The SPP dispersion relation is plotted as the *solid line*

We analyze the grating structure again using FDTD simulations with the models mentioned above [6]. The structure shown in Fig. 1.12a is a grating with 11 uniform periods, truncated at the edges. Such a grating supports at least the first three modes as shown in Fig. 1.12b–d. As we change the period of the grating, the Brillouin zone edges change, and the frequencies of the standing wave modes shift (Fig. 1.12e). By plotting the first-, second-, and third-order frequencies against one, two, and three times the grating **k** vector, respectively, the standing wave points should trace out the SPP dispersion relation. For low grating **k** vectors, the band edge points lie very close to the approximately linear dispersion line. Deviation from the dispersion relationship for higher normalized frequencies is expected as higher order bands have increased bending.

We experimentally investigate the SE enhancement in the metal grating structures. We spincast PbS colloidal QDs on the metallic grating and attempt to measure the effect of the grating. The PL of the bulk QDs is shown in Fig. 1.13a, while the normalized PL intensities of the dots on gratings of various periodicities are shown in Fig. 1.13b. The Purcell factor is defined as in Eq. (1.24). By dividing

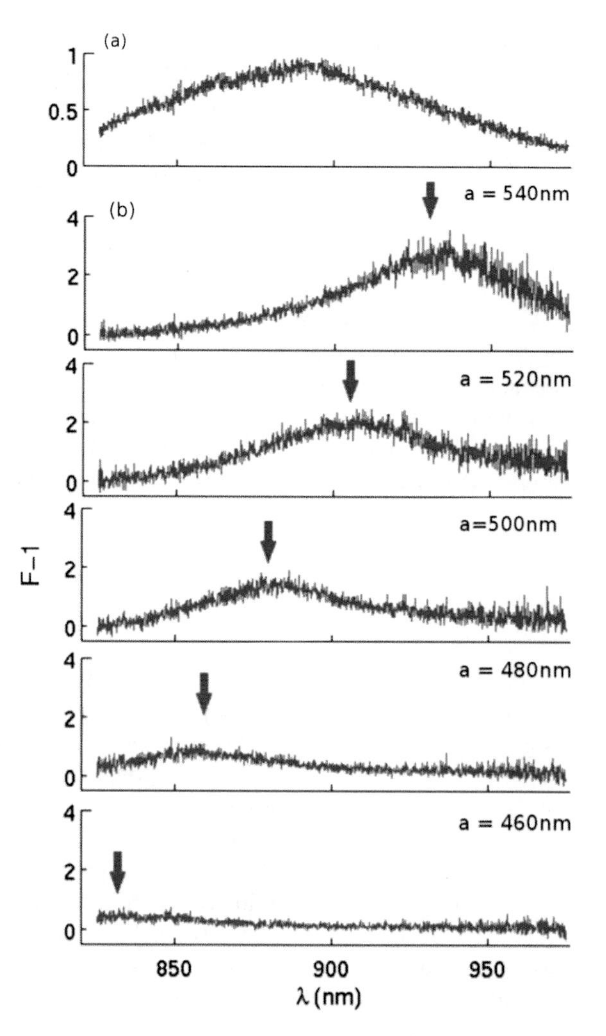

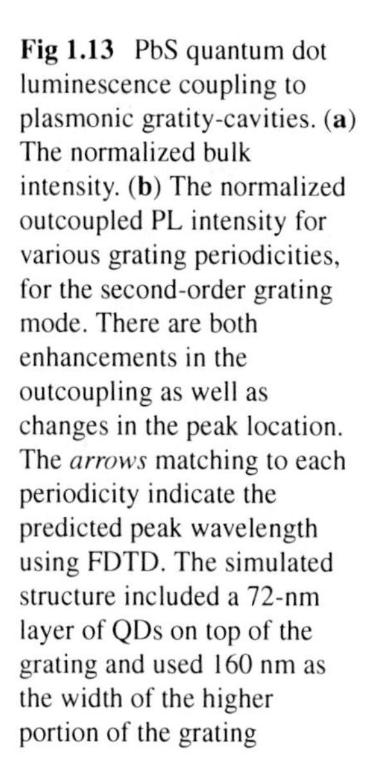
Fig 1.13 PbS quantum dot luminescence coupling to plasmonic gratity-cavities. (**a**) The normalized bulk intensity. (**b**) The normalized outcoupled PL intensity for various grating periodicities, for the second-order grating mode. There are both enhancements in the outcoupling as well as changes in the peak location. The *arrows* matching to each periodicity indicate the predicted peak wavelength using FDTD. The simulated structure included a 72-nm layer of QDs on top of the grating and used 160 nm as the width of the higher portion of the grating

the PL collected from the grating by the bulk PL and normalizing to account for the unenhanced portions of the spectrum, we can obtain $F - 1$, or the outcoupled portion of the enhancement. In this case, there is significant enhancement of PL for grating periodicities of 520 and 540 nm. In addition, the PL of the dots on the grating demonstrates a noticeable shift in frequency, as the PL peak shifts by 100 nm when the grating periodicity decreases by 80 nm. Finally, each of the PL spectra demonstrates a decrease in the linewidth from the bulk PL broadening.

While the SPP grating in air offers a wide range of tuning for wavelength, we must also account how the finite thickness of the quantum dot layer on top of grating affects the mode. In fact, we repeat FDTD simulations for a multilayered device, assume an effective index of 1.8 for the layer of quantum dots. Here, we find a splitting of the modes that is confined differently in the dot layer and air. These two

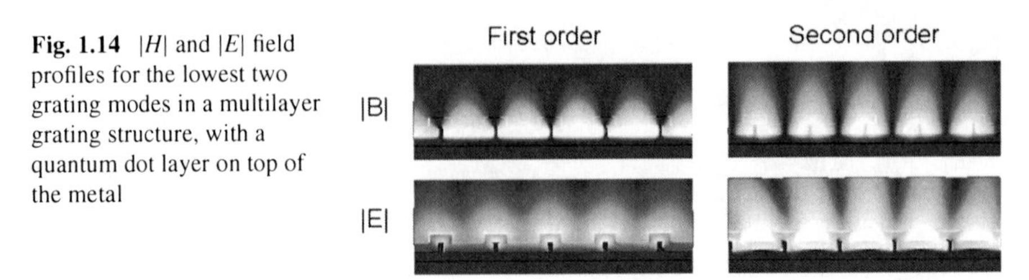

Fig. 1.14 $|H|$ and $|E|$ field profiles for the lowest two grating modes in a multilayer grating structure, with a quantum dot layer on top of the metal

modes form a splitting much like that in a classical one-dimensional DBR stack (Fig. 1.14). The first-order mode has most of the electric field confined to a small volume in the higher index, while the second-order mode electric field is more spatially delocalized and interacts with more of the air. Through trials of varying the thickness of the new layer, we determine with simulations that the dot thickness on top of the metallic grating is approximately 80 nm, which is in agreement with measurements.

4 Conclusions

In conclusion, we have shown theoretically and experimentally that it is possible to control the SE rate of an exciton by its coupling to metal–dielectric surfaces. The design of the structures we discuss in this chapter supports SPP of reduced ohmic loss and small mode volumes that overcome diffraction limit. These advantages make such metal–dielectric structures attractive for novel photoelectric devices. As opposed to waveguides or cavities in dielectric, metal surface naturally confines electromagnetic field without elaborate air bridges or DBR structure, which allows for compact devices. We experimentally show that if the exciton lies close enough to the metal surface, more than 80% of PL is emitted into SPP modes. In addition, we also demonstrate that there is a strong enhancement of emitters coupled to metallic grating-like modes. Both types of devices demonstrate the possibility of efficient coupling between the emitters and SPP waveguides. Finally, we show that defect modes in patterned metallic surfaces can be used to confine SPP modes on the metal–dielectric interface. Such a device is characterized by extremely small mode volumes and high electrical field confinement. We hope that the design principles described in this chapter will help enable improved optoelectronic devices and sensors.

References

1. Ritchie, R.H.: Plasma losses by fast electrons in thin films. Phys. Rev. **106**, 874–881 (1957)
2. Westphalen, M., Kreibig, U., Rostalski, J., Luth, H., Meissner, D.: Metal cluster enhanced organic solar cells. Sol. Energy Mat. Sol. Cells **61**, 97–105 (2000)
3. Derkacs, D., Lim, S.H., Matheu, P., Mar, W., Yu, E.T.: Improved performance of amorphous silicon solar cells via scattering from surface plasmon polaritons in nearby metallic nanoparticles. Appl. Phys. Lett. **89**, 093103 (2006)

4. Maier, S.A.: Plasmonic field enhancement and SERS in the effective mode volume picture. Opt. Express **14**, 1957–1964 (2006)
5. Nie, S., Emory, S.R.: Probing single molecules and single nanoparticles by surface-enhanced Raman scattering. Science **275**, 1102 (1997)
6. Vučković, J., Loncar, M., Scherer, A.: Surface plasmon enhanced light-emitting diode. IEEE J. Quantum Electron **36**, 1131–1144 (2000)
7. Kumar, S., Williams, B.S., Qin, Q., Lee, A.W.M., Hu, Q., Reno, J.L.: Surface-emitting distributed feedback terahertz quantum-cascade lasers in metal-metal waveguides. Opt. Express **15**, 113–123 (2007)
8. Kim, S., Jin, J., Kim, Y.-J., Park, I.-Y., Kim, Y., Kim, S.-W.: High-harmonic generation by resonant plasmon field enhancement. Nature **453**, 757–760 (2008)
9. Purcell, E.M.: Spontaneous emission probabilities at radio frequencies. Phys. Rev. **69**, 681 (1946)
10. Barnes, W.L.: Fluorescence near interfaces: the role of photonic mode density. J. Mod. Opt. **45**, 661–699 (1998)
11. Chance, R.R., Prock, A., Silbey, R.: Lifetime of an emitting molecule near partially reflecting surface. J. Chem. Phys. **60**, 2744–2748 (1974)
12. Chance, R.R., Prock, A., Silbey, R.: Molecular fluorescence and energy transfer near interfaces. Adv. Chem. Phys. **37**, 1–65 (1978)
13. Barnes, W.L., Dereux, A., Ebbesen, T.W.: Surface plasmon subwavelength optics (review). Nature **424**, 824–830 (2003)
14. Iwase, H., Englund, D., Vučković, J.: Spontaneous emission control in high-extraction efficiency plasmonic crystals. Opt. Express **16**, 426–434 (2008)
15. Akimov, A.V., Mukherjee, A., Yu, C.L., Chang, D.E., Zibrov, A.S., Hemmer, P.R., Par, H., Lukin, M.D.: Generation of single optical plasmons in metallic nanowires coupled to quantum dots. Nat. Phys. **450**, 402–406 (2007)
16. Sirtori, C., Gmachl, C., Capasso, F., Faist, J., Sivco, D.L., Hutchinson, A.L., Cho, A.Y.: Long-wavelength ($\lambda \sim 8$–$11.5\ \mu$m) semiconductor lasers with waveguides based on surface plasmons. Opt. Lett. **23**, 1366–1368 (1998)
17. Okamoto, T., H'Dhili, F., Kawata, S.: Towards plasmonic band gap laser. Appl. Phys. Lett. **85**, 3968 (2004)
18. Barnes, W.L., Preist, T.W., Kitson, S.C., Sambles, J.R.: Physical origin of photonic energy gaps in the propagation of surface plasmons on gratings. Phys. Rev. B **54**, 6227–6244 (1996)
19. Chang, D.E., Sorensen, A.S., Hemmer, P.R., Lukin, M.D.: Strong coupling of single emitters to surface plasmons. Phys. Rev. B **76**, 035420 (2007)
20. Gong, Y., Vučković, J.: Design of plasmon cavities for solid-state cavity quantum electrodynamics applications. Appl. Phys. Lett. **90**, 033113 (2007)
21. Kimble, H.J.: Structure and dynamics in cavity quantum electronics. In: Berman, P. (ed.) Cavity Quantum Electrodynamics, pp. 213–219. Academic, San Diego, CA (1994)
22. Englund, D., Faraon, A., Fushman, I., Stoltz, N., Petroff, P., Vučković, J.: Nature **450**, 857 (2007)
23. Iwase, H., Englund, D., Vučković, J.: Analysis of the Purcell effect in photonic and plasmonic crystals with losses. Opt. Express **18**, 16546–16560 (2010)
24. Okamonto, K., Niki, I., Shvartser, A., Narukawa, Y., Mukai, T., Scherer, A.: Surface-plasmon-enhanced light emitters based on InGaN quantum wells. Nat. Mat. **3**, 601–605 (2004)
25. Gontijo, I., Boroditsky, M., Yablonovitch, E., Keller, S., Mishra, U.K., DenBaars, S.P.: Coupling of InGaN quantum-well photoluminescence to silver surface plasmons. Phys. Rev. B **60**, 11564 (1999)
26. Neogi, A., Lee, C., Everitt, H.O., Kuroda, T., Tackeuchi, A., Yablonovitch, E.: Enhancement of spontaneous recombination rate in a quantum well by resonant surface plasmon coupling. Phys. Rev. B **66**, 153305 (2002)
27. Raether, H.: Surface Plasmons on Smooth and Rough Surfaces and on Gratings. Springer, Berlin (1988)

28. Jackson, J.D.: Classical Electrodynamics. Wiley, New York, NY (1998)
29. Scully, M.O., Zubairy, M.S.: Chap. 9. Quantum Optics. Cambridge University Press, London (1997)
30. Manga Rao, V.S.C., Hughes, S.: Single quantum-dot Purcell factor and β factor in a photonic crystal waveguide. Phys. Rev. B **75,** 205437 (2007)
31. Landau, L.D.: Electrodynamics of Continuous Media. Pergamon, New York, NY (1984)
32. Economou, E.N.: Surface plasmons in thin films. Phys. Rev. **182**, 539–554 (1969)
33. Lee, R.K., Xu, Y., Yariv, A.: Modified spontaneous emission from a two-dimensional photonic bandgap crystal slab. J. Opt. Soc. Am. B **17**, 1438 (2000)
34. Vučković, J., Pelton, M., Scherer, A., and Yamamoto, Y.: Optimization of three-dimensional micropost microcavities for cavity quantum electrodynamics. Phys. Rev. A **66**, 023808 (2002)
35. Joannopoulos, J.D., Meade, R.D., Winn, J.N.: Photonic Crystals. Princeton University Press, Princeton, NJ (1995)
36. Boroditsky, M., Vrijen, R., Krauss, T.F., Coccioli, R., Bhat, R., Yablonovitch, E.: Spontaneous emission extraction and Purcell enhancement from thin-film 2-D photonic crystals. J. Lightwave Technol. **17**, 2096–2112 (1999)
37. Plihal, M., Maradudin, A.A.: Photonic band structure of two-dimensional systems: the triangular lattice. Phys. Rev. B **44**, 8565 (1991)
38. Chutinan, A., Ishihara, K., Asano, T., Fujita, M., Noda, S.: Theoretical analysis on light-extraction efficiency of organic light-emitting diodes using FDTD and mode-expansion methods. Org. Electron. **6**, 3–9 (2005)
39. Altug, H., Vučković, J.: Two-dimensional coupled photonic crystal resonator arrays. Appl. Phys. Lett. **84**, 161–163 (2004)
40. Hinds, E.A.: Perturbative cavity quantum electrodynamics. In: Berman, P.R. (eds) Cavity Quantum Electrodynamics. Academic, New York, NY (1994)
41. Coldren, L.A., Corzine, S.W.: Diode Lasers and Photonic Integrated Circuits. Wiley, New York, NY (1995)
42. Kitson, S.C., Barnes, W.L., Sambles, J.R.: Full photonic band gap for surface modes in the visible. Phys. Rev. Lett. **77**, 2670–2673 (1996)
43. Adachi, S.: Physical Properties of III-V Semiconductor Compounds. Wiley, New York, NY (1992)
44. Chang, D.E., Sørenson, A.S., Hemmer, P.R., Lukin. M.D.: Quantum optics with surface plasmons. Phys. Rev. Lett. **97**, 053002 (2006)
45. Weeber, J.-C., Lacroute, Y., Dereux, A., Devaux, E., Ebbesen, T., Girard, C., González, M.U., Baudrion, A.-L.: Near-field characterization of Bragg mirrors engraved in surface plasmon waveguides. Phys. Rev. B **70**, 235406 (2004)
46. Bozhevolnyi, S., Boltasseva, A., Søndergaard, T., Nikolajsen, T., Leosson, K.: Photonic band gap structures for long-range surface plasmon polaritons. Opt. Comm. **250**, 328–333 (2005)
47. Liu, Z.-W., Wei, Q.-H., Zhang, X.: Surface plasmon interference nanolithography. Nano Lett. **5**, 957–961 (2005)
48. Wang, B., Wang, G.P.: Plasmon Bragg reflectors and nanocavities on flat metallic surfaces. Appl. Phys. Lett. **87**, 013107 (2005)
49. Eliseev, P.G., Li, H., Stintz, A., Liu, G.T., Newell, T.C., Malloy, K.J., Lester, L.F.: Transition dipole moment of InAs/InGaAs quantum dots from experiments on ultralow-threshold laser diodes. Appl. Phys. Lett. **77**, 262 (2000)
50. Waks, E., Vučković, J.: Dispersive properties and large Kerr nonlinearities using dipole-induced transparency in a single-sided cavity. Phys. Rev. A **73**, 041803 (2006)

Chapter 2
Surface Plasmon Enhanced Solid-State Light-Emitting Devices

Koichi Okamoto

Abstract A novel method to enhance light emission efficiencies from solid-state materials was developed by the use of surface plasmon (SP). A 17-fold increase in the photoluminescence (PL) intensity along with a 7-fold increase in the internal quantum efficiency (IQE) of light emission from InGaN/GaN quantum wells (QWs) was obtained when nanostructured silver layers were deposited 10 nm above the QWs. A 32-fold increase in the spontaneous emission rate of InGaN/GaN at 440 nm probed by the time-resolved PL measurements was also observed. Likewise, both light emission intensities and rates were enhanced for organic materials, CdSe-based nanocrystals, and also Si/SiO_2 nanostructures. These enhancements should be attributed to the SP coupling. Electron–hole pairs in the materials couple to electron vibrations at the metal surface and produce SPs instead of photons or phonons. This new path increases the spontaneous emission rate and the IQEs. The SP-emitter coupling technique would lead to super bright and high-speed solid-state light-emitting devices that offer realistic alternatives to conventional fluorescent light sources.

Keywords Plasmonics · Surface plasmon · Polaroiton · Light-emitting device · InGaN · Quantum well · CdSe · Quantum dot · Silicon naocrystal

1 Introduction

Conduction electron gas in a metal oscillates collectively and the quantum of this plasma oscillation is called "plasmon." A special plasma oscillation mode called "surface plasmon (SP)" exists at an interface between a metal, which has a negative dielectric constant, and a positive dielectric material [1]. The plasma oscillation frequency (ω_{SP}) of the SP is different to that in the bulk plasmon (ω_P). The charge

K. Okamoto (✉)
Department of Electronic Science and Engineering, Kyoto University, Katsura, Nishikyo-ku, Kyoto 615-8510, Japan; PRESTO, Japan Science and Technology Agency, 4-1-8 Honcho Kawaguchi, Saitama 332-0012, Japan
e-mail: k.okamoto@hy4.ecs.kyoto-u.ac.jp

Z.M. Wang, A. Neogi (eds.), *Nanoscale Photonics and Optoelectronics*, Lecture Notes in Nanoscale Science and Technology 9,
DOI 10.1007/978-1-4419-7587-4_2,

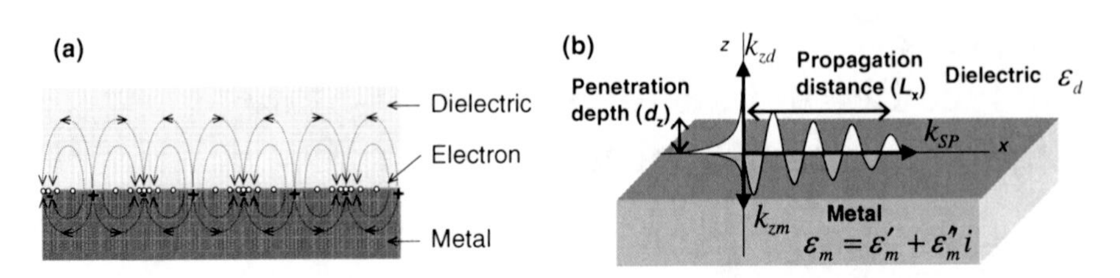

Fig. 2.1 (**a**) Schematic diagram of the surface plasmon (SP) and surface plasmon polariton (SPP) generated at the metal/dielectric interface. (**b**) Propagation along the x axis and penetration along the z axis of the surface plasmon portions

fluctuation of the longitudinal oscillation of the SP, which is localized at the interface, is accompanied by fluctuations of transversal and longitudinal electromagnetic fields, which is called surface plasmon polariton (SPP). Schematic diagram of the SP mode and the SPP mode generated at the metal/dielectric interface is shown in Fig. 2.1a. As the SPP is one of the electromagnetic wave modes, it can interact with light waves at the interface and it brings novel optical properties and functions to materials. The technique of controlling and utilizing the SPP is called "plasmonics" and has attracted much attention with the recent rapid advance of nanotechnology [2–4].

Figure 2.1b shows behaviors of the SPP at a metal/dielectric interface. The wave vector of the SPP (k_{SP}) along the x direction (parallel to the interface) can be written with the following equation when the relative permittivity of the metal is $\varepsilon_1 = \varepsilon'_1 + \varepsilon''_1 i$ and that of the dielectric material is ε_2 [1]:

$$k_{\mathrm{SP}} = \frac{\omega}{c}\sqrt{\frac{\varepsilon'_1 \varepsilon_2}{\varepsilon'_1 + \varepsilon_2}} + \frac{\omega}{c}\left(\frac{\varepsilon'_1 \varepsilon_2}{\varepsilon'_1 + \varepsilon_2}\right)^{\frac{3}{2}} \frac{\varepsilon''_1}{2\varepsilon'^2_1} i, \tag{2.1}$$

where ω and C are the frequency of the SPP and the light velocity in vacuum, respectively. The first term of this equation is known as the dispersion relation of the SPP. Figure 2.2a shows the typical dispersion relations of the SPPs at Al/GaN, Ag/GaN, and Au/GaN interfaces. Usually, the k_{SP} values are much larger than the wave vector of the light wave propagated in the dielectric media, because $\varepsilon'_1 < 0$ at the visible wavelength regions. This fact suggests that the SPP can propagate into nanospaces much smaller than the wavelength. This is one of the most important features of the SPP. This enables us to shrink the sizes of waveguides and optical circuits into nanoscale [5]. k_{SP} becomes infinity when $\varepsilon'_1 + \varepsilon_2 = 0$ and the frequency under this condition is ω_{SP}. The second term of Eq. (2.1) indicates damping of the SPP mode. Figure 2.2b shows the propagation length (L_x) of the SPP at Al/GaN, Ag/GaN, and Au/GaN calculated by Eq. (2.1). This figure suggests that the SPP can propagate to a few tens or a few hundreds of micrometers.

Wave vectors of the SPP (k_{zj}) along the z direction (perpendicular to the interface) in a metal ($j = 1$) or a dielectric material ($j = 2$) are given by [1]

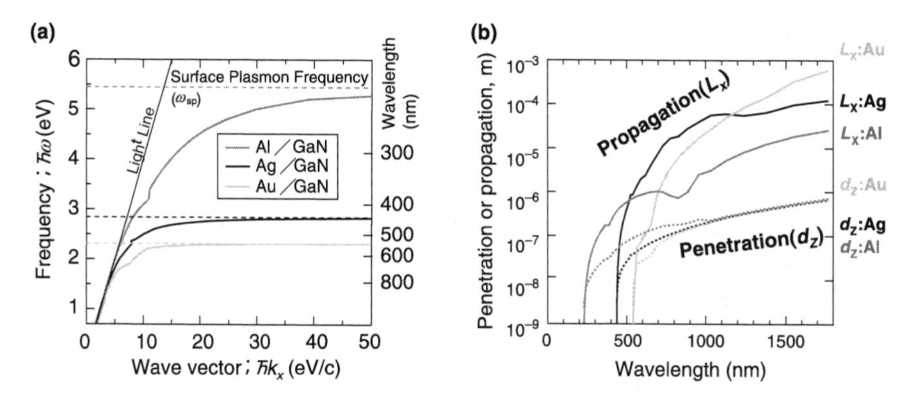

Fig. 2.2 (**a**) Dispersion diagrams of the SPP at Al/GaN, Ag/GaN, and Au/GaN interfaces. (**b**) Penetration depths and propagation distances of the SPP at Al/GaN, Ag/GaN, and Au/GaN interfaces calculated by Eqs. (2.1) and (2.2), respectively

$$k_{\mathrm{z}j} = \sqrt{\varepsilon_j \left(\frac{\omega}{c}\right)^2 - k_{\mathrm{SP}}^2} \qquad (j = 1, 2). \tag{2.2}$$

k_{zj} should be an imaginary number because k_{SP} is larger than the light line. This suggests that the SPP cannot propagate to the z direction but decays exponentially. Figure 2.2b also shows the penetration depth (d_z) of the SPP into GaN at each interface calculated by Eq. (2.2). The d_z values strongly depend on the wavelength but are always much shorter than 1 μm. This means that the electromagnetic fields of the SPP are strongly localized at the interface and it makes giant fields at the interface. This huge field enhancement effect is also one of the most important features of the SPPs. It has been applied to sensors which have high sensitivities based on the surface plasmon resonance (SPR) at the interface [6]. Moreover, the SPPs can be localized into one-dimensional spaces using metal nanowires. Localized surface plasmon (LSP), which is localized into zero-dimensional spaces using metal nanoparticles, is well known for application of the surface enhanced Raman scattering (SERS) [7]. The LSP nanoprobes enable the optical imaging with super resolution and very high sensitivity [8].

Moreover, several future possibilities of plasmonics have been proposed recently, e.g., the plasmonic metamaterials [9] with negative refractive index at visible regions, the plasmonic therapy for cancer, and the optical cloaking technology [10] based on plasmonics. Undoubtedly, plasmonics becomes a key technology at wider fields and would attract much more attention in the near future. Here, I describe one of the new applications of plasmonics, that is, surface plasmon enhanced light emissions. Recently, I and my coworkers developed a novel method to enhance the light emission efficiencies from solid-state materials by the use of surface plasmon. The technique we invented adds new potential to plasmonics.

2 Background of Solid-State Light-Emitting Devices

Since 1993, InGaN quantum wells (QW)-based light-emitting diodes (LEDs) have been continuously improved and commercialized as light sources in the ultraviolet and visible spectral regions [11]. Moreover, white light LEDs, in which a blue LED is combined with a yellow phosphor, have been commercialized and offer a replacement for conventional incandescent and fluorescent light bulbs [12]. However, these devices have not fulfilled their original promise as solid-state replacements for light bulbs as their light emission efficiencies have been limited. The most important requirement for a competitive LED for solid-state lighting is the development of new methods to increase its quantum efficiency of light emission.

The external quantum efficiency (η_{ext}) of light emission from an LED is given by the light extraction efficiency (C'_{ext}) and the internal quantum efficiency (IQE: η_{int}). η_{int} in turn is determined by the ratio of the radiative (k_{rad}) and nonradiative (k_{non}) recombination rates of carriers.

$$\eta_{ext} = C'_{ext} \times \eta_{int} = C'_{ext} \times \frac{k_{rad}}{k_{rad} + k_{non}}. \tag{2.3}$$

Often, k_{non} is faster than k_{rad} at room temperature, resulting in modest η_{int}. There are three methods to increase η_{ext}: (1) increase C'_{ext}, (2) decrease k_{non}, or (3) increase k_{rad}. Previous work has focused on improving C'_{ext} from InGaN LEDs by using the patterned sapphire substrates and mesh electrodes [13]. However, further improvements of extraction of light through these methods are rapidly approaching fundamental limitations. Although much effort has recently been placed into reducing k_{non} by growing higher quality crystals [14], dramatic enhancements of η_{ext} have so far been elusive. On the other hand, there have been very few studies focusing on increasing k_{rad} [15], though that could prove to be most effective for the development of high η_{ext} light emitters. Here, I describe the enhancement of k_{rad} by coupling between surface plasmon and the InGaN QWs. If the plasmon frequency is carefully selected to match the QW emission frequency, the increase of the density of states resulting from the SP dispersion diagram (Fig. 2.2a) can result in large enhancements of the spontaneous emission rate. Therefore, energy coupling between QW and SP as described in this chapter is one of the most promising solutions to increase k_{rad}.

Since 1990, the idea of SP enhanced light emission was proposed and received much attention [16–22]. Vuckovic et al. reported the SP enhanced LED analyzing it both theoretically and experimentally [20]. For InGaN QWs, Gontijo and coworkers reported the coupling of the spontaneous emission from QW into the SP on silver thin film and showed increased absorption of light at the SP frequency [21]. Neogi et al. confirmed that the recombination rate in an InGaN/GaN QW could be significantly enhanced by the time-resolved PL measurement [22]. However, in these early studies for InGaN QWs, light could not be extracted efficiently from the silver/GaN surface. Therefore, the actual enhancements of visible light emissions had not been observed directly before our first report.

3 Surface Plasmon Enhanced Light Emission

Recently, we have reported, for the first time, large photoluminescence (PL) increases from InGaN/GaN QW material coated with metal layers [23]. InGaN/GaN single QW (3 nm) structures were grown on sapphire substrates by a metalorganic chemical vapor deposition (MOCVD), and silver, aluminum, or gold layers (50 nm) were deposited on top of the surfaces of these wafers by a high-vacuum thermal evaporation. The sample structure was shown in Fig. 2.3a. To perform the PL measurements, a cw-InGaN diode laser (406 nm) was used to excite the QWs from the bottom surface of the wafer. The excitation power was 4.5 mW. PL was collected and focused into an optical fiber and subsequently detected with a multichannel spectrometer.

Figure 2.3b shows typical PL spectra from InGaN/GaN QWs separated from Ag, Al, and Au layers by 10 nm GaN spacers. For Ag coatings, the PL peak of the uncoated wafer at 470 nm is normalized to 1, and a 14-fold enhancement in peak PL intensity is observed from the Ag-coated emitter. The PL intensity integrated over the emission spectrum is increased by 17 times, whereas 8-fold peak intensity and 6-fold integrated intensity enhancements are obtained from Al-coated InGaN QW. The PL is not increased after Au coating. A small increase in the PL intensity might be expected after metallization because the metal reflects pump light back through the QW, doubling the effective path of the incident light, but differences between Au and Ag reflectivities at 470 nm cannot explain the large difference in the measured enhancement alone. We believe that these PL enhancements should be attributed to the energy transfer between QWs and SPs. The SPs can increase the density of states and the spontaneous emission rate in the semiconductor and lead to the enhancement of light emission by SP–QW coupling. No such enhancements were obtained from samples coated with Au, as its well-known plasmon resonance occurs only at longer wavelengths.

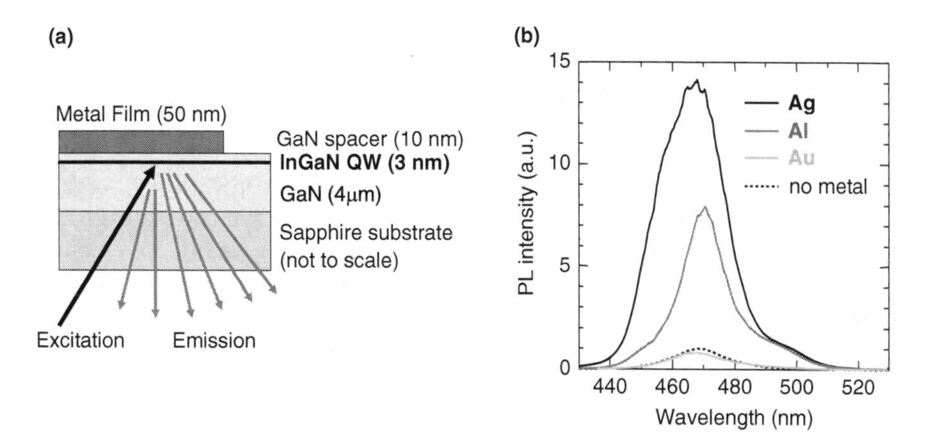

Fig. 2.3 (**a**) Sample structure of InGaN/GaN QW and excitation/emission configuration of PL measurement. (**b**) PL spectra of InGaN/GaN QWs coated with Ag, Al, and Au. The PL peak intensity of uncoated InGaN/GaN QW at 470 nm was normalized to 1

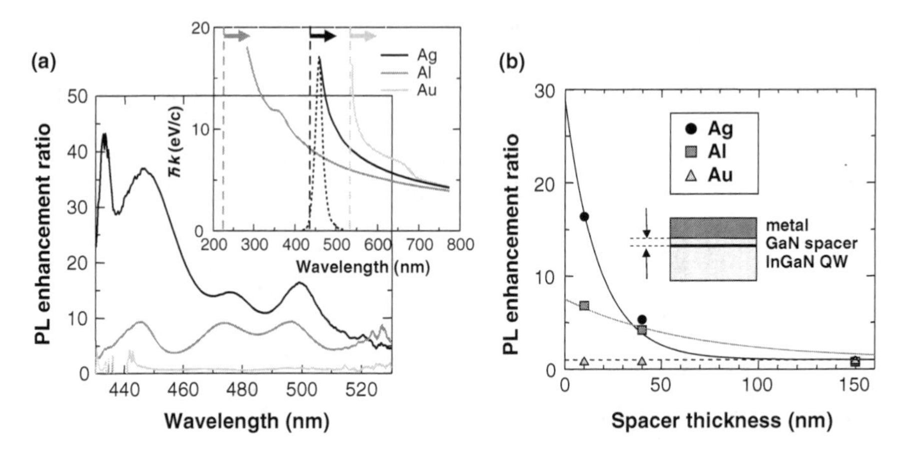

Fig. 2.4 (**a**) PL enhancement ratios at several wavelengths for the same samples as in Fig. 2.3b. (Inset) Dispersion diagrams of surface plasmons generated on Ag/GaN, Al/GaN, and Au/GaN surfaces. (**b**) Integrated PL enhancement ratios for samples with Ag, Al, and Au are plotted against the thicknesses of GaN spacers. The *solid lines* are the calculated values by the penetration depths using Eq. (2.2)

We have several evidences to support the contribution of the SPs to obtained PL enhancements. Figure 2.4a shows the enhancement ratios of PL intensities with metal layers separated from the QWs by 10 nm spacers as a function of wavelength. We find that the enhancement ratio increases at shorter wavelengths for Ag samples, whereas it is independent of wavelength for Al-coated samples. The PL enhancement after coating with Ag and Al can be attributed to strong interaction with SPs. The inset figure shows the dispersion diagrams of SP on metal/GaN surfaces (similar to Fig. 2.2a). The surface plasmon frequency (ω_{SP}) at GaN/Ag is 2.84 eV (437 nm). Thus, Ag is suitable for SP coupling to blue emission, and we attribute the large increases in the PL intensity from Ag-coated samples to such resonant SP excitation. In contrast, ω_{SP} at GaN/Au is 2.462 eV (537 nm), and no measurable enhancement is observed in Au-coated InGaN emitters as the SP and QW energies are not matched. In the case of Al, the ω_{SP} is 5.50 eV (225 nm), and the real part of the dielectric constant is negative over a wide wavelength region for visible light. Thus, a substantial and useful PL enhancement is observed in Al-coated samples, although the energy match is not ideal at 470 nm and a better overlap is expected at shorter wavelengths. The clear correlation between Fig. 2.4a and the dispassion diagrams suggests that the obtained emission enhancement with Ag and Al is due to the SP coupling.

PL intensities of Al- and Ag-coated samples were also found to depend strongly on the distance between QWs and the metal layers, in contrast to Au-coated samples. Figure 2.2b compares integrated PL enhancement ratios for three different GaN spacer thicknesses (10, 40, and 150 nm) for Ag, Al, and Au coatings. Al and Ag samples show exponential decreases in the PL intensity as the spacer thickness is increased, whereas no such reduction was measured in Au-coated QWs. This spacer-layer dependence of the PL enhancement ratios matches our models of SP–QW coupling, as the SPP should be localized at the metal/dielectric interface and exponentially decays with distance from the metal surface. Only electron–hole pairs located within the near-field of the surface can couple to the SPP mode, and this penetration depth (d_z) of the SP fringing field into the semiconductor is given by Eq. (2.2). d_z can be calculated as 18 and 63 nm for Ag and Al, respectively. Figure 2.2b shows a good agreement between these calculated penetration depths (lines) and measured values of the PL enhancement (symbols) for Ag- and Al-coated samples.

4 Surface Plasmon Coupling Mechanism

We propose a possible mechanism of the QW–SP coupling and the light extraction shown in Fig. 2.5 [24]. First, electron–hole pairs are generated in the QW by photo-pumping or electrical pumping. For uncoated samples, these carriers are terminated by the radiative or nonradiative recombination rates, and the IQE is determined by the ratio of these two rates (Eq. (2.3)). When a metal layer is grown within the near-field of the active layer and when the bandgap energy ($\hbar\omega_{BG}$) of InGaN active layer is close to the electron oscillation energy ($\hbar\omega_{SP}$) of SP at the metal/semiconductor surface, the QW energy can transfer to the SP. PL decay rates are enhanced through the QW–SP coupling rate (k_{SPC}), as k_{SPC} values are expected to be very fast.

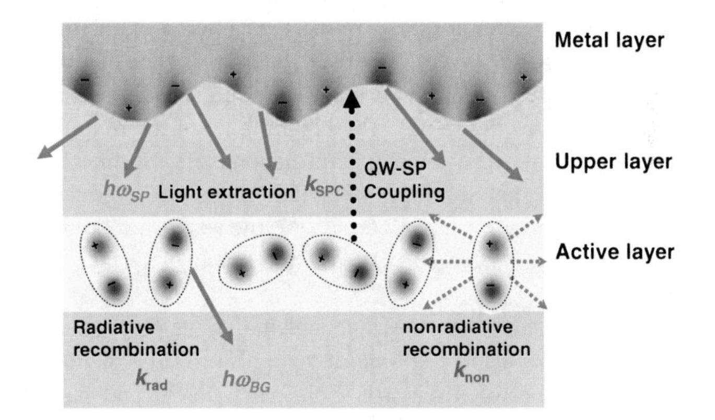

Fig. 2.5 Schematic diagram of the electron–hole recombination and QW–surface plasmon (SP) coupling mechanism

The QW–SP coupling in LED devices may be considered detrimental to the optical efficiency because the SP is a nonradiative wave. If the metal/semiconductor surface were perfectly flat, it would be difficult to extract light from the SPP mode and the SP energy would be thermally dissipated. However, roughness and imperfections in evaporated metal coatings can efficiently scatter SPs as light. However, the SP energy can be extracted as light by providing roughness or nanostructuring

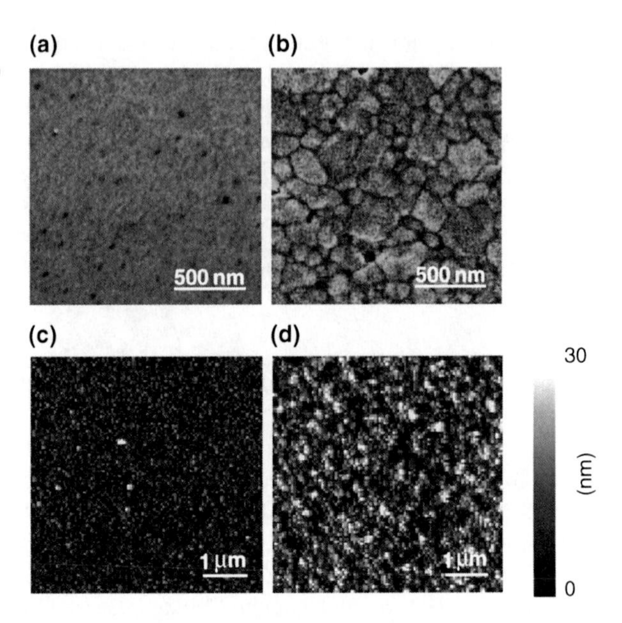

Fig. 2.6 (**a**) SEM image of the uncoated GaN surface. (**b**) SEM image of the 50 nm Ag film evaporated on GaN. (**c**) Topographic image of the uncoated GaN surface. (**d**) Topographic image of a 50-nm-thick Ag film evaporated on GaN. All samples used were coated onto InGaN/GaN QWs with 10 nm GaN spacers

the metal layer. Such roughness allows SPs of high momentum to scatter, lose momentum, and couple to radiated light [25]. The few tens of nanometer-sized roughness in the Ag surface layer can be obtained by controlling the evaporation conditions or by nanofabrication to obtain the high photon extraction efficiencies. Such roughness in the metal layer was observed from higher-magnification scanning electron microscopy (SEM) images of the original GaN surface (Fig. 2.6a) and the Ag-coated surface (Fig. 2.6b). The length scale of the roughness of the Ag surface was determined to be a few hundred nanometers. Similar roughness was also observed from topographic images obtained by shear-force microscopy of the GaN and Ag surfaces shown in Fig. 2.6c, d. We measured a modulation depth of the Ag surface of approximately 30–40 nm while the GaN surface roughness was below 10 nm.

In order to evaluate the SP coupling mechanism that we proposed, we employed a three-dimensional finite-difference time-domain (3D-FDTD) method to represent the coupling processes between electron–hole pairs, SPPs, and photons. To perform 3D-FDTD simulations, we used "Poynting for optics" (Fujitsu Co.) which is known to be very suitable to simulate SP modes [26]. A polarized plane wave with 525 nm wavelength and 1 V/m amplitude was used as a point light source which is an assumption of an electron–hole pair. Figure 2.7 shows the calculated spatial distribution of the electromagnetic field around the Ag/GaN interface. If the point light source was far from the interface, the SPP mode was not excited (Fig. 2.7a). On the other hand, if the point light source was located near the interface, the SPP was well generated and propagated within the interface (Fig. 2.7b). This result suggests that the SPP mode can be generated easily by direct energy transfer from electron–hole pairs without any special structures. Usually, some special configurations are

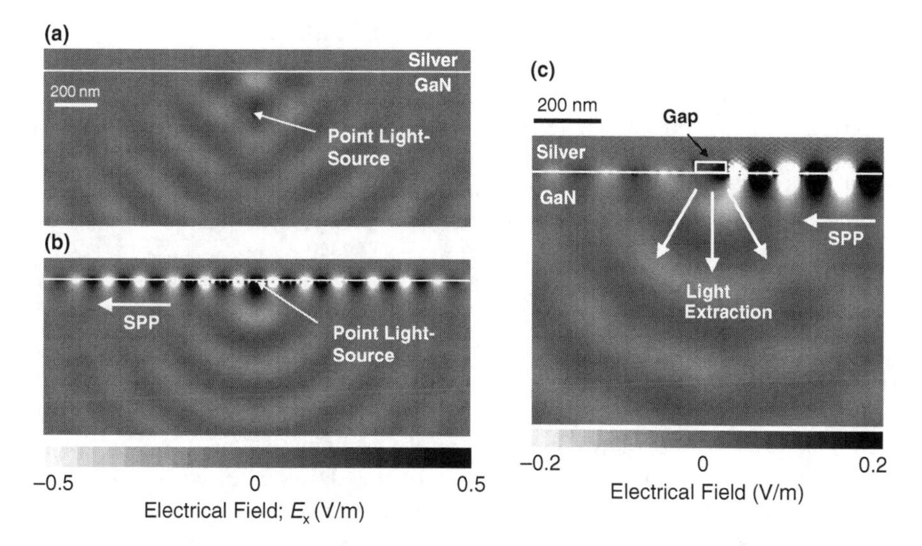

Fig. 2.7 3D-FDTD simulations of generation and light extraction of SPPs. (**a**) Point light source was located at 200 nm below the Ag/GaN interface. (**b**) Point light source was located on the interface. SSP was generated and propagated on the interface. (**c**) Generated SSP was extracted as light at the gap in the metal layer

necessary to generate the SPP mode such as a grating coupler or an attenuated total reflection (ATR) setting to satisfy a phase matting condition between the SPPs and the photons. However, if the light source is located near the metal/dielectric interface within wavelength scale, the SPP mode can be generated regardless of the phase matching condition. The generated SPP mode can be coupled to a photon if there is a nanosized gap structure at the interface (Fig. 2.7c). Then, generated surface plasmon can be extracted from the interface as light and the emission efficiencies should be increased. These calculations support our proposed SP coupling model.

Under the existence of the SP coupling, the enhanced IQE of emission can be described as follows:

$$\eta_{\text{int}}^{*}(\omega) = \frac{k_{\text{rad}}(\omega) + C'_{\text{ext}}(\omega) k_{\text{SPC}}(\omega)}{k_{\text{rad}}(\omega) + k_{\text{non}}(\omega) + k_{\text{SPC}}(\omega)}, \tag{2.4}$$

where $k_{\text{SPC}}(\omega)$ is the SP coupling rate and should be very fast because the density of states of SP modes is much larger than that of the electron–hole pairs in the QW. $C'_{\text{ext}}(\omega)$ is the probability of photon extraction from the SPs energy. $C'_{\text{ext}}(\omega)$ is decided by the ratio of light scattering and dumping of the SPP mode through nonradiative loss. $C'_{\text{ext}}(\omega)$ should depend on the roughness and nanostructure of the metal surface. If the SP coupling rate k_{SPC} is much faster than k_{rad} and k_{non}, the IQE should be dramatically increased.

5 Improvements of IQEs and Emission Rates

Our proposed model suggests that the SP coupling should increase IQEs of emissions. In order to obtain the IQE values to separate the SP enhancement from other possible effects, we have measured the temperature dependence of the PL intensity [23]. Figure 2.8 shows Arrhenius plots of the integrated PL intensities from InGaN QWs separated from Ag and Al films by 10 nm spacers, and compares these to uncoated samples. The IQE values from uncoated QWs were estimated as 6% at room temperature by assuming $\eta_{\text{int}} \sim 100\%$ at 4.2 K. These IQE values increased 6.8 times (to 41%) after Ag coating and 3 times (to 18%) after Al coating, explainable by spontaneous recombination rate enhancements through SP coupling. The 6.8-fold increasing of the IQE means that 6.8-fold improvement of the efficiency of electrically pumped LED devices should be achievable because an IQE is a fundamental property and does not depend on the pumping method. Such improved efficiencies of the white LEDs, in which a blue LED is combined with a yellow phosphor, are expected to be larger than those of current fluorescent lamps or light bulbs.

Quite recently, a few groups reported about the SP enhanced LEDs based on our technique. Yeh et al. reported the SP coupling effect in an InGaN/GaN single-QW LED structure [27]. Their LED structure has a 10 nm p-type AlGaN current blocking layer and a 70 nm p-type GaN layer between the metal surface and the InGaN QW layer. The total distance is 80 nm, which is too far to obtain an effective SP coupling. By this reason, they obtained only 1.5-fold enhancement of the emission. Kwon et al. put metal particle on the InGaN QW layer first, and overgrew a GaN layer above the metal particles [28]. However, a large amount of metal particles were gone by high temperature of the crystal growth and only 3% particles remained. Therefore, they obtained only 1.3-fold enhancement of the emission. These tiny enhancement ratios should not be good enough for device application. Therefore, a highly efficient LED structure based on plasmonics is still not yet achieved.

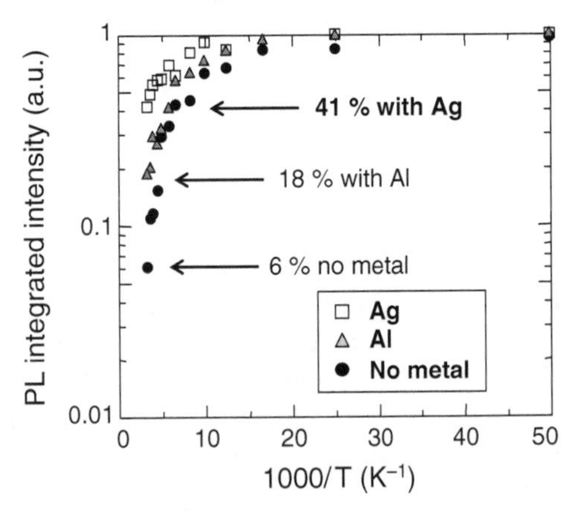

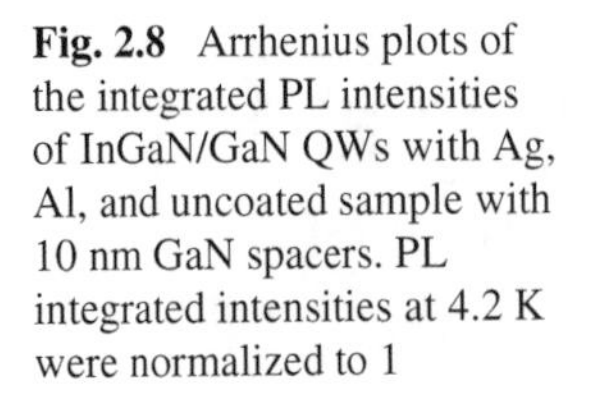
Fig. 2.8 Arrhenius plots of the integrated PL intensities of InGaN/GaN QWs with Ag, Al, and uncoated sample with 10 nm GaN spacers. PL integrated intensities at 4.2 K were normalized to 1

The increased IQE should be due to the enhancement of the spontaneous emission rate. Since the density of states of SP mode is much larger, the QW–SP coupling rate should be very fast, and this new path of a recombination can increase the spontaneous emission rate. We investigated the direct observation of SP-coupled spontaneous emission rate by using the time-resolved PL measurements [24]. To perform time-resolved PL measurements, the frequency-doubled output from a mode-locked Ti:Al_2O_3 laser was used to excite the InGaN QW from the bottom surface of the wafer. The pulse width, wavelength, and repetition rate were chosen as 1.5 ps, 400 nm, and 80 MHz, respectively. A Hamamatsu Photonics C5680 streak camera served as the detector.

Figure 2.9a, b shows the time-resolved PL decay profiles of (a) uncoated and (b) Ag-coated InGaN-GaN QW sample emitters at several wavelengths. All profiles could be fitted to single exponential functions and the spontaneous emission rate (k_{PL}) was obtained. The PL decay profile of each sample was quite different and the k_{PL} values of Ag-coated sample were larger than those of uncoated sample. Also, we found that the decay profiles of the Ag-coated sample strongly depend on the wavelength and become faster at shorter wavelengths, whereas those of the uncoated sample show little spectral dependence. We attribute the increase in both emission intensities and decay rates from Ag-coated samples to the SP coupling.

The original spontaneous emission rate is attributed to the radiative and nonradiative recombination rates of the electron–hole pairs in the QW.

$$k_{PL}(\omega) = k_{rad}(\omega) + k_{non}(\omega). \tag{2.5}$$

By the SP coupling, the spontaneous emission rate should be increased to

$$k^*_{PL}(\omega) = k_{rad}(\omega) + k_{non}(\omega) + k_{SPC}(\omega). \tag{2.6}$$

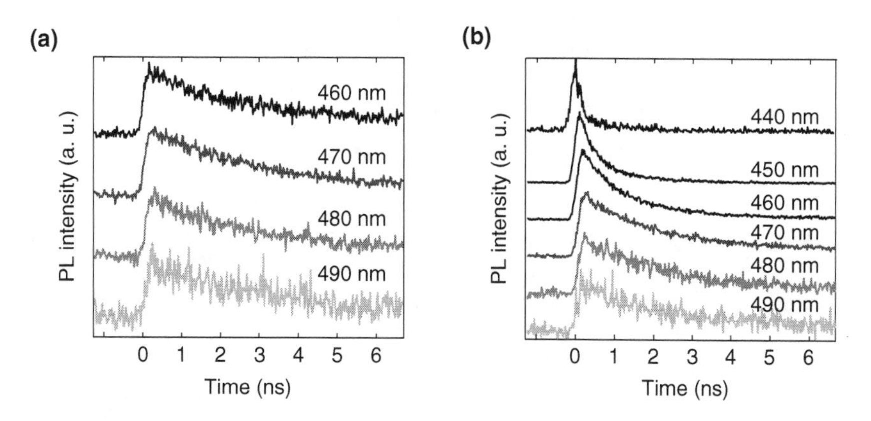

Fig. 2.9 (**a**) Photoluminescence (PL) decay profiles of uncoated InGaN–GaN QW at several wavelengths. (**b**) PL decay profiles of Ag-coated InGaN/GaN QW at several wavelengths. The distance between the Ag layers and QWs was 10 nm

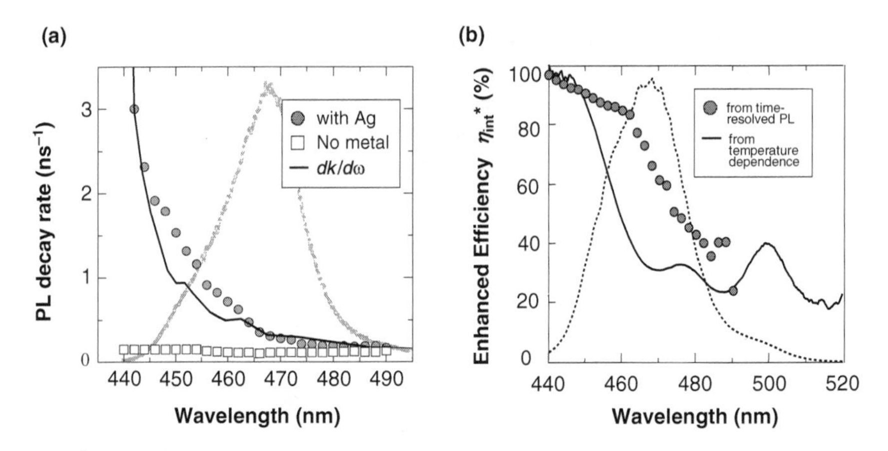

Fig. 2.10 (**a**) The spontaneous emission rates of InGaN/GaN with/without silver layer plotted against wavelength. The *solid line* is dk/dω of the SPP mode at the silver/GaN interface obtained by the dispersion curve (Fig. 2.2a). The *dashed line* is the emission spectrum. (**b**) Wavelength-dependent internal quantum efficiencies of the InGaN/GaN with Ag layer with 10 nm GaN spacers estimated by the ratios of the emission rates (*marks*) and the temperature dependence of the PL intensities (*line*). The *dotted black line* is the PL spectrum of the same sample

The observed $k_{\rm PL}$ and $k^*_{\rm PL}$ values were plotted against wavelength in Fig. 2.10a. The emission rates of Ag-coated sample were much faster than those of the uncoated sample and strongly depend on the wavelength. This difference becomes dramatically larger at the shorter wavelength region. The spontaneous emission rate into the SPP mode (SP coupling rate) depends on the density of states of the SPP by Fermi's golden rule [16, 17]. The density of states of the SPP mode is proportional to $\mathrm{d}k/\mathrm{d}\omega$ which can be obtained by the dispersion curve. $\mathrm{d}k/\mathrm{d}\omega$ is also plotted in Fig. 2.10a as the solid line. The SP coupling rate should be almost equal to the PL decay rate with Ag layers because those values were much larger than the values of the PL decay rate without Ag. Figure 2.10a shows that the wavelength dependence of the SP coupling rates is similar to that of $\mathrm{d}k/\mathrm{d}\omega$.

Figure 2.10b shows the enhanced IQEs ($\eta^*_{\rm int}$) estimated by the ratios between $k_{\rm PL}$ and $k^*_{\rm PL}$ with Eq. (2.4) under $C'_{\rm ext}(\omega) = 1$. The SP coupling becomes remarkable when the energy is near to the SP frequency described in Fig. 2.2a as 2.84 eV (437 nm). At this shorter wavelength region, the SP coupling rates are much faster than the radiative or nonradiative recombination rates of electron–hole pairs ($k_{\rm SPC} >> k_{\rm rad} + k_{\rm non}$), and the $\eta^*_{\rm int}$ values are reached to almost 100%. Wavelength-dependent $\eta^*_{\rm int}$ values were estimated also from the temperature dependence of the PL intensities (Fig. 2.8) and plotted in Fig. 2.10b (solid line). Both the data show similar behavior. The discrepancy of each data should be due to the light extraction probability from the SPP. $\eta^*_{\rm int}$ estimated by the temperature-dependent measurements of the PL intensities should include the damping energy loss of the SPP. The important fact is that both values are reached to almost 100% at the shorter wavelength region. This suggests one of the most important advantages of the SP coupling technique to enhance the emission efficiencies. If we can control the SP frequency and obtain the best matching condition between the emission wavelength and the SP frequency, we can increase both the η^*_{int} and C'_{ext} to 100% at any wavelength. It is perfect efficiency and would bring full color devices and natural white LEDs. Tuning of SP coupling should be available by choosing the appropriate metal, metal mixture alloy, multiple layers, or nanostructures. For example, we could improve the green emission of InGaN by fabricating the nano-grating structures of gold layer by E-beam lithography and Ar ion milling [29]. A theoretical study was also reported by Paiella to tune the SP frequency by using metallo-dielectric multiple layers [30].

6 Applications for Organic Light-Emitting Materials

The most important advantage of the SP coupling technique is that the technique can be applied not only to InGaN-based materials but also to various materials. Therefore, we have used this technique for various other light-emitting materials. For example, polymers, appropriately doped with dye molecules, emitting in the visible spectrum provide stable sources of light for displays and illumination sources at a significantly lower cost than semiconductors. Organic light-emitting diodes (OLEDs) have become widely available and are used for replacing inorganic light-emitting diodes as they are less expensive and provide many opportunities with regard to structural placement. Despite the tremendous promise for efficient solid-state lighting offered by such organic light emitters, the road toward spectrally broad white light polymer emitters still holds many design challenges. Thus, it is of both commercial and scientific interest to improve the IQEs of the polymer dyes within such light emitters, as well as to increase the light extraction efficiencies from such organic films. Here, we focus on enhancing the light emission efficiency from organic thin films by using the SP coupling [31].

The experimental setup used to measure our samples is shown in Fig. 2.11a. Dye polymer solution was prepared by dissolving common laser dye molecules of Coumarin 460 in chlorobenzene. This laser dye emits blue light at 460 nm with UV excitation. Then, 2% polymethylmethacrylate (PMMA) was added to the mixture as a host matrix to obtain a 20 mM/L solution of the dye doped polymer solution. Only half of each substrate was metallized, enabling the rapid comparison between polymer emission on top of metal layers with polymer deposited on quartz. After the metallization step, the dye doped PMMA layers were spun onto both gold and silver substrates to obtain layer thicknesses of ~200 nm.

Figure 2.11b shows typical PL spectra of Coumarin 460 on Ag, Au, and bare quartz substrate. While the Au assisted in reflecting the pump laser, the surface plasmons did not seem to couple to the emission wavelength of Coumarin 460 to offer any measurable enhancement. However, we do observe an 11-fold enhancement of the emission light from the Coumarin doped PMMA on silver due the coupling of the surface plasmons generated on the Ag film as the plasmon resonance frequency

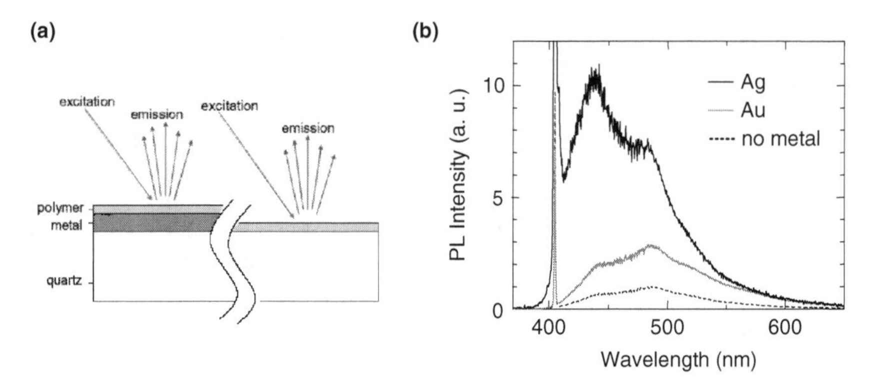

Fig. 2.11 (**a**) Sample structure of dye doped polymer with both pump light and emission light configurations. (**b**) PL spectra of Coumarin 460 on Ag, Au, and quartz. The PL peak intensity of Coumarin 460 on quartz was normalized to 1

closely matches the emission frequency of the dye. Indeed, the dielectric constants for Ag match well with the emission wavelength of Coumarin 460, and if the data with the Coumarin 460 PL intensity normalized to 1. While reflection can be used to account for some of the increased brightness, only the SP coupling can explain the enhancement measured.

Likewise, we obtained obvious enhancements of both PL intensities and emission rates for three conjugated polymers: polyfluorenes (PF)-cyanophenylene(CNP) (1:1), PF-CNP (3:1), and polyfluorenes(PF)-triphenylamine(TPA)-quinoline(Q) [32]. These polymers have been used for OLEDs actually as high-efficient light-emitting materials [33, 34].

7 Applications for CdSe-Based Quantum Dots

CdSe-based quantum dot (QD) nanocrystals are also very promising materials for light-emitting sources. CdSe-based QD nanocrystals possess a number of advantageous features and have been used in LEDs [35, 36] and as biological fluorescent labels [37, 38]. However, their light emission efficiencies are still substantially lower than those of fluorescent tubes. Therefore, we investigate the direct observation of SP-coupled spontaneous emission from CdSe-based QDs [39]. CdSe-based QDs were purchased from Evident Technologies. These QDs have an emission peak around 620 nm and a crystal diameter of approximately 5 nm. The toluene solutions of the QDs were dispersed on quartz substrates. After the solutions evaporated, a monolayer of the QD nanocrystals remained on the substrates. The half parts of the quartz surface were covered by a 50 nm gold layer by thermal evaporation. The sample structure is shown in Fig. 2.12a.

We used two types of nanocrystals: one was naked CdSe nanocrystals and other was CdSe core with ZnS shells (CdSe/ZnS). The IQE of naked CdSe ($\sim$2%) was

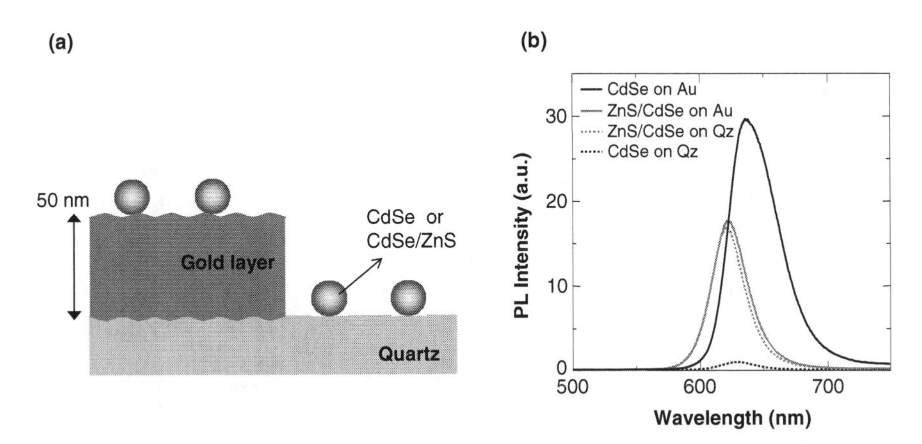

Fig. 2.12 (**a**) Sample structures of CdSe nanocrystals on Au-coated quartz chips. (**b**) PL spectra for CdSe and CdSe/ZnS nanocrystals on Au and quartz (Qz)

well increased for CdSe/ZnS structure (~40%) because generated carriers can be well confined into core/shell structures. Figure 2.12b shows PL spectra of naked CdSe and CdSe/ZnS on gold layers and quartz substrate. A dramatic enhancement in the PL intensity from the QDs on gold layer was very clearly observed for naked CdSe. When the PL peak of the QDs on quartz was normalized to 1, a 30-fold increase of PL intensity was observed. On the other hand, the enhancement of PL intensity of CdSe/ZnS was not remarkable compared with the result of naked CdSe without shells. This fact indicates the merit and demerit of the SP coupling technique for enhancing light emission. The SP coupling increases IQE values by enhancement of spontaneous emission rates. The SP coupling condition is decided by the matching of energies between the SP frequency and the emission wavelength. Thus, the enhancement condition does not depend on the intrinsic IQE values of materials. This feature suggests that the SP-coupling technique is very effective for increasing the emission efficiency of materials with low intrinsic efficiency like naked CdSe, but not so effective for high-efficiency materials like CdSe/ZnS, which were used in this study.

The SP enhanced luminescence of CdSe QDs has been reported by a few groups. Kulakovich et al. [40] reported 5-fold enhancement of the PL intensity for CdSe/ZnS QD and gold colloids. Song et al. [41] achieved ~50-fold enhancement by using CdSe/ZnS QDs and nanoperiodic silver arrays fabricated by electron-beam lithography. Grycczynski et al. [42] reported a well-polarized, directional, and photostable SP coupling emission by using CdSe/ZnS QDs on SiO_2/silver thin layers. Compared with these reports, our setup is much simpler and easier. We used naked CdSe and an evaporated gold layer, but we still obtained remarkable enhancement (30-fold). Special geometry or nanoperiodic structures are not necessary for our setup.

8 Applications for Silicon-Based Nanocrystals

The SP coupling technique can be applied to materials that suffer from low emission efficiencies, which include the indirect semiconductor. Usually, the emission efficiencies of such indirect semiconductors are quite low but it is possible to enhance these efficiencies to values as large as those available from direct compound semiconductors by SP enhancement. Accordingly, we tried to enhance emissions from silicon-based semiconductors.

Silicon photonics has attracted a great deal of attention in this decade and is expected as a light-emitting material alternative to compound semiconductors. Several nanostructures such as porous silicon [43], nanocrystals [44], quantum wells [45], and nanowires [46] were fabricated to obtain bright emissions from Si. We tried to enhance emission from Si nanocrystals in SiO_2 media with gold thin layers [47]. Silicon nanocrystal QDs were prepared by reactive thermal evaporation of SiO powders in an oxygen atmosphere under vacuum. After rapid thermal annealing, size-controlled Si nanocrystals ($\sim$3 nm diameter) were formed in SiO_2 by phase separation. This technique was developed by Zacharias et al. and the details have already been published [48]. Metal thin layers (50 nm) were prepared by thermal evaporation. Figure 2.13a shows a sample structure and Fig. 2.13b shows the PL spectra for Si nanocrystals. A 70-fold large PL enhancement was observed with gold coating at the wavelength region longer than 650 nm, whereas only 2-fold enhancement was obtained from aluminum-coated sample. This should be reasonable because the calculated dispersion diagram of the SP at Au/SiO_2 interface suggests that the SP coupling must be effective at a longer wavelength region than 600 nm. On the other hand, the SP at Al/SiO_2 is not effective around this wavelength region. It should be effective at much shorter wavelength region.

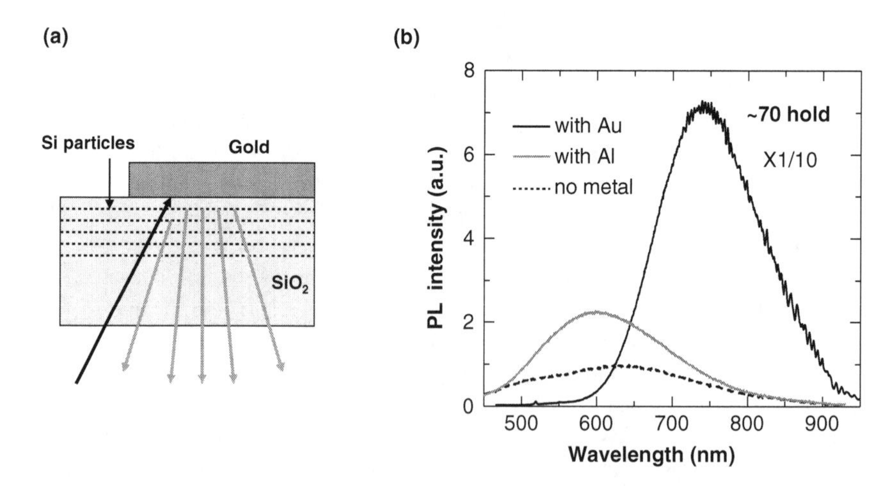

Fig. 2.13 (**a**) Sample structure of Si nanoparticles dispersed in SiO_2 media and excitation/emission configuration of PL measurement. (**b**) PL spectra of Si/SiO_2 with Au, Al, and no metal layer

After our first report of the SP coupling technique, it was already applied to Si nanocrystals and similar enhancements have been reported at room temperature [49, 50]. We measured temperature dependence of PL intensities to estimate the enhanced IQE values [47]. The IQE value from uncoated Si/SiO_2 was estimated as 6% at room temperature by this assumption. The IQE value increased to 36% after Au coating, explainable by spontaneous emission rate enhancements through SP coupling. This value is as large as that of a compound semiconductor with direct transition. However, the emission intensity of Si/SiO_2 was still much weaker than that of InGaN/GaN or CdSe/ZnS with the same IQE value. It was reported that the emission lifetimes of Si/SiO_2 were usually very long (∼ms) even though some of Si nanocrystals have very high IQE values (>50%) [51–53]. The SP coupling can enhance the emission rate, but the enhanced emissions still have long lifetimes with millisecond scale [49, 50]. These lifetimes are 1,000 times longer than those of InGaN/GaN or CdSe/ZnS which has similar IQE values. The slower emission rates should be the reason for weak emission intensities of Si/SiO_2. The excitation densities of nanocrystals become saturated easily and this brings poor carrier injection efficiencies in spite of their high IQE values. Due to this reason, so far, silicon-based materials are still not useful for light-emitting materials. We believe that both emission rates and excitation densities of Si nanocrystal can be increased by optimizing the SP coupling condition and it would bring super bright silicon LEDs, which could be very cheap to make, easy to process.

9 Conclusions

We conclude that the SP enhancement of PL intensities of light emitters is a very promising method for developing highly efficient LEDs. We have directly measured significant enhancements of IQE and the spontaneous recombination rate. Even when using unpatterned metal layers, the SP energy can be extracted by the submicron scale roughness on the metal surface. A possible mechanism of the QW–SP coupling and emission enhancement has been proposed and highly efficient light emission is predicted for optically as well as electrically pumped light emitters because the mechanism should not be related to the pumping method. Enhanced spontaneous emission should also be very useful for high-speed light-emitting devices for the development of communication technology and optical computing. Moreover, similar plasmonic design should also be applicable to devices based on nonlinear optical materials, photo detectors, waveguides, optical modulators, plasmonic metamaterials, and other optical and electric devices. This technique is very simple and easy, and moreover, can be applicable to various materials that suffer from low quantum efficiencies. So far as we think, the SP coupling may be the only technique with a big possibility of developing the super bright light-emitting devices by the use of the silicon-based semiconductors. We believe that the QW–SP coupling technique would bring super bright plasmonic LEDs, which become the dominant white light source and serve as an alternative to conventional fluorescent tubes.

Acknowledgments The author wishes to thank Professor Y. Kawakami (Kyoto University) and Professor A. Scherer (Caltech) for valuable discussions and support. The author also thanks Mr. A. Shavartser, Dr. T. D. Neal, and Dr. S. Vyawahare for collaboration and help. InGaN/GaN materials were provided by Mr. I. Niki, Dr. Y. Narukawa, and Dr. T. Mukai (Nichia Co.). A part of this study was supported by the Precursory Research for Embryonic Science and Technology (PRESTO), Japan Science and Technology Agency (JST).

References

1. Raether, H.: Surface Plasmons on Smooth and Rough Surfaces and on Gratings. Springer, Berlin (1988)
2. Shalaev, V.M., Kawata, S.: Nanophotonics with Surface Plasmons. Elsevier, Amsterdam (2007)
3. Brongersma, M.L., Kik, P.G.: Surface Plasmon Nanophotonics. Springer, Berlin (2007)
4. Maier, S.A.: Plasmonics: Fundamentals and Applications. Springer, Berlin (2007)
5. Barnes, W. L., Dereux, A., Ebbesen, T.W.: Surface plasmon subwavelength optics. Nature **424**, 824–830 (2003)
6. Dostalek, J., Homola, J., Jiang, S., Ladd, J.: Surface Plasmon Resonance Based Sensors. Springer, Berlin (2006)
7. Fleischmann, M., Hendra, P.J., McQuillan, A.J.: Raman spectra of pyridine adsorbed at a silver electrode. Chem. Phys. Lett. **26**, 163–166 (1974)
8. Kawata, S.: Near-Field Optics and Surface Plasmon Polaritons. Springer, Berlin (2001)
9. Fang, N., Lee, H., Sun, C., Zhang, X.: Sub-diffraction-limited optical imaging with a silver superlens. Science **308**, 534–537 (2005)
10. Atwater, H.A.: The promise of plasmonics. Scientific American, pp. 56–63 (April 2007)
11. Nakamura, S., Mukai, T., Senoh, M., Nagahama, S., Iwasa, N.: P-GaN/N-InGaN/N-GaN double-heterostructure blue-light-emitting diodes, Jpn. J. Appl. Phys. **32**, L8 (1993)
12. Nakamura, S., Fasol, G.: The Blue Laser Diode: GaN Based Light Emitting Diode and Lasers. Springer, Berlin (1997)
13. Yamada, M., Mitani, T., Narukawa, Y., Shioji, S., Niki, I., Sonobe, S., Deguchi, K., Sano, M., Mukai, T.: InGaN-based near-ultraviolet and blue-light-emitting diodes with high external quantum efficiency using a patterned sapphire substrate and a mesh electrode, Jpn. J. Appl. Phys. **41**, L1431–L1433 (2002)
14. Mukai, T., Takekawa, K., Nakamura, S.: InGaN-based blue light-emitting diodes grown on epitaxially laterally overgrown GaN substrates, Jpn. J. Appl. Phys. **37**, L839–L841 (1998)
15. Walterelt, P., Brandt, O., Trampert, A., Grahn, H.T., Menniger, J., Ramsteiner, M., Reiche, M., Ploog, K.H.: Nitride semiconductors free of electrostatic fields for efficient white light-emitting diodes. Nature **406**, 865–868 (2000)
16. Köck, A., Gornik, E., Hauser, M., Beinstingl, M.: Strongly directional emission from AlGaAs/GaAs light-emitting diode. Appl. Phys. Lett. **57**, 2327–2329 (1990)
17. Hecker, N.E., Hopfel, R.A., Sawaki, N., Maier, T., Strasser, G.: Surface plasmon-enhanced photoluminescence from a single quantum well. Appl. Phys. Lett. **75**, 1577–1579 (1999)
18. Barnes, W.L.: Electromagnetic crystals for surface plasmon polaritons and the extraction of light from emissive devices. J. Light. Tech. **17**, 2170–2182 (1999)
19. Hobson, P.A., Wedge, S., Wasey, J.A.E., Sage, I., Barnes, W.L.: Surface plasmon mediated emission from organic light emitting diodes. Adv. Mater. **14**, 1393–1396 (2002)
20. Vuckovic, J., Loncar, M., Scherer, A.: Surface plasmon enhanced light-emitting diode. IEEE J. Quantum Electron **36**, 1131–1144 (2000)
21. Gontijo, I., Borodisky, M., Yablonvitch, E., Keller, S., Mishra, U.K., DenBaars, S.P.: Coupling of InGaN quantum-well photoluminescence to silver surface plasmons. Phys. Rev. B **60**, 11564–11567 (1999)

22. Neogi, A., Lee, C.-W., Everitt, H.O., Kuroda, T., Tackeuchi, A., Yablonvitch, E.: Enhancement of spontaneous recombination rate in a quantum well by resonant surface plasmon coupling. Phys. Rev. B **66**, 153305 (2002)
23. Okamoto, K., Niki, I., Shvartser, A., Narukawa, Y., Mukai, T., Scherer, A.: Surface-plasmon-enhanced light emitters based on InGaN quantum wells. Nat. Mater. **3**, 601–605 (2004)
24. Okamoto, K., Niki, I., Narukawa, Y., Mukai, T., Kawakami, Y., Scherer, A.: Surface plasmon enhanced spontaneous emission rate of InGaN/GaN quantum wells probed by time-resolved photoluminescence spectroscopy. Appl. Phys. Lett. **97**, 071102 (2005)
25. Barnes, W.L.: Light-emitting devices: Turning the tables on surface plasmons. Nat. Mater. **3**, 588–589 (2004)
26. Nomura, W., Ohtsu, M., Yatsui, T.: Nanodot coupler with a surface plasmon polariton condenser for optical far/near-field conversion. Appl. Phys. Lett. **86**, 181108 (2005)
27. Yeh, D.-M., Huang, C.-F., Chen, C.-Y., Lu, Y.-C., Yanga, C.C.: Surface plasmon coupling effect in an InGaN/GaN single-quantum-well light-emitting diode. Appl. Phys. Lett. **91**, 171103 (2007)
28. Kwon, M.-K., Kim, J.-Y., Kim, B.-H., Park, I.-K., Cho, C.-Y., Byeon, C.C., Park, S.-J.: Surface-plasmon-enhanced light-emitting diodes. Adv. Mater. **20**, 1253–1257 (2008)
29. Okamoto, K., Niki, I., Shvartser, A., Maltezos, G., Narukawa, Y., Mukai, T., Kawakami, Y., Scherer, A.: Surface plasmon enhanced bright light emission from InGaN/GaN. Phys. Stat. Sol. (a) **204**, 2103–2107 (2007)
30. Paiella, R.: Tunable surface plasmons in coupled metallo-dielectric multiple layers for light-emission efficiency enhancement. Appl. Phys. Lett. **87**, 111104 (2005)
31. Neal, T.D., Okamoto, K., Scherer, A.: Surface plasmon enhanced emission from dye doped polymer layers. Opt. Express **13**, 5522–5527 (2005)
32. Neal, T.D., Okamoto, K., Scherer, A., Liu, M.S., Jen, A.K-Y.: Time-resolved photoluminescence spectroscopy of surface-plasmon-enhanced light emission from conjugate polymers. Appl. Phys. Lett. **89**, 221106 (2006)
33. Liu, M.S., Jiang, X., Herguth, P., Jen, A.K.-Y.: Efficient cyano-containing electron-transporting polymers for light-emitting diodes. Chem. Mater. **13**, 3820–3822 (2001)
34. Shu, C.-F., Dodda, R., Wu, F.-I., Liu, M.S., Jen, A.K.-Y.: Highly efficient blue-light-emitting diodes from polyfluorene containing bipolar pendant groups. Macromolecules **36**, 6698–6703 (2003)
35. Colvin, V.L., Schlamp, M.C., Ailvisatos, A.P.: Light-emitting diodes made from cadmium selenide nanocrystals and a semiconducting polymer. Nature **370**, 354–357 (1994)
36. Dabbousi, B.O., Bawendi, M.G., Onitsuka, O., Rubner, M.F.: Electroluminescence from CdSe quantum-dot/polymer composites. Appl. Phys. Lett. **66**, 3116–3118 (1995)
37. Bruchez, M. Jr., Moronne, M., Gin, P., Weiss, S., Alivisatos, A.P.: Semiconductor nanocrystals as fluorescent biological labels. Science **281**, 2013–2016 (1998)
38. Chan, W., Nie, S.: Quantum dot bioconjugates for ultrasensitive nonisotopic detection. Science **281**, 2016–2018 (1998)
39. Okamoto, K., Vyawahare, S., Scherer, A.: Surface plasmon enhanced bright emission from CdSe quantum dots nanocrystal. J. Opt. Soc. Am. B **23**, 1674–1678 (2006)
40. Kulakovich, O., Strekal, N., Yaroshevich, A., Maskevich, S., Gaponenko, S., Nabiev, I., Woggon, U., Artemyev, M.: Enhanced luminescence of CdSe quantum dots on gold colloids. Nano Lett. **2**, 1449–1452 (2002)
41. Song, J.-H., Atay, T., Shi, S., Urabe, H., Nurmikko, A.V.: Large enhancement of fluorescence efficiency from CdSe/ZnS quantum dots induced by resonant coupling to spatially controlled surface plasmons. Nano Lett. **5**, 1227–1561 (2005)
42. Gryczynski, I., Malicka, J., Jiang, W., Fischer, H., Chan, W.C.W., Gryczynski, Z., Grudzinski, W., Lakowicz, J.R.: Surface-plasmon-coupled emission of quantum dots. J. Phys. Chem. B **109**, 1088–1093 (2005)
43. Cullis, A.G., Canham, L.T.: Visible light emission due to quantum size effects in highly porous crystalline silicon. Nature **353**, 335–338 (1991)

44. Brongersma, M.L., Polman, A., Min, K.S., Boer, E., Tambo, T., Atwater, H.A.: Tuning the emission wavelength of Si nanocrystals in SiO_2 by oxidation. Appl. Phys. Lett. **72**, 2577–2579 (1998)
45. Steigmeier, E.F., Morf, R., Grützmacher, D., Auderset, H., Delley, B., Wessicken, R.: Light emission from a silicon quantum well. Appl. Phys. Lett. **69**, 4165–4167 (1996)
46. Kanemitsu, Y., Sato, H., Nihonyanagi, S., Hirai, Y.: Efficient radiative recombination of indirect excitons in silicon nanowires. Phys. Stat. Sol. (a) **190**, 755–758 (2002)
47. Okamoto, K., Scherer, A., Kawakami, Y.: Surface plasmon enhanced light emission from semiconductor materials. Phys. Stat. Sol. c, **5**, 2822–2824 (2008)
48. Zacharias, M., Heitmann, J., Scholz, R., Kahler, U., Schmidt, M., Bläsing, J.: Size-controlled highly luminescent silicon nanocrystals: A SiO/SiO2 superlattice approach. Appl. Phys. Lett. **80**, 661–663 (2002)
49. Biteen, J.S., Pacifici, D., Lewis, N.S., Atwater, H.A.: Enhanced radiative emission rate and quantum efficiency in coupled silicon nanocrystal-nanostructured gold emitters. Nano Lett. **5**, 1768–1773 (2005)
50. Kalkman, J., Gersen, H., Kuipers, L., Polman, A.: Excitation of surface plasmons at a SiO_2/Ag interface by silicon quantum dots: Experiment and theory. Phys. Rev. B **73**, 075317 (2006)
51. Brongersma, M.L., Kik, P.G., Polman, A., Min, K.S., Atwater, H.A.: Size-dependent electron-hole exchange interaction in Si nanocrystals. Appl. Phys. Lett. **76**, 351–353 (2000)
52. Kovalev, D., Heckler, H., Polisski, G., Koch, F.: Optical properties of Silicon nanocrystals. Phys. Stat. Sol. (b) **215**, 871–932 (1999)
53. Timoshenko, V.Yu, Lisachenko, M.G., Shalygina, O.A., Kamenev, B.V., Zhigunov, D.M., Teterukov, S.A., Kashkarov, P.K., Heitmann, J., Schmidt, M., Zacharias, M.: Comparative study of photoluminescence of undoped and erbium-doped size-controlled nanocrystalline Si/SiO_2 multilayered structures. J. Appl. Phys. **96**, 2254–2260 (2004)

Chapter 3
Polariton Devices Based on Wide Bandgap Semiconductor Microcavities

Ryoko Shimada, Ümit Özgür, and Hadis Morkoç

Abstract Cavity polaritons which are the elementary optical excitations in semiconductor microcavities may be viewed as a superposition of excitons and cavity photons. The major feature of cavity polariton technology centers on large and unique optical nonlinearities which would lead to a new class of optical devices such as polariton lasers exhibiting very low thresholds and polariton parametric amplifiers with ultrafast response. Among the wide bandgap semiconductors, GaN and ZnO are promising candidates for low-threshold polariton lasers operating at room temperature because of their large oscillator strengths and large exciton binding energies, particularly ZnO with its unmatched exciton binding energy of 60 meV. In this chapter, the recent progress on polariton devices based on wide bandgap semiconductor microcavities is reviewed.

Keywords Cavity polaritons · Strong coupling regime · Wide bandgap semiconductors · Microcavities · Polariton lasing

1 Introduction

Planar semiconductor microcavities (MCs) have attracted a good deal of attention owing to their potential to enhance and control the interaction between photons and excitons, which leads to cavity polaritons [1, 2]. The control of the aforementioned interaction is expected to lead to the realization of coherent optical sources such as polariton lasers, which are based on Bose–Einstein condensation (BEC) due to collective interaction of cavity polaritons with photon modes. In contrast with the bulk polaritons, cavity polaritons are half-light and half-matter entities, having quasi two-dimensional nature with a finite energy at zero wavevector, $\mathbf{k} = 0$, and are characterized by a very small in-plane effective mass. These characteristics lead

---

R. Shimada (✉)
Department of Mathematical and Physical Sciences, Japan Women's University, 2-8-1 Mejirodai, Bunkyo-ku, Tokyo, 112-8681, Japan
e-mail: shimadar@fc.jwu.ac.jp

Z.M. Wang, A. Neogi (eds.), *Nanoscale Photonics and Optoelectronics*, Lecture Notes in Nanoscale Science and Technology 9,
DOI 10.1007/978-1-4419-7587-4_3,

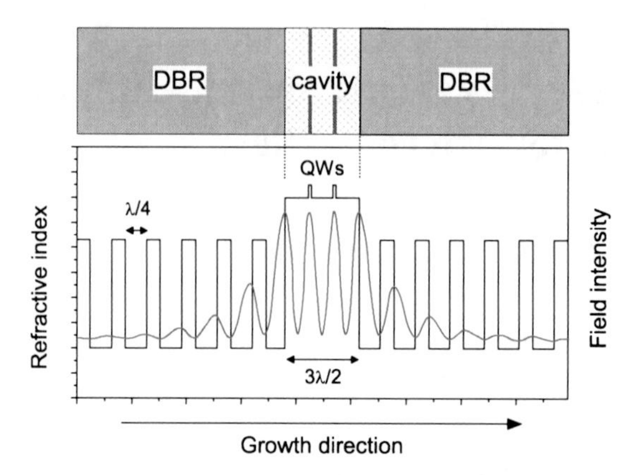

Fig. 3.1 The *top panel* shows the schematic of a typical microcavity (MC) structure. The cavity layer having quantum wells (two in the particular example above) is sandwiched between two distributed Bragg reflectors. The *bottom panel* shows the refractive index profile and the cavity mode electric field intensity throughout the MC. Quantum wells should be embedded at the antinodes of the cavity mode electric field to provide the strongest coupling to light

to bosonic effects in MCs that cannot be achieved in bulk material. In particular, the large occupation number and BEC at the lower polariton branch (LPB) can be accessible at densities well below the onset of exciton bleaching. This can potentially pave the way for ultra-low-threshold polariton lasers. This feature is markedly different from conventional lasers. Lasing in conventional lasers occurs upon population inversion which requires substantial pumping/carrier injection. In polariton lasers, however, the lasing condition is uniquely dependent only on the lifetime of the lower polariton ground state. This is expected to lead to extremely low-threshold lasers, even when compared to the conventional vertical cavity surface emitting lasers (VCSELs).

Planar MCs whose optical length ($m\lambda/2$, m being an integer) is a half-integer multiple of the optical transition wavelength (λ) are highly suited for the manipulation of cavity polaritons. As shown in Fig. 3.1 the MCs are simply composed of an active cavity medium, which can be a semiconductor bulk layer, a number of quantum wells, or a combination of both sandwiched between two highly reflective mirrors. Multiple layers of dielectrics or semiconductors in the form of distributed Bragg reflectors (DBRs) are used as high-reflectivity mirrors. Higher mirror reflectivity results in higher cavity quality factor, and therefore, a better confined cavity mode and a stronger photon–exciton coupling.

1.1 Distributed Bragg Reflectors

Planar MCs require high-reflectivity mirrors which are formed by DBRs composed of alternating $\lambda/4$ stacks of semiconductor and/or dielectric materials with high refractive index contrast. The advantages of DBRs compared with other conventional metallic mirrors include much lower loss and tunability of the high-reflectivity bands (stopbands) to the desired wavelength region by adjusting the thickness of the stacks. The peak reflectivity of DBR is given by [3]

$$R = \left(\frac{1 - (n_{\text{ext}}/n_{\text{c}}) (n_{\text{L}}/n_{\text{H}})^{2N}}{1 + (n_{\text{ext}}/n_{\text{c}}) (n_{\text{L}}/n_{\text{H}})^{2N}} \right)^2, \tag{3.1}$$

where n_{L}, n_{H}, n_{c}, and n_{ext} are the refractive indices of the low and high index layers, the cavity material, and the external medium, respectively, and N is the number of stacked pairs. If the refractive index difference between the layers of the DBR stack is relatively small, a large number of pairs are needed for high reflectivity. For large N, this equation can be simplified to

$$R = \tanh^2 \left(N \ln \frac{n_{\text{L}}}{n_{\text{H}}} + \frac{1}{2} \ln \frac{n_{\text{ext}}}{n_{\text{c}}} \right). \tag{3.2}$$

The peak reflectivity approaches unity when N increases. The stopband width $\Delta\lambda$ is approximately given by

$$\Delta\lambda = \frac{2\lambda \left(n_{\text{H}} - n_{\text{L}} \right)}{\pi n_{\text{eff}}}, \tag{3.3}$$

where n_{eff} is the effective refractive index of the DBR. It is clear that a smaller refractive index contrast results in a narrower stopband. DBRs can provide high reflectivity over any desired wavelength region by properly tailoring the layer thickness. Unlike ordinary metallic mirrors, there is a strong phase dispersion in DBRs. A phase shift of π occurs at the center of the stopband. The dispersion in the center of the stopband is independent of the number of layers and it depends only on the refractive index difference between the layers making up the stack. The reflection occurs gradually when the wave penetrates the DBR. The reflected wave has phase delay of $\exp(-\text{i}\ 2k\ L_{\text{DBR}})$ relative to the incident wave, where L_{DBR} stands for the penetration depth into the DBR at cavity wavelength λ_{c} and is given by [4]

$$L_{\text{DBR}} = \frac{\lambda_{\text{c}}}{2n_{\text{c}}} \frac{n_{\text{L}} n_{\text{H}}}{n_{\text{H}} - n_{\text{L}}}. \tag{3.4}$$

Therefore, an effective value L_{eff} as defined below needs to be used for the cavity length:

$$L_{\text{eff}} = L_{\text{c}} + L_{\text{DBR}} \tag{3.5}$$

1.2 Cavity Polaritons

Bulk polaritons and cavity polaritons differ drastically in nature. In bulk materials, photon dispersion is linear ($\omega = kc/n$, where n is the refractive index of the

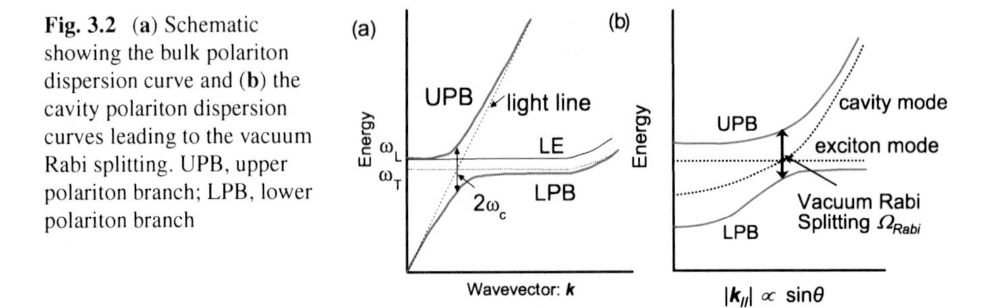

Fig. 3.2 (**a**) Schematic showing the bulk polariton dispersion curve and (**b**) the cavity polariton dispersion curves leading to the vacuum Rabi splitting. UPB, upper polariton branch; LPB, lower polariton branch

materials). Therefore, no photons and the polariton states in the lower polariton branch (LPB) exist at $\mathbf{k} = 0$ as shown in Fig. 3.2a. The polaritons accumulate at the anticrossing point in LPB, the so-called bottleneck region, from where light is usually emitted [5]. It is difficult to observe the strong coupling regime in bulk crystals because the strength of the exciton–photon coupling is orders of magnitude smaller than in MCs. Moreover, in the relaxation bottleneck of bulk materials, polaritons cannot decay directly into external photons [5].

In MC structures on the other hand, photons are quantized in the growth (vertical) direction, while the in-plane photon states are not affected. As a result, the dispersion of cavity photons is strongly modified relative to that of photons in free space. The exciton states are also quantized along the growth direction with a continuum of the in-plane excitonic wavevector states. The coupling modes between photons and excitons form the new quantized states and are called the cavity polaritons. The energy relaxation in MCs is strongly modified relative to that in uncoupled excitons, as illustrated in the schematic cavity polariton dispersion curve of Fig. 3.2b. Unlike the bulk case, the LPB in the cavity polariton dispersion curve has a nonzero minimum at wavevector $\mathbf{k} = 0$. Notable is that the cavity polaritons have very small effective mass around $\mathbf{k} = 0$. Polariton relaxation is very fast due to scattering with acoustic phonons, but polariton relaxation rate becomes slower around the anticrossing point according to the so-called bottleneck effect [6–8]. The major issue in the bottleneck region is that an acoustic phonon cannot assist in polariton relaxation due to small in-plane wavevectors. Therefore, polaritons accumulate in this bottleneck region. This bottleneck effect prevents polaritons from rapid relaxation down to the ground states at $\mathbf{k} = 0$, which would be a major issue for realization of polariton lasers.

The energy splitting between the two coupled modes (lower and upper polariton branches) is called the vacuum Rabi splitting (Ω_{Rabi}) by analogy to the atom–cavity coupling in atomic physics. When Ω_{Rabi} is significantly greater than both the exciton and the cavity linewidths, the system is said to be in the strong coupling regime. Ω_{Rabi} in the case of single quantum well (SQW) can be expressed as [4]

$$\Omega_{\mathrm{Rabi}} = \sqrt{\omega_0 \bar{\omega}_{\mathrm{LT}} \left(\frac{L_{\mathrm{w}}}{L_{\mathrm{eff}}}\right)}, \tag{3.6}$$

where $\bar{\omega}_{\mathrm{LT}} = 2\Gamma_0/kL_{\mathrm{w}}$ is the effective longitudinal–transverse (LT) splitting for the SQW, L_{w} is the well width, and $\hbar\Gamma_0$ is the radiative linewidth of a free exciton at $k_x = 0$ in an SQW and is given in terms of the exciton oscillator strength per unit area f_{ex} by [4]

$$\hbar\Gamma_0 = \frac{\hbar}{4\pi\varepsilon_0}\frac{\pi}{n_{\mathrm{c}}}\frac{e^2}{m_{\mathrm{e}}c}f_{\mathrm{ex}}. \tag{3.7}$$

Ω_{Rabi} in an SQW is reduced by $2\sqrt{L_w/L_{\mathrm{eff}}}$ compared to that in bulk ($\sqrt{2\omega_0\omega_{\mathrm{LT}}}$). This is due to the fact that the excitons can interact with the electromagnetic field only in the QW region along the growth direction. In the case of multiple QWs (MQW), which are located at the electric field antinodes, Ω_{Rabi} is given by

$$\Omega_{\mathrm{Rabi}} \approx 2\hbar\left(\frac{2\Gamma_0 c N_{\mathrm{QW}}}{n_{\mathrm{c}}L_{\mathrm{eff}}}\right)^2, \tag{3.8}$$

where N_{QW} is the number of QWs in the cavity.

1.3 Polariton Lasing

The concept of polariton laser, which was first pointed out by Imamoğlu et al. [9], relates the conditions for the onset of equilibrium and nonequilibrium exciton–polariton condensations. Figure 3.3 illustrates some characteristic features of polariton lasers in comparison with those of BEC of excitons and photon lasers [9]. When a thermal equilibrium reservoir and a photon character of exciton–polaritons vanish, the system tends to become BEC of excitons. In the opposite limit of a nonequilibrium reservoir and a vanishing exciton character, the polariton laser becomes

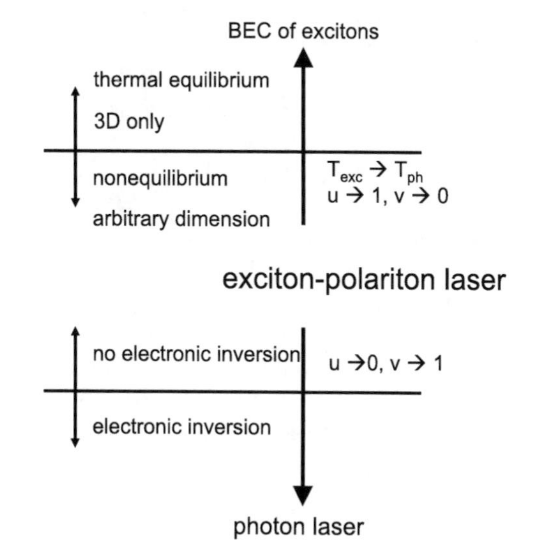

Fig. 3.3 Diagram comparing the exciton-polariton laser to the more familiar concepts of Bose–Einstein condensate of excitons and photon lasers. The annihilation operator for the exciton–polariton quasiparticles is given by $\hat{p} = u\hat{C} + v\hat{a}$, where u and v determine the exciton and photon character of the polariton, respectively. [Reprinted with permission from Ref. 9. Copyright 1996 by the American Physical Society]

indistinguishable from a photon laser. These features of cavity polaritons can be suitable for attaining an exciton–polariton condensate and hopefully emission of coherent light. Conventional lasers require population inversion for lasing, while in polariton lasers only the lifetime of the ground state does matter. As mentioned before, polariton relaxation is fairly fast, but it becomes increasingly slower upon approaching the bottleneck region, preventing smooth relaxation of polaritons from decaying down to their ground states. To overcome this problem and to realize room temperature polariton lasers, it is necessary to find out materials which allow fast polariton relaxation with thermally stable excitons.

Following the proposal of this conceptual scheme of polariton lasers [9], the polariton-stimulated scattering was experimentally observed in resonantly pumped GaAs-based MCs by Savvidis et al. [10]. In that experiment, a 100 μm diameter region on the sample surface was excited resonantly by a pump pulse having a power up to 2 mW and a bandwidth of 1.5 meV. At the same time, a broadband 100 fs probe pulse (<0.3 mW) was focused to a spot size of 50 μm on the sample surface at normal incidence. The response from the lowest energy polariton states was measured by collecting the back reflected probe as shown in Fig. 3.4. The spectra of the probe without the pump and with cross-circularly polarized pump show two polariton dips. In the spectrum of cocircularly polarized pump case, however, an extremely strong and sharp emission peak was observed to emerge at an energy slightly higher than that of the lower energy polariton dip. The two polariton dips were still visible, showing that this nonlinear feature is an outcome of the strong coupling regime. Enormous single-pass optical gains approaching 100 were observed at 4 K. This result clearly demonstrated that polaritons behave as bosons. Moreover, ultrafast polariton parametric amplification was soon observed up to 120 K in GaAs and 220 K in CdTe-based MCs by Saba et al. [11]. The cut-off temperature for the amplification was observed to depend on the exciton binding energy. At the moment, much attention has been generated to develop MCs based on wide bandgap semiconductors, such as GaN and ZnO, owing to their large exciton binding energies and oscillator strengths in order to explore the room temperature polariton nonlinearities.

Polariton lasers, however, are different from polariton parametric amplifiers in that the excitation is made nonresonantly and optically/electrically. Polaritons relax along LPB and scatter to the lower energy states as shown in Fig. 3.5a. Polariton emission is coherent, monochromatic, and spontaneous; therefore, population inversion is not required. As a result, polariton lasers have extremely low threshold and can even be thresholdless. The main issue with regard to polariton lasing is how we can effectively achieve polariton condensates in MCs in spite of the very short lifetime of polaritons. The lifetime of polariton in MCs is shorter or comparable to their energy relaxation time due to the leakage of their photonic components from MCs. In addition, the bottleneck effect in LPB disturbs the relaxation toward the ground state. The formation of condensate within the polariton lifetime is the only way, albeit hard, to achieve polariton lasing.

To overcome this major obstacle, Malpuech et al. proposed to utilize the interactions between polariton and free electrons [12]. They investigated by simulation of

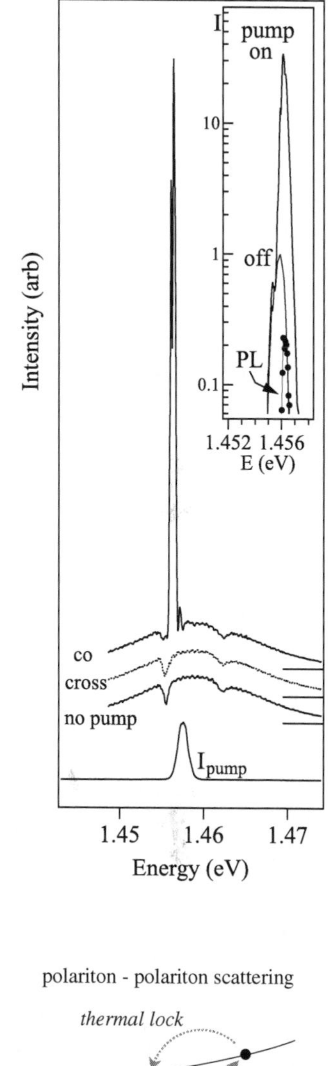

Fig. 3.4 Reflected probe spectra at $\tau = 0$ ps for pump off, co-, and cross-circularly polarized pumped cases. Spectral oscillations are caused by interference between reflections from the front and back surfaces of the sample. Pump spectrum is shown with the lower trace. Inset: Reflected narrow band probe spectra at $\tau = 0$ ps, with pump pulse on/off, together with pump PL without the probe pulse (*solid circles*). [Reprinted with permission from Ref. 10. Copyright 2000 by the American Physical Society]

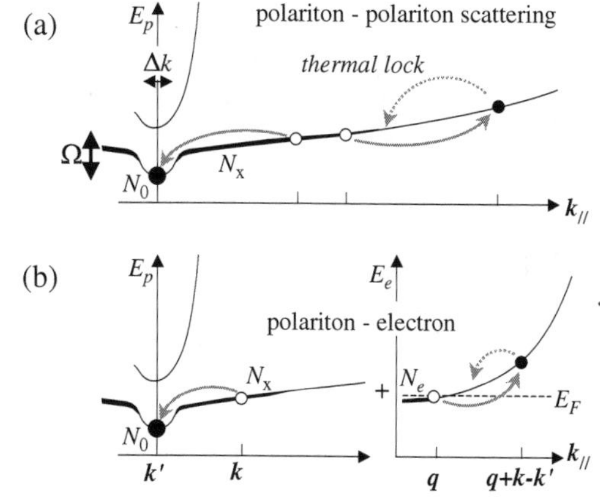

Fig. 3.5 (**a**) Schematic dispersion relations of polariton–polariton scattering. (**b**) Schematic showing the polariton–electron scattering process. The *right panel* shows the free exciton dispersion. [Reprinted with permission from Ref. 12. Copyright 2002 by the American Physical Society]

the possibility for polaritons being efficiently scattered from the bottleneck region to the ground state by introducing free electrons in MCs. Figure 3.5b shows the schematic polariton–electron scattering process. According to the aforementioned simulation, the cavity polaritons which are excited nonresonantly in continuous regime are accumulated near the bottleneck region and then efficiently scattered by the free electrons down to the final emitting state with a transition time of a few picoseconds. This idea might pave the way to a new generation of low-threshold devices.

2 Experimental Studies on Wide Bandgap Semiconductor Microcavities

Since the first observation of the vacuum Rabi splitting in GaAs-based QW-MCs [13], the strong coupling of light with excitons in semiconductor MCs has attracted a great deal of interest for fundamental studies of exciton–polariton BEC in solid state as well as promising applications to very low-threshold vertical cavity lasers. So far, however, cavity polariton and BEC research based on GaAs-based MCs are observable only at very low temperatures because of the slow relaxation of cavity polariton due to the bottleneck effect at LPB [8]. Very recently, polariton light-emitting diode (LED) [14] and electroluminescence from polariton state in the strong coupling regime [15] were reported. Yet, these experiments were conducted in the 4–10 K temperature range, still with the low temperature limitation.

It is only natural then that wide bandgap semiconductors such as GaN- and ZnO-based MCs come to attract increasing attention for room temperature polariton devices, such as polariton lasers, polariton LEDs, and polariton parametric amplifiers owing to their large exciton binding energies and oscillator strengths. The GaN technology is now reasonably well developed and the observation of the strong coupling regime in GaN MCs has been reported by a number of research groups [16–23]. In principle, ZnO is even more attractive as it has a much larger exciton binding energy of 60 meV as compared to 26 meV for GaN. However, the ZnO technology is not so well developed as yet as compared to GaN. In the following section, we review the state-of-the-art experimental research activities on GaN- and ZnO-based MCs.

2.1 GaN-Based Microcavities

GaN-based MCs provide a number of challenging topics for investigating cavity polariton physics and polariton-based devices operating at room temperature due to the large exciton binding energy of bulk GaN ($\sim$26 meV), which can be enhanced in QW structures (>40 meV in narrow QWs [24]), and strong coupling to the light field at room temperature. In vertical cavity devices, the reflectivity of top and bottom DBRs must be sufficiently high in order to minimize optical loss for lasing because the short gain region leads to low gain. In fact the bottom DBR should have a high reflectivity as close to 100% as possible, while the top one is made to have slightly lower reflectivity to match with the optical power desired for extraction. A more precise detailed account of fabrication technique of nitride-based DBRs can be found in Ref. [25]. The precise thickness control is required for high-reflectivity DBRs, exacerbating the situation for short wavelengths which is the realm of nitride-based lasers. Even small errors in the thickness of each layer in the stack can easily lead to large shifts in the central wavelength of the reflection band. It is also imperative to use materials, which are practically loss-free at the operation wavelengths, in the reflector region.

The refractive index contrast is the highest in AlN/GaN among all the III–V compounds. As a result, high reflectivity is accomplished with a reduced number of periods. However, the lattice mismatch between the two materials, which is near 4%, introduces defects which also propagate. The large lattice mismatch between AlN and GaN leads to a built-in strain in vertical structures and even to notorious crack formation. Despite the cracking problem, there are several reports on AlN/GaN DBRs [26–28] indicating promising future applications. To overcome crack formation, GaN/AlGaN DBRs consisting of low Al concentration (<40%) are reported to be useful by several groups [29–33]. Some 60 pairs amounting to a total thickness of 4$\sim$5 μm are then needed because of low Al composition, and thus reduced refractive index contrast, Δn, among the layers forming the stack.

In order to reduce cracks and decrease the built-in strain in vertical structures, two kinds of approaches have been employed, namely, AlN/AlGaN DBRs using an AlN buffer layer [34, 35] and AlGaN/AlGaN DBRs using an AlN interlayer grown on 1 μm thick GaN buffer layers [36]. A 25-pair AlGaN/AlGaN DBR intended for the spectral region around 350 nm grown on 100 μm thick GaN template by plasma-assisted molecular beam epitaxy (MBE) has been reported by Mitrofanov et al. [37]. Good structural quality of the DBR layers was maintained by compensating the compressive and tensile stresses in each $\lambda/4$ pair. This crack-free DBR provided a 26 nm wide stopband centered at 347 nm with a reflectivity above 99%. Another notable example is lattice-matched $Al_{0.82}In_{0.18}N$/GaN DBRs reported by Carlin et al. [38, 39]. The lattice-matched $Al_{0.85}In_{0.15}N/Al_{0.2}Ga_{0.8}N$ DBRs have been deposited on nearly strain-free $Al_{0.2}Ga_{0.8}N$ templates to avoid strain-induced structural degradation and cracking. These completely crack-free DBRs exhibited a reflectivity higher than 99% at a wavelength around 340 nm and a stopband width of 20 nm [40].

A realistic model of the room temperature polariton laser was proposed for a GaN MC by Malpuech et al. [24]. The model structure was a $3\lambda/2$ MC which consisted of a cavity layer with 4-monolayer-thick 9 QWs between $Al_{0.2}Ga_{0.8}N/Al_{0.9}Ga_{0.1}N$ DBRs, 11 pairs on the top and 14 pairs at the bottom. The critical temperature of BEC of cavity polaritons was predicted to be 460 K with a room temperature polariton lasing threshold as small as 100 mW.

The first experimental results of the strong coupling regime in GaN-based MCs have been reported by Antoine-Vincent et al. [16]. The MCs are composed of a $\lambda/2$ GaN active layer grown on a three-layer Bragg stack consisting of a 2λ–$\lambda/4$

layer of AlN, a 2λ–λ/4 layer of $Al_{0.2}Ga_{0.9}N$, and a λ/4 layer of AlN grown on Si (111) for the bottom reflector and a SiO_2/Si_3N_4 top DBR. Although the MC finesse was low, a vacuum Rabi splitting of 31 meV was obtained at 5 K from the angle-resolved reflectivity measurements. Room temperature cavity polaritons have been observed in an InGaN/InGaN QW MC [17]. The MCs fabricated by a wafer-bonding technique was composed of $In_{0.15}Ga_{0.85}N/In_{0.02}Ga_{0.98}N$ QWs embedded in a GaN-based cavity layer sandwiched between two SiO_2/ZrO_2 DBRs. The anticrossing behavior was observed by angle-resolved reflectivity measurements showing the vacuum Rabi splitting of 6 meV. By increasing the number of QWs from 3 to 10, the vacuum Rabi splitting was increased to 17 meV. An impediment for strong coupling regime in this particular InGaN QW-MC was a low finesse cavity and/or large inhomogeneous broadening of the QW emission [18].

Bulk GaN-based MCs were further studied for polariton emission in the strong coupling regime [19–22]. In a bulk GaN MC with lattice-matched AlInN/(Al)GaN DBRs, a strong bottleneck effect was observed at room temperature by photoluminescence (PL) measurements [21]. In an attempt to use ubiquitous Si substrates, bulk GaN MCs with 10 pair $AlN/Al_{0.2}Ga_{0.8}N$ DBR have been grown directly on Si (111) [19, 20]. A vacuum Rabi splitting of approximately 50 meV was observed up to room temperature by angle-resolved reflectivity and PL measurements. A vacuum Rabi splitting of 43 meV in GaN hybrid MCs in the strong coupling regime was reported by Alyamani et al. [22], although the highest cavity quality value (*Q*-value) was fairly low, about 160 at best. A $GaN/Al_{0.2}Ga_{0.8}N$ QW-MC with a sharper linewidth enabled to observe cavity polaritons at room temperature using angle-resolved PL [23]. A vacuum Rabi splitting Ω of 30 meV was observed as shown in Fig. 3.6. The exciton oscillator strength was estimated to be $\sim 3 \times 10^{13}$ cm^{-2}/QW.

The first observation of polariton lasing in bulk GaN-based MCs at room temperature has been reported by Christpoulos et al. [41]. Under similar experimental geometry to that described in Ref. [10], a 150 fs tunable pump pulse light was

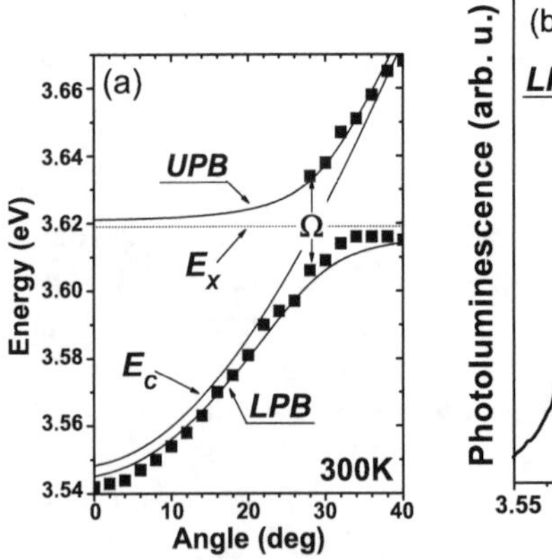

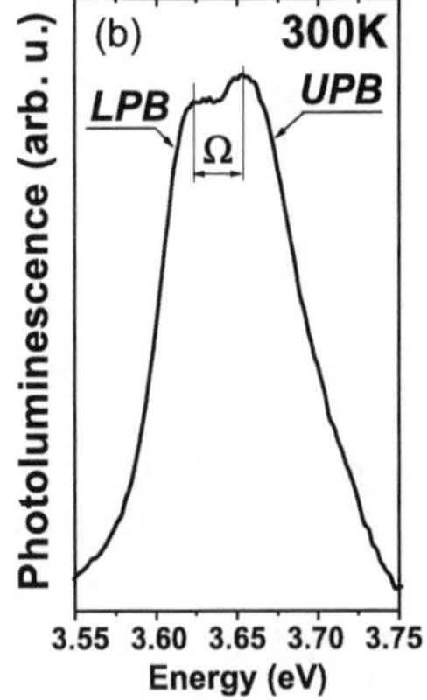

Fig. 3.6 (**a**) Experimental dispersion curves (*black squares*) and fits of the lower (LPB) and upper (UPB) polariton branches. The cavity mode energy (E_c) and the uncoupled exciton energy (E_X) are also shown. (**b**) PL spectra of the 3λ/2 AlGaN MC containing six QWs close to resonance. [Reprinted with permission from Ref. 23. Copyright 2006 by the American Institute of Physics]

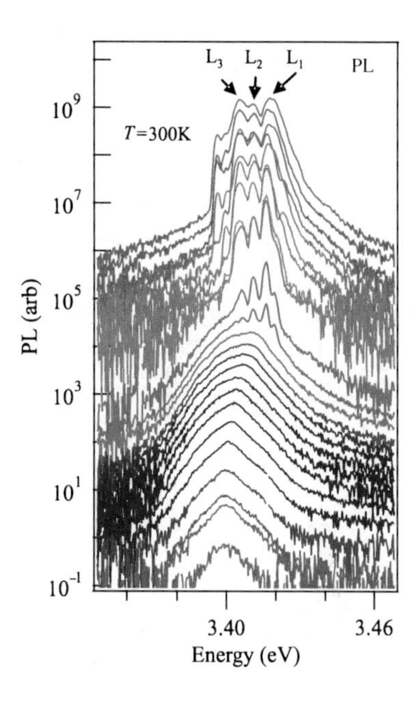

Fig. 3.7 Emission spectra for pump powers ranging from 20 μW to 2 mW at 0°. Spectra integrated over 10 ms show multiple emission lines (L_1, L_2, and L_3). [Reprinted with permission from Ref. 41. Copyright 2007 by the American Physical Society]

focused on a 60 μm diameter spot on the top DBR to excite the sample non-resonantly. The emission was collected over an angular range of ±3°. Figure 3.7 shows the integrated output intensity collected at normal direction (0°) as a function of pump intensity ranging from 20 μW to 2 mW. A clear nonlinear behavior was observed at ∼365 nm, with an increase by a factor over 10^3 at the critical threshold of around $I_{th} = 1.0$ mW. The emission peak energy is sensitive to the spatial alignment on the sample, and spectra integrated over 10 ms show multiple line emissions (L_1, L_2, and L_3 in Fig. 3.7). The blue shift in the emission spectra with increasing pumping power was attributed to polariton–polariton interaction. The observed coherent emission threshold is one order of magnitude smaller than that in previously reported nitride-based VCSELs [42].

2.2 ZnO-Based Microcavities

To reiterate, ZnO is a wide bandgap semiconductor having a large exciton binding energy (60 meV), much larger than that of GaN (23 meV), and a large oscillator strength, and therefore, is a potential candidate as GaN for the realization of room temperature polariton devices.

The most adopted structure for the observation of polariton lasing is a model ZnO-based MC proposed by Zamfirescu et al. [43]. The structure consists of a λ-thick ZnO cavity layer sandwiched between ZnO/$Zn_{0.7}Mg_{0.3}O$ DBRs having 14 and 15 pairs of λ/4 thick layers on the top and at the bottom, respectively. Figure 3.8a shows the calculated angle-resolved reflectivity spectra. Amazingly, the vacuum

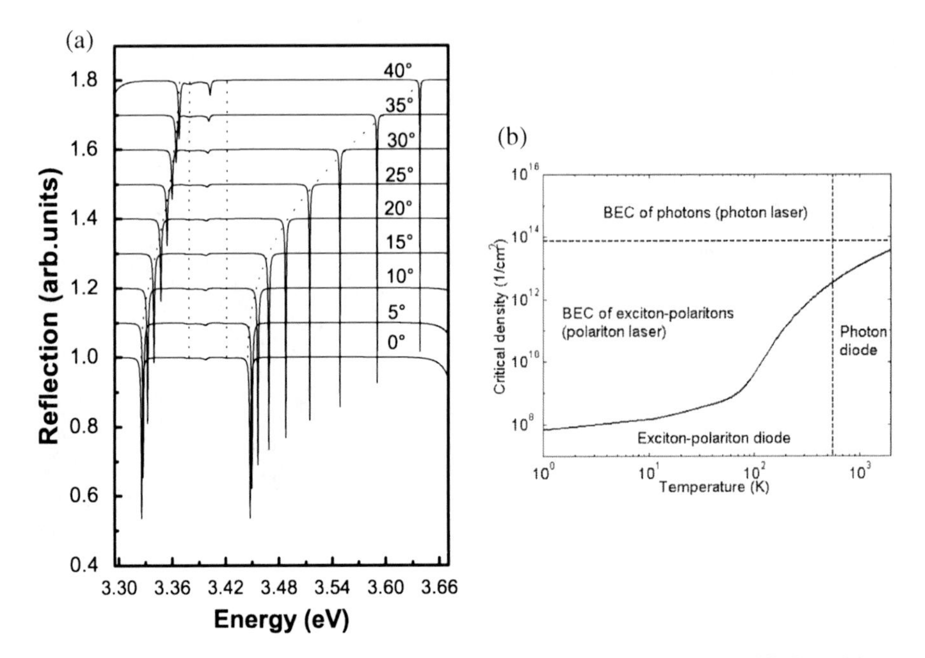

Fig. 3.8 (**a**) The calculated angle-resolved reflectivity spectra of a ZnO-based MC. *Dotted lines* indicate the cavity eigenmodes. (**b**) Exciton–polariton phase diagram in the ZnO-based MC studied. The *solid line* shows the polariton critical density versus lattice temperature. The *vertical dashed line* shows the exciton thermal dissociation limit. The *horizontal dashed line* shows the Mott transition for excitons. [Reprinted with permission from Ref. 43. Copyright 2002 by the American Physical Society]

Rabi splitting is as large as 120 meV in this model ZnO-based MC, suggesting room temperature polariton lasing to be possible. According to calculations, the BEC phase is formed in a wide temperature range with pumping power limited by a critical temperature as shown in Fig. 3.8b. The critical temperature 560 K is determined by the exciton dissociation energy and would be the highest for BEC if this model is truly applicable. From this phase diagram, a lasing threshold was estimated to be 2 mW at room temperature. Another model MC consisting of λ-thick ZnO sandwiched between seven and eight pairs of ZrO_2/SiO_2 top and bottom DBRs has been proposed by Chichibu et al. [44]. It was suggested that the vacuum Rabi splitting increased by up to 191 meV.

There is a limited number of experimental reports on planar ZnO-based MCs [45, 46]. Schmidt-Grund et al. recently reported the results on ZnO-based resonators grown by pulsed laser deposition (PLD) [45]. Their MCs consisted of $\lambda/2$-thick ZnO cavity layers sandwiched between ZrO_2/MgO DBRs having a Q-value of $\sim$80. A vacuum Rabi splitting of some 50 meV was obtained at $\phi = 60°$ from the angle-resolved reflectivity and PL measurements. This value was lower than expected [43, 44] due to the large inhomogeneous broadening of the excitons.

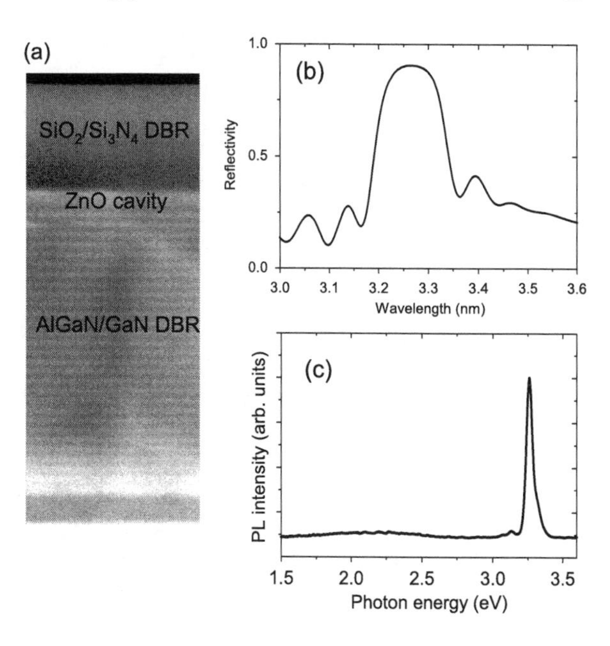

Fig. 3.9 (**a**) Cross-sectional SEM image of a ZnO-based hybrid MC. (**b**) Reflectivity spectrum of a 29-pair $Al_{0.5}Ga_{0.5}N$/GaN DBR showing a peak reflectivity of ~90% and a stopband width of ~150 meV. (**c**) Room temperature PL of a ZnO half cavity (without the top SiO_2/SiN_x DBR) (after Shimada et al. [Ref. 46])

More recently, cavity polaritons in bulk ZnO-based hybrid MCs has been reported by Shimada et al. [46]. The structure shown in Fig. 3.9a consists of a λ-thick ZnO cavity layer sandwiched between a 29-pair $Al_{0.5}Ga_{0.5}N$/GaN bottom DBR and an 8-pair SiO_2/SiN_x top DBR. The bottom $Al_{0.5}Ga_{0.5}N$/GaN DBR was directly grown on a 200 nm-thick AlN buffer layer on a (0001) sapphire substrate by low-pressure metalorganic chemical vapor deposition (MOCVD). The AlN buffer was chosen to avoid cracking due to the built-in strain caused by lattice mismatch. The Al composition in the AlGaN layer is nearly 50% as determined from X-ray diffraction (XRD) measurements, which also revealed clear interference fringes indicative of smooth interfaces between layers. The $Al_{0.5}Ga_{0.5}N$/GaN pair layer thickness was determined to be 77 nm from both XRD measurements and cross-sectional scanning electron microscopy (SEM) images. As shown in Fig. 3.9b, the measured peak reflectivity and the stopband width are ~90% and ~150 meV, respectively, for the 29-pair $Al_{0.5}Ga_{0.5}N$/GaN DBR. A λ-thick (optical thickness being ~160 nm for $\lambda \approx 380$ nm in air) ZnO cavity layer was grown on the bottom $Al_{0.5}Ga_{0.5}N$/GaN DBR by plasma-assisted molecular beam epitaxy (MBE). The photoluminescence (PL) spectrum at room temperature is shown in Fig. 3.9c. The deep level emission band of the ZnO cavity layer is nearly fully suppressed indicating good material quality of the sample.

The top dielectric DBR which consisted of $\lambda/4$-thick SiO_2 and SiN_x layers was deposited on the ZnO half-MC by ultra-high vacuum remote plasma enhanced chemical vapor deposition (UHV-RPECVD) to complete the MC structure. The optical reflectivity and PL spectra from the full-MC (including the top SiO_2/SiN_x DBR) at near-normal incidence are shown in Fig. 3.10. A Q-value of ~100 was obtained for the full MC from both the optical reflectivity and the PL spectra.

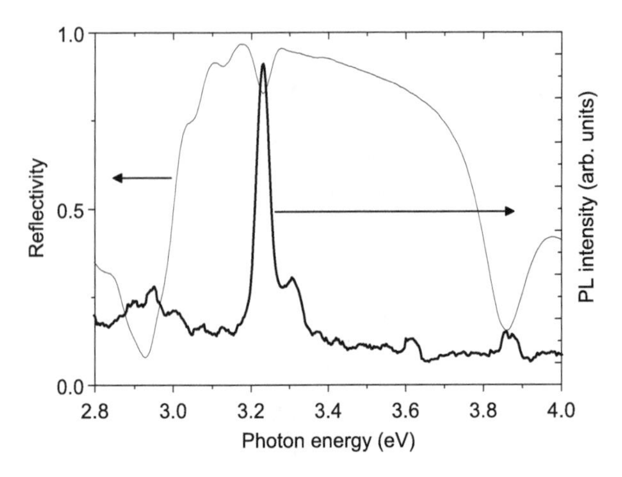

Fig. 3.10 Room temperature reflectivity and PL spectra of a ZnO-based hybrid MC (after Shimada et al. [Ref. 46])

This low Q-value is due to the optical loss attributed mainly to the cracks in the bottom DBR. A reflectivity dip is observed at 3.23 eV which coincides with the PL data corresponding to the energy of the lower polariton mode for the ZnO-based hybrid MC. The weak emission peak at 3.30 eV corresponds to the upper polariton mode.

Angle-resolved measurements are conventional means to trace the cavity polariton modes in MCs without changing the position or the temperature. By using the above-mentioned ZnO MC samples, the angle-resolved PL spectra were measured at room temperature over the range 0º–40º using 325 nm excitation light from a He–Cd laser. The results are shown in Fig. 3.11a, where the dotted line indicates the uncoupled exciton mode while the solid lines are guide to the eye. As the angle increases, the lower polariton mode approaches the uncoupled exciton mode, while the upper polariton mode is dispersed from the exciton mode toward the cavity mode. These mode positions are plotted as a function of angle in Fig. 3.11b, indicating a typical anticrossing behavior between the cavity mode and the exciton mode when the cavity mode crosses the exciton. Since the stopband width of the bottom DBR is narrow (~150 meV) due to relatively low refractive index contrast in semiconductor DBR layers, the upper polariton features are not clear at large angles, making it difficult to observe a clear anticrossing behavior. In addition, the relaxation processes at the lower polariton branch and thermalization issues due to the large vacuum Rabi splitting might be also responsible for poor resolution of the upper polariton branch. Yet, the anticrossing behavior is clearly seen in Fig. 3.11b and confirms the strong coupling regime in ZnO-based hybrid MCs. At the resonant condition of $\theta = 22°$, the vacuum Rabi splitting is estimated to be ~50 meV. This value is larger than that in bulk-GaN hybrid MCs [19, 20, 22], but is far below 120 meV predicted by Zamfirescu et al. [43]. The reason for such a large discrepancy is not clear at present,

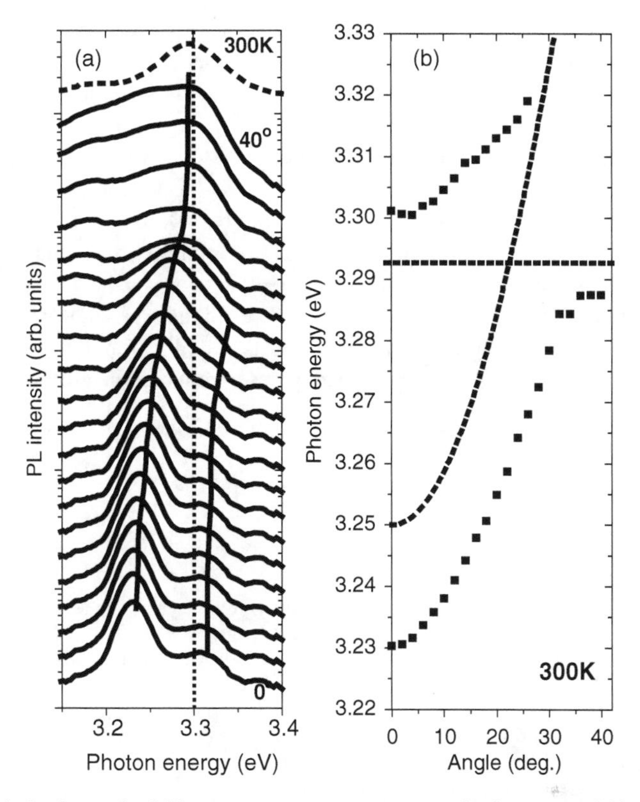

Fig. 3.11 (**a**) Angle-resolved PL spectra at room temperature in the range of 0°–40° for a λ-thick ZnO hybrid MC. The *dotted line* is the exciton mode and the *solid lines* are guide to the eye. (**b**) Experimental dispersion curves for upper and lower cavity polariton modes. The *dotted curves* represent the uncoupled cavity and exciton modes. [Reprinted with permission from Ref. 46. Copyright 2008 by the American Institute of Physics]

but partly might be attributed to the experimental problems such as the inhomogeneous broadening in the ZnO cavity layer and the low Q-value in the MC. Since polariton lasing depends on the formation of a BEC at the lower energy trap states in the lower polariton branch, high-resolution spectroscopy of the lower polariton branch is imperative for the development of the polariton laser in the strong coupling regime.

3 Conclusions

In this chapter, we have reviewed the basic microcavity physics and state-of-the-art technologies of wide bandgap semiconductor-based polariton devices. In GaN-based MCs, vacuum Rabi splittings of 30–43 meV have been reported at room temperature in the strong coupling regime. Moreover, room temperature low-threshold transition to a coherent polariton state have been demonstrated in bulk-GaN MCs in the strong coupling regime under nonresonant pulsed optical pumping. In ZnO-based MCs, a vacuum Rabi splitting as large as ~50 meV was observed at room temperature, pending further improvements for the realization of polariton lasing. Meanwhile, electroluminescence from GaAs-based MCs in the strong coupling regime [15] and also GaAs-based polariton light-emitting diodes [14] have been reported at low temperatures. Based on these encouraging advances, further efforts should pave the way for attaining high-quality GaN- and ZnO-based MCs, which have great potential/advantages in the realm of polariton devices operating at room temperature because of their large exciton binding energies and oscillator strengths. The focus now shifts to realization of room temperature polariton devices based on these wide bandgap semiconductor microcavities.

Acknowledgments This work was supported by Air Force Office Scientific Research under the direction of Drs. K. Reinhardt and D. Silversmith.

References

1. Skolnick, M.S., Fisher, T.A., Whittaker, D.M.: Strong coupling phenomena in quantum microcavity structures. Semicond. Sci. Technol. **13**, 645–669 (1998)
2. Kavokin, A., Malpuech, G.: Cavity Polaritons, Elsevier, Amsterdam (2003)
3. Yeh, P.: Optical Waves in Layered Media. Wiley, New York (1988)
4. Savona, V., Andreani, L.C., Schwendimann, P., Quattropani, A.: Quantum well excitons in semiconductor microcavities: Unified treatment of weak and strong coupling regimes. Solid. State Commun. **93**, 733–739 (1995)
5. Toyozawa, Y.: On the dynamical behavior of an exciton. Suppl. Prog. Theor. Phys. **12**, 111–140 (1959)
6. Tassone, F., Piermarocchi, C., Savona, V., Quattropani, A., Schwendimann, P.: Bottleneck effects in the relaxation and photoluminescence of microcavity polaritons. Phys. Rev. B **56**, 7554–7563 (1997)
7. Tassone, F., Yamamoto, Y.: Exciton-exciton scattering dynamics in a semiconductor microcavity and stimulated scattering into polaritons. Phys. Rev. B **59**, 10830–10842 (1999)
8. Tartakovskii, A.I., Emam-Ismail, M., Stevenson, R.M., Skolnick, M.S., Astratov, V.N., Whittaker, D.M., Baumberg J.J., Roberts, J.S.: Relaxation bottleneck and its suppression in semiconductor microcavities. Phys. Rev. B **62**, R2283–R2286 (2000)
9. Imamoğlu, A., Ram, R.J., Pau, S., Yamamoto, Y.: Nonequilibrium condensates and lasers without inversion: Exciton-polariton lasers. Phys. Rev. A **53**, 4250–4253 (1996)
10. Savvidis, P.G., Baumberg, J.J., Stevenson, R.M., Skolnick, M.S., Whittaker, D.W., Roberts, J.S.: Angle-resonant stimulated polariton amplifier. Phys. Rev. Lett. **84**, 1547–1550 (2000)
11. Saba, M., Ciuti, C, Bloch, J., Thierry-Mieg, V., André, R., Dang, L.S., Kundermann, S., Mura, A., Bongiovanni, G., Staehli, J.L., Deveaud, B.: High-temperature ultrafast polariton parametric amplification in semiconductor microcavities. Nature. **414**, 731–735 (2001)
12. Malpuech, G., Kavokin, A., Di Carlo, A., Baumberg, J.J.: Polariton lasing by exciton-electron scattering in semiconductor microcavities. Phys. Rev. B **65**, 153310 (2002)
13. Weisbuch, C., Nishioka, M., Ishikawa, A., Arakawa, Y.: Observation of the coupled exciton-photon mode splitting in a semiconductor quantum microcavity. Phys. Rev. Lett. **69**, 3314–3317 (1992)
14. Bajoni, D., Semenova, E., Lamaître, A., Bouchoule, S., Wertz, E., Senellart, P., Bloch, J.: Polariton light-emitting diode in a GaAs-based microcavity. Phys. Rev. B **77**, 113303 (2008)

15. Khalifa, A.A., Love, A.P.D., Krizhanovskii, D.N., Skolnick, M.S., Roberts, J.S.: Electroluminescence emission from polariton states in GaAs-based semiconductor microcavities. Appl. Phys. Lett. **92**, 061107 (2008)
16. Antoine-Vincent, N., Natali, F., Byrne, D., Vasson, A., Disseix, P., Leymarie, J., Leroux, M., Semond, F., Massies, J.: Observation of Rabi splitting in a bulk GaN microcavity grown on silicon. Phys. Rev. B **68**, 153313 (2003)
17. Tawara, T., Gotoh, H., Akasaka, T., Kobayashi, N., Saitoh, T.: Cavity polaritons in InGaN microcavities at room temperature. Phys. Rev. Lett. **92**, 256402 (2004)
18. Christmann, G., Butté, R., Feltin, E., Carlin, J.–F., Grandjean, N.: Impact of inhomogeneous excitonic broadening on the strong exciton-photon coupling in quantum well nitride microcavities. Phys. Rev. B **73**, 153305 (2006)
19. Sellers, I.R., Semond, F., Leroux, M., Massies, J., Disseix, P., Henneghien, A.L., Leymarie, J., Vasson, A.: Strong coupling of light with *A* and *B* excitons in GaN microcavities grown on silicon. Phys. Rev. B **73**, 033304 (2006)
20. Sellers, I.R., Semond, F., Leroux, M., Massies, J., Zamfirescu, M., Stokker-Cheregi, F., Gurioli, M., Vinattieri, A., Colocci, M., Tahraoui, A., Khalifa, A.A.: Polariton emission and reflectivity in GaN microcavities as a function of angle and temperature. Phys. Rev. B **74**, 193308 (2006)
21. Butté, R., Christmann, G., Feltin, E., Carlin, J.–F., Mosca, M., Ilegems, M., Grandjean, N.: Room-temperature polariton luminescence from a bulk GaN microcavity. Phys. Rev. B **73**, 033315 (2006)
22. Alyamani, A., Sanvitto, D., Khalifa, A.A., Skolnick, M.S., Wang, T., Ranalli, F., Parbrook, P.J., Tahraoui, A., Airey, R.: GaN hybrid microcavities in the strong coupling regime grown by metal-organic chemical vapor deposition on sapphire substrates. J. Appl. Phys. **101**, 093110 (2007)
23. Feltin, E., Christmann G., Butté, R., Carlin, J.–F., Mosca, M., Grandjean, N.: Room temperature polariton luminescence from a GaN/AlGaN quantum well microcavity. Appl. Phys. Lett. **89**, 071107 (2006)
24. Malpuech, G., Carlo, A.D., Kavokin, A., Baumberg, J.J., Zamfirescu, M., Lugli, P.: Room-temperature polariton lasers based on GaN microcavities. Appl. Phys. Lett. **81**, 412–414 (2002)
25. Butté, R., Feltin, E., Dorsaz, J., Christmann, G., Carlin, J.–F, Grandjean, N., Ilegems, M.: Recent progress in the growth of highly reflective nitride-based distributed Bragg reflectors and their use in microcavities. Jpn. J. Appl. Phys. **44**, 7207–7216 (2005)
26. Ng, H.M., Doppalapudi, D., Iliopoulos, E., Moustakas, T.D.: Distributed Bragg reflectors based on AlN/GaN multilayers. Appl. Phys. Lett. **74**, 1036–1038 (1999)
27. Ng, H.M., Moustakas, T.D., Chu, S.N.G.: High reflectivity and broad bandwidth AlN/GaN distributed Bragg reflectors grown by molecular-beam epitaxy. Appl. Phys. Lett. **76**, 2818–2820 (2000)
28. Ive, T., Brandt, O., Kostial, H., Hesjedal, T., Ramsteiner, M., Ploog, K.H.: Crack-free and conductive Si-doped AlN/GaN distributed Bragg reflectors grown on 6H-SiC (0001). Appl. Phys. Lett. **85**, 1970–1972 (2004)
29. Khan, M.A., Kuznia, J.N., Van Hove, J.M., Olson, D.T.: Reflective filters based on single-crystal $GaN/Al_xGa_{1-x}N$ multilayers deposited using low-pressure metalorganic chemical vapor deposition. Appl. Phys. Lett. **59**, 1449–1451 (1991)
30. Someya, T., Arakawa, Y.: Highly reflective $GaN/Al_{0.34}Ga_{0.66}N$ quarter-wave reflectors grown by metal organic chemical vapor deposition. Appl. Phys. Lett. **73**, 3653–3655 (1998)
31. Waldrip, K.E., Han, J., Figiel, J.J., Zhou, H., Makarona, E., Nurmikko, A.V.: Stress engineering during metalorganic chemical vapor deposition of AlGaN/GaN distributed Bragg reflectors. Appl. Phys. Lett. **78**, 3205–3207 (2001)
32. Nakata, N., Ishikawa, H., Egawa, T., Jimbo, T., Umeno, M.: MOCVD growth of high reflective GaN/AlGaN distributed Bragg reflectors. J. Cryst. Growth. **237–239**, 961–967 (2002)

33. Langer, R., Barski, A., Simon, J., Pelekanos, N.T., Konovalov, O., André, R.,Dang, L.S.: High-reflectivity GaN/GaAlN Bragg mirrors at blue/green wavelengths grown by molecular beam epitaxy. Appl. Phys. Lett. **74**, 3610–3612 (1999)
34. Bhattacharyya, A., Iyer, S., Iliopoulos, E., Sampath, A.V., Cabalu, J., Moustakas, T. D., Friel, I.: High reflectivity and crack-free AlGaN/AlN ultraviolet distributed Bragg reflectors. J. Vac. Sci. Technol. B **20**, 1229–1233 (2002)
35. Natali, F., Antoine-Vincent, N., Semond, F., Byrne, D., Hirsch, L., Barrière, A.S., Leroux, M., Massies, J., Leymarie, J., Jpn. J.: AlN/AlGaN Bragg-Reflectors for UV Spectral Range Grown by Molecular Beam Epitaxy on Si(111). Appl. Phys. **41**, L1140–L1142 (2002)
36. Wang, T., Lynch, R.J., Parbrook, P.J., Butté, R., Alyamani, A., Sanvitto, D., Whittaker, D.M., Skolnick, M.S.: High-reflectivity $Al_xGa_{1-x}N/Al_yGa_{1-y}N$ distributed Bragg reflectors with peak wavelength around 350 nm. Appl. Phys. Lett. **85**, 43–45 (2004)
37. Mitrofanov, O., Schmult, S., Manfra, M.J., Siegrist, T., Weimann, N.G. and Sergent, A.M., Molnar, R.J.: High-reflectivity ultraviolet AlGaN/AlGaN distributed Bragg reflectors. Appl. Phys. Lett. **88**, 171101 (2006)
38. Carlin, J.-F., Ilegems, M.: High-quality AlInN for high index contrast Bragg mirrors lattice matched to GaN. Appl. Phys. Lett. **83**, 668–670 (2003)
39. Carlin, J.-F., Dorsaz, J., Feltin, E., Butté, R., Grandjean, N., Ilegems, M.: Crack-free fully epitaxial nitride microcavity using highly reflective AlInN/GaN Bragg mirrors. Appl. Phys. Lett. **86**, 031107 (2005)
40. Feltin, E., Carlin, J.-F., Dorsaz, J., Christmann G., Butté, R., Laügt, M., Ilegems, M., Grandjean, N.: Crack-free highly reflective AlInN/AlGaN Bragg mirrors for UV applications. Appl. Phys. Lett. **88**, 051108 (2006)
41. Christopoulos, S., Baldassarri Höger Von Högersthal, G., Grundy, A.J.D., Lagoudakis P.G., Kavokin, A.V., Baumberg, J.J., Christmann, G., Butté, R., Feltin, E., Carlin, J.–F., Grandjean, N.: Room-temperature polariton lasing in semiconductor microcavities. Phys. Rev. Lett. **98**, 126405 (2007)
42. Zhou, H., Diagne, M., Makarona, E., Nurmikko, A.V., Han, J., Waldrip, K.E.,Figiel, J.J.: Near ultraviolet optically pumped vertical cavity laser. Electron. Lett. **36**, 1777–1779 (2000)
43. Zamfirescu, M., Kavokin, A., Gil, B., Malpuech, G., Kaliteevski, M.: ZnO as a material mostly adapted for the realization of room-temperature polariton lasers. Phys. Rev. B **65**, 161205 (2002)
44. Chichibu, S.F., Uedono, A., Tsukazaki, A., Onuma, T., Zamfirescu, M., Ohtomo, A., Kavokin, A., Cantwell, G., Litton, C.W., Sota, T., Kawasaki, M.: Exciton-polariton spectra and limiting factors for the room-temperature photoluminescence efficiency in ZnO. Semicond. Sci. Technol. **20**, S67–S77 (2005)
45. Schmidt-Grund, R., Rheinländer, B., Czekalla, C., Benndorf, G., Hochmut, H, Rahm, A., Lorenz, M., Grundmann, M.: ZnO based planar and micropillar resonators. Superlattices Microstruct. **41**, 360–363 (2007)
46. Shimada, R., Xie, J., Avrutin, V., Özgür, Ü., Morkoç, H.: Cavity polaritons in ZnO-based hybrid microcavities. Appl. Phys. Lett. **92**, 011127 (2008)

Chapter 4
Search for Negative Refraction in the Visible Region of Light by Fluorescent Microscopy of Quantum Dots Infiltrated into Regular and Inverse Synthetic Opals

R. Moussa, A. Kuznetsov, E. Neiser, and A.A. Zakhidov

Abstract In this chapter, regular and inverse synthetic opals are examined experimentally by infiltrating them with CdSe quantum dots (QDs). Confocal microscopy measurements in which we track the infiltration of QDs inside the regular and inverse opals show indications of focusing of light emitted by QDs, which can be due to negative refraction occurring at the opal–glass interface. The formation of a focus can be an indication of the left-handed behavior of these synthetic opals in the [111] direction in its higher photonic band, above the photonic band gap (PBG). This result can be very promising because, until now, left-handed behavior has not been demonstrated in 3D photonic crystals in the visible region of light. This result was made possible due to the use of infiltrated QDs as internal light sources inside the porous photonic crystal, which appears to be a very useful technique for the study of other negative-index materials (NIM) effects.

Keywords Negative refraction · Opals · Photonic crystals · Left-handed materials

Recent experimental and theoretical results [1–3] have confirmed the existence of negative refraction in a specific type of material known as left-handed material (LHM). This material exhibits unusual properties such as negative refractive index; antiparallel wave vector, **k**, and Poynting vector, **S**; and antiparallel phase, $\mathbf{V_p}$, and group, $\mathbf{V_g}$, velocities. This phenomenon was predicted years ago but has since attracted little notice [4]. However, the recent publication by Pendry [5] in which he suggested that a slab of material with a negative permittivity and negative permeability can be used as a perfect lens has renewed interest in this subject. Meanwhile, other authors related this phenomenon to the negative group velocity of the system and demonstrated it [6]. Many recently published papers [7–10] have studied negative refraction and focused on the superlensing phenomenon or on the diffraction limit problem. The key feature of these phenomena is the excitation of surface waves and the amplification of the evanescent waves inside the LH slab. We

R. Moussa (✉)
UTD-Nanotech Institute, University of Texas at Dallas, Richardson, TX 75083, USA
e-mail: rabia.moussa@utdallas.edu

Z.M. Wang, A. Neogi (eds.), *Nanoscale Photonics and Optoelectronics*,
Lecture Notes in Nanoscale Science and Technology 9,
DOI 10.1007/978-1-4419-7587-4_4,

have succeeded in exciting surface waves in two-dimensional (2D) PC structures in the microwave regime. However, applications in nanoscale technology require another procedure especially when it comes to extracting light from optical devices. Quantum dots (QDs) seem to be a solution [11]. Indeed, QDs can be used to imitate a point source. Many research groups [12–14] have focused on 1D and 2D PCs based on periodic dielectric planes or rods. Our research focuses mainly on the propagation of light and characterization of 3D photonic crystals. There have been some attempts to study negative refraction in 3D PC, but this chapter examines new aspects of 3D NIM behavior [15]. It is important to notice that up to date there are no PCs operating as left-handed materials in the optical range and all attempts of building artificial material and metamaterial structures face the same major problems, namely, the high absorption and the fragility of materials in the absence of strong and robust fabrication techniques that build structures in the nanoscale or even the micrometer scale. The PCs used in this study are fabricated by the self-assembly of silica and polystyrene spheres [16–18]. As the PCs are self-assembled, many of the problems encountered with more traditional nanoscale fabrication techniques are bypassed. By using these self-assembled arrays infiltrated with QDs, one can easily track the propagation of light and demonstrate the material's ability to have negative refraction.

While many studies have been done using external sources of light to investigate light propagation of PCs, very little work has been done using embedded sources [19]. Our approach involves using CdSe QDs that act as quasi-point sources that can insert themselves into the PC through Brownian motion. Some groups have already studied the QD effect in PCs [20, 21]. However, left-handed behavior has not been examined using this technique. The theoretical part of an embedded point source inside the LHM was extensively studied in a recent paper [22].

In this chapter, we will experimentally demonstrate that like a point source a QD can be embedded in a PC and following it inside the material will allow us to check the path of the scattered light and determine the rightness (whether it is right-hand or left-hand material) of the sample.

1 Experimental Details

The 3D opal photonic crystals used in this experiment are prepared by self-assembly of monodispersed silica particles in a closely packed arrangement of SiO_2 spheres. To prepare the samples we use sedimentation of silica particles from solution [23]. Monodispersed suspension of the particles is being deposited by sedimentation on a polished silicon substrate during several weeks. After removal from the liquid, the deposit is sintered in order to reinforce it. Then it is polished with fine sandpaper. Grown photonic crystals show perfect short-range order on a length scale of tens of microns while the domain structure is prevalent over a long range. An SEM image of the photonic crystal made of 820 nm silica spheres is shown in Fig. 4.1.

Meanwhile, the inverse opals are prepared. In general, inverse structures are prepared in the following way: first, a template is infiltrated with a solid material and then the template is removed by etching. To make the inverse opals studied in this

Fig. 4.1 SEM image of the photonic crystal made of 820 nm silica spheres

chapter, we used direct silica opals as a template. The direct opals were heated to 100°C in vacuum oven to remove all the excess oxygen from them. They were infiltrated later on with optical glue with a refractive index $n = 1.6$. To ensure that all air voids are filled with the material, the infiltration was done in vacuum oven heated to 100°C for 24 h. After the infiltration, the samples were etched in 32% HF solution diluted in a proportion of 1 to 6. An SEM image of the inverse opal made of silica spheres is shown in Fig. 4.2.

Three microscopes have been used to image our opal PC. First, the Cytoviva microscope is used which allows us to see real-time images of QDs simultaneously in fluorescent and dark-field modes. To this microscope a highly sensitive camera

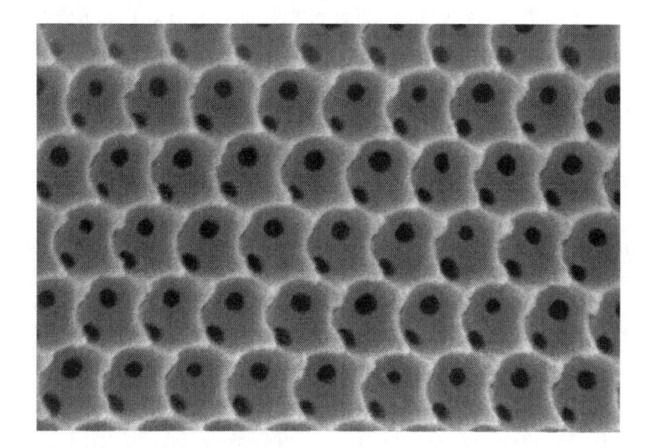

Fig. 4.2 SEM image of the inverted opal

was attached that detects concentrations of QDs as low as 5×10^{-5} mg/ml. The Cytoviva microscope is the first step because it is a good indicator of where to locate aggregations of QDs inside the sample. In order to see inside the PC, the sample needs to be transparent. Drops of specific index-matching oil were used not just to make our samples transparent but also to spread the light equally on the surface. Images were taken under different light intensities to get glimpses of the structure of the opal and the QDs.

The second microscope was the Leica microscope. It is possible to investigate the propagation of light inside our samples by using confocal imaging that this microscope allows. By changing the focus distance and collecting only the focused light, confocal microscopes scan through the sample. Special software is then used to convert consecutive series of images into a three-dimensional (3D) image. This technique requires the samples to be transparent, so the photonic crystals were immersed in oil as for the Cytoviva imaging. Clear images of the light distribution can be taken even by scanning 100 μm into the sample.

The third microscope used is the Nikon confocal microscope. The illumination in the Nikon microscope is achieved by scanning one or more focused beams of light, usually from a laser or arc-discharge source, across the sample. This point of illumination is brought to focus in the sample by the objective lens and laterally scanned using a scanning device under computer control. The sequences of points of light from the sample are detected by a photomultiplier tube (PMT) through a pinhole, and the output from the PMT is built into an image and displayed by the computer. Regardless of the sample preparation employed, a primary benefit of the manner in which confocal microscopy is carried out is the flexibility in image display and analysis that results from the simultaneous collection of multiple images, in digital form, into a computer.

Since we know from the simulation results [22] what to expect, we started investigating experimentally the opal PC system. One of the main difficulties that we faced while doing the experimental measurement was to determine how deep the infiltrated QDs go inside the opal. Indeed, in the first samples we noticed that most of the QDs were concentrating at the interface and that made our conclusion concerning the nature of the focus not as conclusive because it could be just a result of a rough interface. Therefore, we decided first to investigate several samples with different concentration of QDs and analyze them. In the result plotted in Fig. 4.3, three different samples were examined. The three samples were treated the same way and were cut so that the [111] face was the main surface. The samples were then polished to remove any large scratches before being soaked in solutions of CdSe QDs in toluene. Several concentrations of QD solutions have been used to explore the effects of solution concentration on sample infiltration. Two concentrations that gave the best results were solutions with concentrations of 8×10^{-4} mg CdSe QD/ml toluene which will be referred to as the low high concentration (LH) and 8×10^{-5} mg QD/ml toluene which will be referred to as the medium concentration (M). The samples were soaked in these solutions for two and a half days and then imaged within a day of being removed from the solutions. There were three samples studied. The first sample consists of 600 nm opal size infiltrated in LH QDs

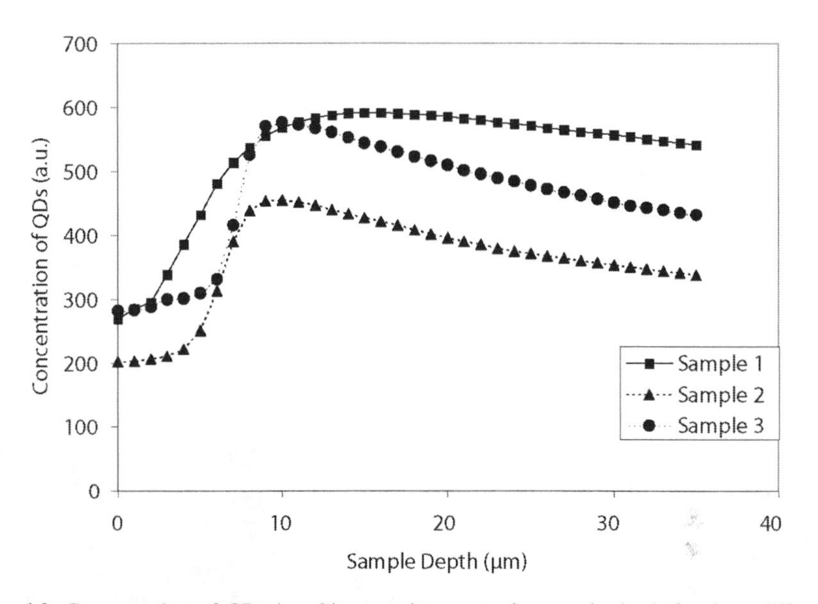

Fig. 4.3 Concentration of QDs in arbitrary units versus the sample depth for three different samples

solution of 599 nm wavelength, the second consists of 800 nm opal size infiltrated in LH QDs solution of 599 nm wavelength, and the third consists of 800 nm opal size infiltrated in M QDs solution of 612 nm wavelength. Before imaging, the samples were polished to remove large surface scratches and excess QD build up on the surface of the opal with 1.0 μm polishing paper. After polishing, the samples were placed on a glass slide and then immersed in ethanol to make the opal transparent to allow imaging inside the opal. The microscope used was an Olympus FluoView 300 with a 60× water immersion objective lens. Imaging started at the bottom of the opal, the top of the glass slide, and proceeded into the opal. The images were taken in 1.0 μm steps for 35 μm.

Figure 4.3 shows the concentration of QDs versus the sample depth for the three samples. The opal–glass interface is shown to be between 0 and 10 μm. One possibility for the uncertainty in determining the interface is change of focus distance when switching from optical to computer scanning. The initial increase in these curves is mainly due to the high concentration of QDs close to the interface. In the case of these samples, this interface is roughly around 10 μm. The concentration starts to decrease afterward except for sample 1 in which it stays quite high deep inside the sample. As Fig. 4.3 shows it clearly, the concentration of QDs remains considerable even after 35 μm and actually the depth would be greater if the ethanol had not dried out leaving the sample opaque and thus too dark to image. It is important to remind the reader that despite the differing size of the opals used in the three samples, the wavelength of the infiltrated QDs and the concentration of the QDs solution, the tendency for high concentration of QDs close to the interface and gradual decrease away from the interface remains the same for all three samples.

Additionally, the knowledge of how deep the QDs can go inside the opal is valuable information knowing that currently scientists in the field are trying to determine the optical properties of materials by the different responses they get from the QDs embedded inside these materials. So, the more these QDs travel inside the samples, the better information we can collect from the samples. Therefore, results of Fig. 4.3 are important for many fields.

Now that we are more confident that the QDs can indeed go as deep as 30 or 40 μm inside the opal, it is equally important to track the path of these QDs. Before proceeding to the result that tracks the propagation of the QDs inside the opal and close to the interface with the glass, we ensure with the dark-field/fluorescent "Cytoviva" microscope that this aggregation of QDs is present and localize its position for better confocal imaging.

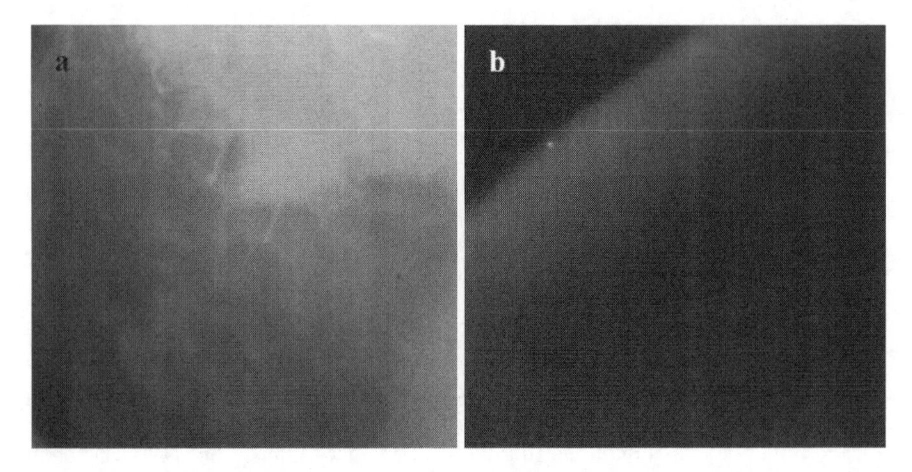

Fig. 4.4 Cytoviva images of the red opal with blue QDs. In (**a**) the opal is seen under fully optical light and in (**b**) it is seen under fully fluorescent light

The sample examined in this case is a red opal with blue CdSe QDs (481 nm) with high (H) concentration (8×10^{-2} mg/ml). QDs have a bright luminescence when seen with this kind of microscope. Figure 4.4 is a Cytoviva image of the opal infiltrated with blue QDs made using the 10× magnifying objective lens. The left picture (Fig. 4.4a) is an optical image of the sample and the right one (Fig. 4.4b) is an image in fluorescent mode. As it is clearly seen from Fig. 4.4a, the sample structure is in the direction [111] and this is shown through the honeycomb shape as indicated in Fig. 4.4a. The right image shows the bright light emanating from the aggregation of QDs under the fluorescent light. Notice that Fig. 4.4b is the top left corner of Fig. 4.4a in which we found this high concentration of QDs. The real-time images from the Cytoviva microscope are good indicators of where to focus during the confocal experiment.

Indeed, the sample shown in Fig. 4.4 was examined under the Leica confocal microscope exactly at the top left side and the generated movie can be provided as a supplementary material. Two frames of this movie are shown in Fig. 4.5. Figure 4.5a

Fig. 4.5 Two snapshots of the two generated movies: (**a**) The snapshot for the whole surface of the imaged red opal with blue QDs and (**b**) the snapshot for the top left part of the surface of the imaged red opal with blue QDs

displays a snapshot of the movie for its entire imaged surface and Fig. 4.5b shows the zoomed part of the top left of the imaged surface. It is important to mention that with this kind of confocal microscope setup, the sample is kept transparent by using specific Leica oil for the duration of all imaging time. The 10× magnifying objective lens was used. The excitation laser was chosen to be 456 nm. The movie was generated using the z scan feature with 1.5 μm step. Thus, the main experimental result is shown on the frames as well as in the movie in which the high concentration of QDs close to the opal–glass interface is observed; however, the very important one is the one situated at the top left as indicated by the Cytoviva image as well as the Leica confocal microscope (Figs. 4.4b and 4.5b). It is well known that if a light is coming from a negative medium and propagating into a positive one, it gets bent in the negative direction giving rise to a focus as was discussed theoretically in [22].

In Fig. 4.5a, different aggregations of the QDs close to the opal–glass interface are seen but the most important one is the aggregation close to the top left and this is why we focus on that huge aggregation in Fig. 4.5b that gives rise to the nice focus observed in the snapshot as well as in the movie, the supplementary material for this chapter.

What we have seen in this result is what is expected from blue QDs that had an excitation frequency right on the band that leads to the negative refraction. A nice focus is being formed above from the interface. Such a result is very important because it leads us to the conclusion that this focus might be due to the negative refraction in nanoscale materials, namely, the opal PCs.

Another red opal was infiltrated with H concentration blue QDs and the results are shown in Fig. 4.6. The images were done using the Leica confocal microscope. A sequence of 60 consecutive scans (2 μm step) was taken and processed into 3D picture. The top view shows the honeycomb structure of the opal infiltrated with 481 nm CdSe QDs. Due to the existence of lots of defects and emptiness between the domains of opals, most of the QDs are concentrated on boundaries of these

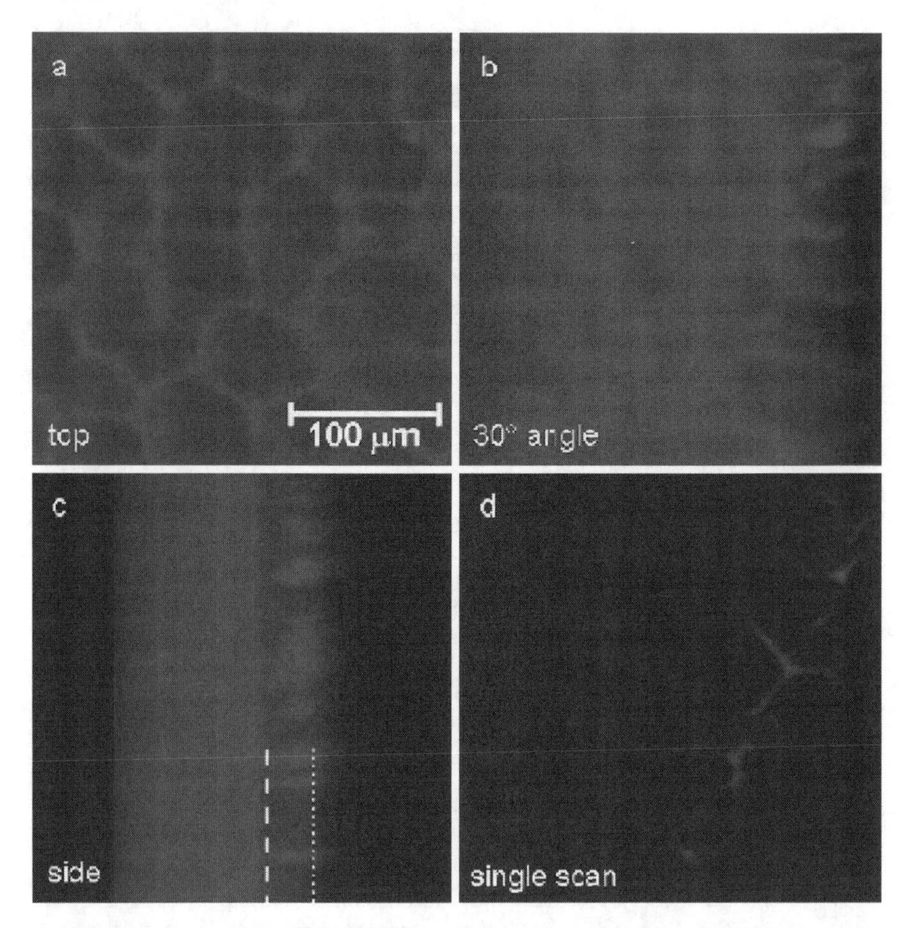

Fig. 4.6 3D reconstructed image of light propagation in red opal infiltrated with 481 nm QDs. (**a**) The top view of the opal. (**b**) 30° inclined from top view. (**c**) The side view of the opal where the dashed line shows opal–glass interface with opal being on the left side and the dotted line is where the single scan was taken. (**d**) The single scan at 30 μm above the surface of the opal

domains (see Fig. 4.6a). The 30° inclined from top-view image shows the focusing and the imaging on the glass side and particularly at the top right of the picture (Fig. 4.6b). The side view (Fig. 4.6c) is clearer and shows the nice focusing and the imaged picture on the glass side. Notice that the opal–glass interface is shown with the dotted line in Fig. 4.6c.

In contrast with the result of the previous figure, this sample shows more homogeneous concentration of the QDs at the opal–glass interface that lead to a nice imaging and beautiful focus in the other side. We succeeded in imaging a single scan of the top view of the glass side plotted in Fig. 4.6d and despite the fact that this image was taken 30 μm above the surface of the opal (the dotted line in Fig. 4.6c), it is clearly seen that an image is being formed on the glass side. It seems that focusing is what is happening at the opal–glass interface. Again, in the glass side, we could recognize the honeycomb shape of the QDs localized primarily on the boundaries of the opal's domains that were translated and refocused in the glass.

It has been shown that the phenomenon of the negative refraction can be seen easily in strong modulation photonic crystal structures in which an effective refractive index can be attributed. Such a strong modulation has a chance to be present not in regular opals but in inverted opals in which the infiltrating material can play an important role and be a key factor for the negative refraction phenomenon. However, one has to choose carefully the infiltrating materials that lead to high refractive index contrast as well as the transparency needed for the material. It is equally important that the material should be nonabsorbing at the band gap frequencies in order to avoid complications arising from electronic transitions. With such criteria in mind we infiltrate the opal with optical glue with a corresponding refractive index of $n = 1.6$. The refractive index is not as high as we would like it to be but with this material we succeeded in having inverted transparent opals with very successful etching phase process.

The inverted red opals were infiltrated later on with blue QDs (481 nm wavelength) in a solution of 8×10^{-4} mg CdSe QD/ml toluene. The samples were soaked in these solutions for 1 day and then imaged within the first few days of being removed from the solutions. Contrary to the regular opals, inverted opals and especially the one infiltrated with optical glue were very soft and curved. Therefore, we escaped the sanding part due to the softness and fragility of the samples. The samples were imaged using the Leica confocal microscope. Figure 4.7 shows the imaged inverted red opal in the [111] direction with its honeycomb structure. The size of the sample was 1 mm by 1 mm and the magnifying lens used was 10×. The sample was not polished; therefore, large surface scratches and excess QD build up on the surface of the inverted opal are seen, namely, the two diagonal lines due to the cuts operated on the sample. In comparison with the regular opal, the inverted one attracts more QDs around the domains and actually those domains of 30–40 microns each looks beautiful in inverted opals because the QDs surround them and shape them beautifully. Thus, the QDs infiltration technique can be used as a test to measure the quality of the infiltration of the inverted opal especially if we know that dissolving the silica spheres and filling the surrounding space with high refractive index material is never a 100% fully done process.

Fig. 4.7 3D reconstructed image of light propagation in red inverted opal with optical glue and infiltrated with 481 nm blue QDs

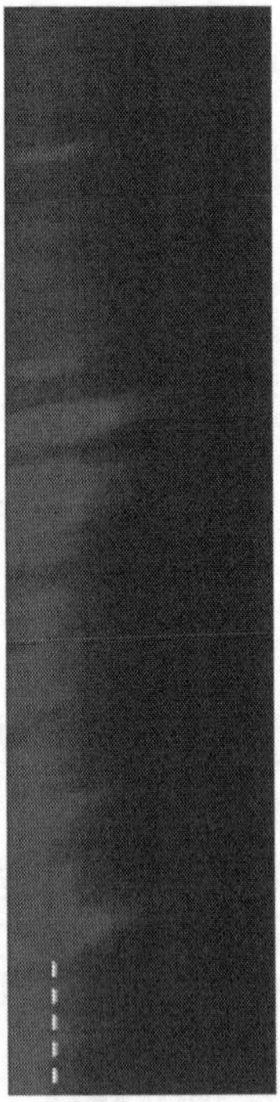

Fig. 4.8 3D reconstructed image of light propagation in inverted red opal with optical glue ($n = 1.6$) infiltrated later on with 481 nm QDs. The *dashed line* indicates roughly the interface between the inverted opal and the glass

The second step was to check with these samples the possibility of focusing. For that, several samples were checked and the result in Fig. 4.8 corresponds to the red inverted opal infiltrated with optical glue and later on with 481 nm CdSe blue QDs at a concentration of 8×10^{-4} mg CdSe QD/ml toluene. Left for a day and removed and imaged right after, the image shows some indication of a converging and close shape form of the focus. The dashed line in the figure approximately indicates the interface between the inverted opal and the glass. In the glass part, many close shape forms are observed along the entire interface indicating focus formation. It is clear from previous works [24] and from the pioneer paper of Notomi [25] that achieving a focus due to a negative refraction can happen in strong modulation structures. However, this phenomenon appears to be occurring for weak modulation in our case. The question that needs to be addressed in future work then becomes: Is this focus due to negative refraction inside the opal or is it due to some other mechanism involving system anisotropy or some other complicated physical phenomenon?

Thus, we have shown in this chapter that infiltrating PC with QDs can be a very useful tool not only to examine optical properties of PCs in photonic band region, as some groups do, but also to follow the dispersion of these photons emitted by QDs inside PC structure and determine whether the right-handed or the left-handed material is used.

Although our experiments have not undoubtedly proved the existence of negative refraction in the upper bands in opals, due to the uncertainty of the photon pathways in defective opal structures, we have demonstrated that there is a clear imaging of QD aggregates accumulated in defective parts with honeycomb shapes. To assign this type of focusing to focusing by negatively refracting beam, we are creating defect-free opal films for future experiments.

In conclusion, we succeeded to show experimentally a strong indication of the formation of focusing at the interface between opal and glass and between inverted opal and glass. The infiltration technique of QDs inside a PC appears to be a useful technique and could be the experimental replacement for the point source. Such an interesting result is very promising knowing that up to date the left-handed behavior is not demonstrated yet at the nanoscale range. Contrary to what was believed before that the infiltrated QDs are mostly very close to the interface, our results demonstrate that the high concentration for a well-prepared sample can be found within a reasonable distance from the interface and decay in a smooth way up to 35 or 40 μm, which opens the door for many physical applications and many perspective uses of this technique for further investigation of complicated structures such as opals.

References

1. Shelby R.A, Smith, D.R., Schultz S.: Experimental verification of a negative index of refraction. Science **292**, 77 (2001)
2. Foteinopoulou, S., Economu, E.N., Soukoulis, C.M.: Refraction in media with a negative refractive index. Phys. Rev. Lett. **90**, 107402 (2003)

3. Smith, D.R., Padilla, W.J., Vier, D.C., Nemat-Nasser, S.C., Schultz, S.: Composite medium with simultaneously negative permeability and permittivity. Phys. Rev. Lett. **84**, 4184 (2000)
4. Veselago, V.G.: The electrodynamics of substances with simultaneously negative values of ε and μ. Sov. Phys. Usp. **10**(4), 509–514 (1968)
5. Pendry, J.B.: Negative refraction makes a perfect lens. Phys. Rev. Lett. **85**, 3966 (2000)
6. Agranovich, V., Shen, Y.R., Baughman, R., Zakhidov, A.: Linear and nonlinear wave propagation in negative refraction metamaterials. Phys. Rev. B **69**, 165112 (2004)
7. Luo, C., Johnson, S.G., Joannopoulos, J.D., Pendry, J.B.: All-angle negative refraction without negative effective index. Phys. Rev. B **65**, 201104® (2002)
8. Cubukcu, E., Aydin, K., Ozbay, E., Foteinopoulou, S., Soukoulis, C.M.: Subwavelength resolution in a two-dimensional photonic-crystal-based superlens. Phys. Rev. Lett. **91**, 207401 (2003)
9. Cubukcu, E., Aydin, K., Ozbay, E., Foteinopoulou, S., Soukoulis, C.M.: Electromagnetic waves negative refraction by photonic crystals. Nature **423**, 604 (2003)
10. Parimi, P.V., Lu, W.T., Vodo, P., Sridhar, S.: Imaging by flat lens using negative refraction. Nature **426**, 404 (2003)
11. Schuster, R., Barth, M., Gruber, A., Cichos, F.: Defocused wide field fluorescence imaging of single CdSe/ZnS quantum dots. Chem. Phys. Lett. **413**, 280 (2005)
12. Vinogradov, A.P., Dorofeenko, A.V., Erokhin, S.G., Inoue, M., Lisyansky, A.A., Merzlikin, A.M., Granovsky, A.B.: Surface state peculiarities in one-dimensional photonic crystal interfaces. Phys. Rev. B **74**, 045128 (2006)
13. Moussa, R., Foteinopoulou, S., Soukoulis, C.M.: Delay-time investigation of electromagnetic waves through homogeneous medium and photonic crystal left-handed materials. Appl. Phys. Lett. **85**, 1125 (2004)
14. Sigalas, M., Soukoulis, C.M., Economou, E.N., Chan, C.T., Ho, K.M.: Photonic band gaps and defects in two dimensions: Studies of the transmission coefficient. Phys. Rev. B **48**, 14121, (1993)
15. Ao, X., He, S.: Adaptive-optics ultrahigh-resolution optical coherence tomography. Optic. Lett. **29**, 2542 (2004)
16. Zakhidov, A.A., Baughman, R.H., Iqbal, Z., Cui, C., Khayrullin, I., Dantas, S., Marti, J., Ralchenko, V.: Carbon structures with three-dimensional periodicity at optical wavelengths. Science **282**, 897 (1998)
17. Aliev, A.E., Zakhidov, A.A., Baughman, R.H.: Chalcogenide inverted opal photonic crystal as infrared pigments. Int. J. Nanosci. **5**(1), 157–172 (2006)
18. Shkunov, M.N., Vardeny, Z.V., DeLong, M.C., Polson, R.C., Zakhidov, A.A., Baughman, R.H.: Tunable, gap-state lasing in switchable directions for opal photonic crystals. Adv. Funct. Mater. **12**, 21 (2002)
19. Berrier, A., Mulot, M., Swillo, M., Qiu, M., Thylén, L., Talneau, A., Anand, S.: Negative refraction at infrared wavelengths in a two-dimensional photonic crystal. Phys. Rev. Lett. **93**, 073902 (2004)
20. Barth, M., Schuster, R., Gruber, A., Cichos, F.: Imaging single quantum dots in three-dimensional photonic crystals. Phys. Rev. Lett. **96**, 243902 (2006)
21. Strauf, S.: Self-tuned quantum dot gain in photonic crystal lasers. Phys. Rev. Lett. **96**, 127404 (2006)
22. Moussa, R., Kuznetsov, A., Neiser, E., Roberson, A.L., Zakhidov, A.A.: Negative refraction in the visible spectrum in photonic crystals: search for focusing by fluorescent quantum dots inside synthetic opals J. Nanophoton., **4**, 043503 (2010)
23. Lopez, C., Vazquez, L., Meseguer, F., Mayoral, R., Ocana, M.: Photonic crystal made by close packing SiO2 submicron spheres. Superlattice. Microst. **22**(3), 399 (1997)
24. Ren, K., Li, Z.Y., Ren, X., Feng, S., Cheng, B., Zhang, D.: Three-dimensional light focusing in inverse opal photonic crystals, Phys. Rev. B **75**, 115108 (2007)
25. Notomi, M.: Theory of light propagation in strongly modulated photonic crystals: Refractionlike behavior in the vicinity of the photonic band gap, Phys. Rev. B **62**, 10696 (2000)

Chapter 5
Self-Assembled Guanosine-Based Nanoscale Molecular Photonic Devices

Jianyou Li, Hadis Morkoç, and Arup Neogi

Abstract The semiconductor industry has seen a remarkable miniaturization trend, driven by innovations in nanofabrication and nanoscale characterization [1]. Semiconductor technology can currently manufacture devices with feature size less than 100 nm. A modern microprocessor can have more than 500 million transistors. Electronic integrated circuits are inherently single-channel connected device arrays within a two-dimensional printed circuit board [2]. By further shrinking transistor size, one approaches the technical, physical, and economical limits, which will be reached within a few years [3]. At the same time, the insulating layer is also getting thinner leading to an enhancement in the current leakage and resulting in short circuit [4]. Manufacture cost increases drastically with further size reduction. As this trend is likely to yield faster and compact electronic and photonic devices, the size of microelectronic circuit components will soon need to reach the scale of atoms or molecules [1]. Those limitations and application requirements inspired extensive research aimed at developing new materials, device concepts, and fabrication approaches that may enable the integrated devices to overcome the limitations of the conventional microelectronic technology. This will require a conceptual design of new device structures beyond CMOS technology which may require alternative materials to overcome these limits. Hybrid organic–inorganic system is one of the alternative solutions.

Keywords self-assembled guanosine · wide-bandgap semiconductor · photonic crystal

1 Introduction

The semiconductor industry has seen a remarkable miniaturization trend, driven by innovations in nanofabrication and nanoscale characterization [1]. Semiconductor technology can currently manufacture devices with feature size less than 100 nm.

J. Li (✉)
Department of Physics, University of North Texas, Denton, TX, USA
e-mail: lijianyou@hotmail.com

Z.M. Wang, A. Neogi (eds.), *Nanoscale Photonics and Optoelectronics*,
Lecture Notes in Nanoscale Science and Technology 9,
DOI 10.1007/978-1-4419-7587-4_5,

A modern microprocessor can have more than 500 million transistors. Electronic integrated circuits are inherently single-channel connected device arrays within a two-dimensional printed circuit board [2]. By further shrinking transistor size, one approaches the technical, physical, and economical limits, which will be reached within a few years [3]. At the same time, the insulating layer is also getting thinner leading to an enhancement in the current leakage and resulting in short circuit [4]. Manufacture cost increases drastically with further size reduction. As this trend is likely to yield faster and compact electronic and photonic devices, the size of microelectronic circuit components will soon need to reach the scale of atoms or molecules [1]. Those limitations and application requirements inspired extensive research aimed at developing new materials, device concepts, and fabrication approaches that may enable the integrated devices to overcome the limitations of the conventional microelectronic technology. This will require a conceptual design of new device structures beyond CMOS technology which may require alternative materials to overcome these limits. Hybrid organic–inorganic system is one of the alternative solutions.

Hybrid organic–inorganic materials can be integrated into one nanoscale composite [5] and harness the properties of both organic and inorganic materials. An atom or a molecule is much smaller than any of the smallest electronic components produced by current semiconductor technology. Moreover, it is easier to replicate a molecular structure in a cost-effective way for mass production of devices. Thus, the dramatic reduction in size, and ease of manufacturing large numbers of devices, can be obtained by the field of molecular electronics facilitated by organic materials. Organic materials are structurally flexible and can have highly efficient luminescence with large degree of polarizability as the molecules generally interact weakly by hydrogen bond and van der Waals force. The manufacture cost of the hybrid devices based on organic materials can be low because they can be mass produced. The electronic properties of the materials can also be tuned. Inorganic

Fig. 5.1 Carbon nanotube–C60–carbon nanotube molecular junction

material molecules typically interact strongly by covalent and ionic bond. So, inorganic materials can have high electrical mobility, a wide range of band gaps and dielectric constants, and substantial mechanical and thermal stability [6]. By appropriate design, the hybrid system can have properties that either component does not exhibit or even have more significant properties than that of either component [7]. During the past years, diode and transistor based on carbon nanotube, DNA, and other materials [8–12] have already been fabricated (Fig. 5.1).

The choice of an appropriate hybrid molecular electronic material system for multifunctional capability is a crucial parameter due to the technical difficulty of controlling the electric contact between molecules and conventional inorganic materials.

2 Photonic Crystals for the Ultraviolet–Visible Region

In the late 1980s, Yablonovitch and John hypothesized that artificial crystals constructed from periodic arrays of dielectric media have the ability to selectively prohibit the propagation of light [13, 14]. Artificial crystals, constructed from a spatially alternating array of materials with differing indices of refraction, have been fabricated with the demonstrated ability to inhibit the propagation of electromagnetic waves of a specific frequency [15, 16]. These crystals, termed photonic band gap (PBG) materials or photonic crystals (PCs), reflect light with a wavelength comparable to the length scale of their dielectric contrast and can be used to create filters, mirrors, resonant cavities, waveguides, optical fibers, and low-threshold lasers. These PBG materials need to have:

- spatial dielectric modulation with periodic length scales comparable to the wavelength of light to be forbidden,
- contrast of indices of refraction of at least 2 for three-dimensional PCs and even more for achieving the forbidden gap at shorter wavelengths (visible or UV), and
- optical transparency in the relevant wavelength range.

PCs that have been fabricated in 1, 2, or 3 dimensions comprise a completely new class of materials with extraordinary optical properties [17–19]. The photonic band structure of PCs can exhibit multiple photonic band gaps, large dispersion, and strong anisotropy in the allowed bands. By exploiting these properties together with nonlinearity, a number of next-generation optoelectronic devices such as waveguides, lasers, modulator, switches, and second-harmonic (SH) generators can be envisaged.

In another analogy to an electronic insulator, photonic crystals can have an optical *band gap* that gives them the ability to control visible light. They can trap, reflect, and guide light in 3D around sharp angles with 100% transmission and it is these properties that have made photonic crystals crucial for the development of high-density integrated optical circuits for optical communications. This has led to the thrust in the fabrication of photonic crystals in the optical communication wavelength regime ranging from 1.3 to 1.55 μm [16, 20]. Although high-refractive-index photonic crystals have been reported in the literature for some time with compelling evidence of total photonic band gaps mainly in the infrared regime, the same has not yet been reported for lower dielectric materials. The lower dielectric materials give much narrower band gaps that are difficult to realize in practice. The refractive index contrast requirement can be less than 2 for 2D PCs, but the band gap is too narrow for practical purposes.

For the control of UV or visible light using photonic crystals, the fabrication of convenient structures poses two problems:

- Materials – to avoid light absorption by the photonic crystal, wide-gap semiconductors, as for example GaN, ZnO, or insulators such as diamonds must be used. Unfortunately, these materials have weak dielectric constants, which reduce the photonic band gap width necessitating novel etching or machining techniques.
- Scale – photonic crystals must also have a period smaller than the wavelength of light to be controlled. This leads to the second problem, due to the technological challenge involved in the nanoscale lithography and etching of material structure.

In practice, it has proved extremely difficult to make 3D photonic crystals on the scale of a micron. Recent improvement in fabrication technology including nanolithography, wafer fusion, etc. has led to the realization of 3D photonic crystal [20, 21]. Some groups have thereby employed a "bottom-up" approach with spontaneous self-organization routes to inverse opal structures or organic molecules, cast from colloidal-sphere templates, but in crystals with a usefully large volume, this can lead to fractures and stacking faults. Organic block copolymers [22] and cholesteric liquid crystals have been used to develop photonic crystals [23]. The organic molecules also have a low refractive index, resulting in a narrow band gap, as a value of n greater than 2 is necessary to create a complete photonic band gap. One early solution to this problem was to use the epoxy material as a template for inorganic structures with larger refractive indices. However, this templating method can lead to shrinkage-induced fractures in the photonic crystals produced. Recently, a hybrid organic–inorganic material system has been proposed [24]. This approach circumvents the problems of template synthesis and allows for the control of the refractive index of a photonic crystal via the variation of the inorganic component of the organic–inorganic composite.

Another attractive alternative to these fabrication techniques also exists in systems that self-assemble, such as those created from DNA junctions. Sauer et al. have designed two- and three-dimensional photonic crystals based on DNA lattices that function either by themselves or as a support scaffolding for the attachment of inorganic colloidal particles [25]. Due to the small dimensions of the DNA arrays, these crystals can reflect light in the UV and soft X-ray portions of the electromagnetic spectrum. Since the DNA lattice can be altered on a nanometer length scale, the width and frequency position of the band gap can be selectively engineered. DNA lattice-based arrays offer a promising new fabrication method for photonic crystals operating at short wavelengths. The modeled DNA-based PCs had very narrow photonic band gaps, making it extremely difficult to measure the transmission or reflection properties without near-field optical techniques. The present status of fabrication techniques leading to the realization of PCs in various wavelength regimes extending from microwave to the soft X-ray regime is shown in Fig. 5.2.

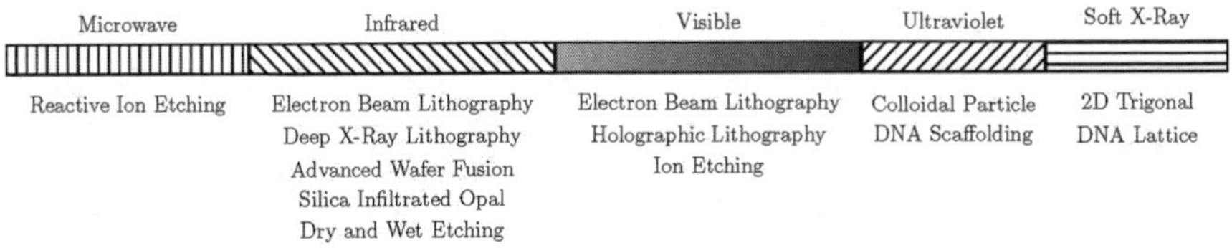

Microwave	Infrared	Visible	Ultraviolet	Soft X-Ray
Reactive Ion Etching	Electron Beam Lithography Deep X-Ray Lithography Advanced Wafer Fusion Silica Infiltrated Opal Dry and Wet Etching	Electron Beam Lithography Holographic Lithography Ion Etching	Colloidal Particle DNA Scaffolding	2D Trigonal DNA Lattice

Fig. 5.2 Present status of wavelength range covered by photonic crystals and related fabrication techniques

Though theoretical simulation has predicted a novel scheme for achieving photonic band gap in the UV–soft X-ray regime, the actual realization of these structures will be extremely challenging and is yet to be realized. Moreover, photonic crystals based solely on pure DNA lattices or conventional polymers are formed of linear chains with an inversion symmetry resulting in low-optical nonlinearity and a low-optical damage threshold. These material systems are therefore not preferred for nonlinear or high-power optoelectronic applications. To overcome the low-dielectric contrast in organic semiconductors, the use of organic semiconductors coupled with high-dielectric-constant inorganic compounds has been proposed. Thus, the future of using hybrid organic–inorganic material system depends on the development of materials with larger inorganic content and higher refractive index [26]. The field is wide open to developments in the synthesis of composite materials that can meet this objective. In this chapter, we present novel hybrid photonic crystals by employing a novel material system consisting of a DNA base encapsulated within highly polar GaN nanoscale confined structures. This approach will result in the development of integrated nanophotonic and biomolecular devices based on hybrid organic–inorganic semiconductors.

2.1 Material System for UV–Visible Photonic Crystals

Among the variety of optical materials, only those with a refractive index roughly greater than 2.0 are capable of supporting a photonic band gap. For fully functional optoelectronics, the III–V semiconductors, such as GaAs, or GaN, will ultimately be preferred because they combine both optical and electronic function. Although there has been considerable progress with other substances, such as TiO2 and silicon, the III–V semiconductors remain the preeminent materials of choice.

2.2 GaN-Based Photonic Crystals

Among III–V semiconductors available for the fabrication of UV–visible photonic band gap structures, nitride-based semiconductors are essential compounds due to their transparency in the visible wavelength regime. Gallium nitride (GaN) is often referred to as the "final frontier of semiconductors" for its physical attributes, its primary performance capabilities, as well as the placement of gallium and nitrogen at the extremes of the periodic table. GaN, with a very wide band gap and a correspondingly high band gap energy of 3.4 eV, has physical properties which translate to fundamental performance advances in the area of high-power, high-frequency power transistors for RF transmission applications, and visible and UV light-emitting diodes. Despite the large inbuilt strain in the nitride system and a high dislocation density, due to the lack of a native or lattice-matched substrate, nitride semiconductors have remarkable optical properties. Due to the anisotropy in its wurtzite crystal structure, GaN semiconductor-based PCs will be attractive for fabricating nonlinear optical devices for optical wavelength conversion from ultraviolet to infrared [27].

A graphite lattice of dielectric rods in an air background has provided one of the most promising two-dimensional (2D) photonic PCs, and recent studies have shown that PBGs are obtained in the near-infrared range for high air filling factors using GaAs ($\varepsilon = 13.6$) [28]. Control of light in the visible wavelength using GaN PCs was proposed using graphite lattices with relatively larger dimensions compared to triangular lattices [29]. Although the optical nonlinearity (such as the second-order coefficient of GaN) [30] is smaller than that of materials such as $LiNbO_3$ or GaAs [31], GaN possesses several advantages:

(i) a wide useful spectral transmission range, from the electronic band gap at 365 nm [32] to the single phonon Reststrahlen absorption band at 13.5 μm [33];
(ii) a mechanical ruggedness;
(iii) a high optical damage threshold; and
(iv) a high internal efficiency leading to a low recombination velocity despite high dislocation density.

A 20-fold enhancement of light extraction by PCs using optical pumping in the nitride material was achieved at 475 nm [34]. Recently, a triangular PC lattice structure was fabricated in GaN using e-beam lithography and ICP dry etching to achieve power enhancement in the ultraviolet regime at 333 nm [35]. However, there are no reports on the fabrication of GaN using graphite lattice structures, which has been predicted [29] to be more efficient compared to the triangular structure fabricated by the Kansas group [34, 35]. The recent success in the nanofabrication of GaN-based PCs has opened up the possibilities of developing new material system for bioconjugation using hybrid organic molecules that can be coupled to GaN material system for the development of biomolecular or biophotonic sensors.

One of the main drawbacks of the nitride system is its relatively low dielectric constant ($\varepsilon = 8$) in the ultraviolet regime, which is further reduced in the infrared regime and results in relatively smaller spatial dimensions. As a result, efforts have been directed toward optimizing the design of the GaN-based structure with larger periodic dimensions, which can be achieved by conventional etching or semiconductor processing techniques. Using organic semiconductors with lower dielectric constant is an effective way to design photonic crystals for the UV–visible regime.

2.3 Diamond-Based Photonic Crystals

Diamond is the hardest known material, has the lowest coefficient of thermal expansion, is chemically inert and wear resistant, offers low friction, has high thermal conductivity, and is electrically insulating and optically transparent from the ultraviolet (UV) to the far infrared (IR). Given these many notable properties, diamond already finds use in many diverse applications including, of course, its use as a precious gem, but also as a heat sink, as an abrasive, and as inserts and/or wear-resistant coatings for cutting tools. Obviously, given its many unique properties, it is possible to envisage many other potential applications for diamond as an engineering material, but progress in implementing many such ideas has been hampered by the comparative scarcity of natural diamond. The so-called industrial diamond has been synthesized commercially for over 30 years using "high-pressure high-temperature (HPHT) techniques," in which diamond is crystallized from metal-solvated carbon at $P \sim$ 50–100 kbar and $T \sim$ 1,800–2,300 K.

The ideal structure for photonic band gaps must recreate, at the optical wavelength scale, the beautiful valence bond structure of diamond crystals at the atomic scale. Diamond-like connectivity or geometry in photonic crystals has provided the widest photonic band gaps observed to date even for relatively low refractive index contrast. Diamonds with a band gap of 5.45 eV are transparent at wavelengths below the GaN band gap energy with a higher refractive index coefficient. This facilitates the fabrication of UV wavelength photonic crystals. Diamonds also have better thermal, optical, and electrical (mobility) properties compared to GaN. The difficulty in processing diamond however restricts the wider application of this system for optoelectronic devices. The availability of focus-ion beam nano-machining opens up the option for realizing photonic crystal using diamond films.

3 Self-assembled Guanine-Based Oligonucleotide Molecules

One of the fundamental goals in the interdisciplinary field of biophotonics or bioelectronics is to realize a material system for nanoscale devices in which a few or a single biomolecule can be used to transfer and process an electronic or optical signal. Among biomolecules, DNA or oligonucleotides has a fundamental role in biological processes. The combination of molecular biology (for engineering DNA with the desired functional and/or self-assembling properties) and nanotechnology (for device fabrication) thus becomes the tool to realize a new class of nanophotonic elements.

DNA has been one of the most investigated class of biomolecules, leading to a somewhat controversial description of its electrical properties, and, hence, of its potential for electronic applications. Depending on the interconnection mechanism (chemical bonding of the DNA on a metal by a selected sequence of oligonucleotides [36], mechanical contact with a gold interdigitated patterns [37], or single DNA molecule immobilized in a metal contact [38]), the DNA molecules have been found to be conductive, nonconductive, or rectifying. Recently, self-assembled DNA bases on graphite for adenine [39] and for guanine [40] have been achieved. In particular, the self-organization of organic molecules on flat surfaces gives structures with a high degree of order, thereby opening a wide range of applications in electronic [41] and optical devices [42]. The spontaneous self-assembly of small molecules from solution directly onto solid surfaces has been used to design two-dimensional organized structures [40, 43, 44]. The hydrogen-bonded networks that can be formed between DNA bases provide the potential for the fabrication of a hybrid molecular electronic material system. While considerable work has been done on unsubstituted bases [36, 37], the self-assembly of ribose-functionalized bases has been recently utilized for the development of molecular electronic devices [45, 46].

Among the conjugated bases forming DNA, the guanine (Fig. 5.3a) represents a versatile molecule that, depending on the environment, can undergo different self-assembly pathways. In the presence of cations, it is notorious for its propensity to associate into planar tetrameric nanostructures, namely G-quartets (G4 in Fig. 5.3b), which are a thermodynamically stable architecture consisting of a cyclic array of four guanines joined by hydrogen bondings [47, 48]. In these structures, each base behaves equally both as donor and as acceptor in four H bonds with its neighbors. By varying the ratio of the concentration of cation vs. molecule, one can build up supramolecular structures composed of stacks of guanine quartets; this approach allows the development of octamers, quadruplexes, or even polymeric species (Fig. 5.3b) [49, 50]. In this regard, by simply casting a guanine solution containing $\sim 10^{-2}$ M cations like K^+ or Na^+, Henderson and coworkers have grown dry tubular nanostructures, known as G wires, lying flat at a surface [51]. By tuning the chemical composition of a guanosine derivative, it is possible to form G4 also in the absence of a templating metal cation [52]. Alternatively, in the absence of cations, guanine derivatives have usually been found to self-assemble into hydrogen-bonded networks [37, 40]. It has been demonstrated that the physisorption of a lipophilic deoxyguanosine derivative on differently treated mica surfaces can be driven toward distinct supramolecular species consisting either of layers of G4 or hydrogen-bonded nanoribbons.

Our group has recently investigated the effect of self-assembly on a highly polar semiconductor surface such as GaN [53, 54]. GaN is a highly polar semiconductor with a large strain-induced polarization field developed due to the noncentrosymmetric nature of the crystalline symmetry and the absence of a naturally available

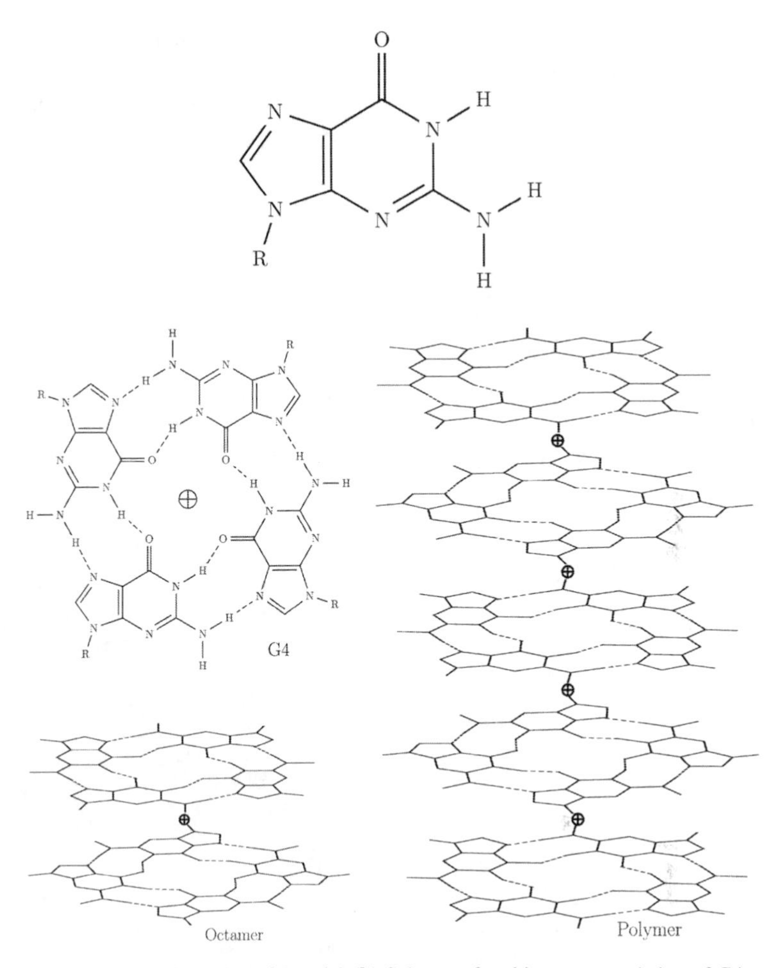

Fig. 5.3 (**a**) Chemical formulae of 1 and 2 (**b**) Scheme of architecture consisting of G4. quartet: the four guanines are joined by hydrogen bonding with a central cavity that can be filled by a cation such as K^+. Stacked species such as octamers or more complex polymers can also be formed

substrate for the epitaxial growth of these materials [55]. The surface of GaN layers can be tailored to have positive or negative polarity via Ga or N termination of the cap layer. The existence of polarization charges at the surface and interface results in domain formation of electronic charges as observed by electron-force microscopy [56]. The formation of domains with a specific polarity has been utilized for the enhancement of the self-assembly behavior of guanosine molecules by suitable modification of the side chains.

The primary choice of a deoxyguanosine–GaN interface is due to the similarity in their optical properties. By analogy to solid-state physics, these biomolecular systems have been termed as self-assembled guanosine crystals (SAGCs) [44, 46, 57]. The mobility of charge carriers for electronic transport in the deoxyguanosine film depends on the ordering of the molecules. The self-assembly process resulting in the formation of a biological semiconductor occurs only at a specific concentration of the solution and great care has to be taken in order to control the solid-state assembly of the molecules. The self-assembled deoxyguanosine films behave like wide-gap semiconductors, with energy gap in the range of 3–3.5 eV and electron effective mass $m_e > 2m_0$ (similar to GaN).

The energy gap onset (3.4 eV) demonstrated by current–voltage measurement [53, 58, 59] demonstrates the band-like description of this particular biomolecular material at the optimum concentration, and its similarities to other wide-band-gap materials, such as GaN, ZnO, and CdSe, with near UV absorption onset (Fig. 5.4a). This band gap energy range of the SAGC molecule is resonant to the exciton energy or the free carrier absorption band edge of GaN substrate, which facilitates energy transfer of the radiative component of the electron–hole recombination energy from the GaN semiconducting layer to the biomolecular layer via resonant

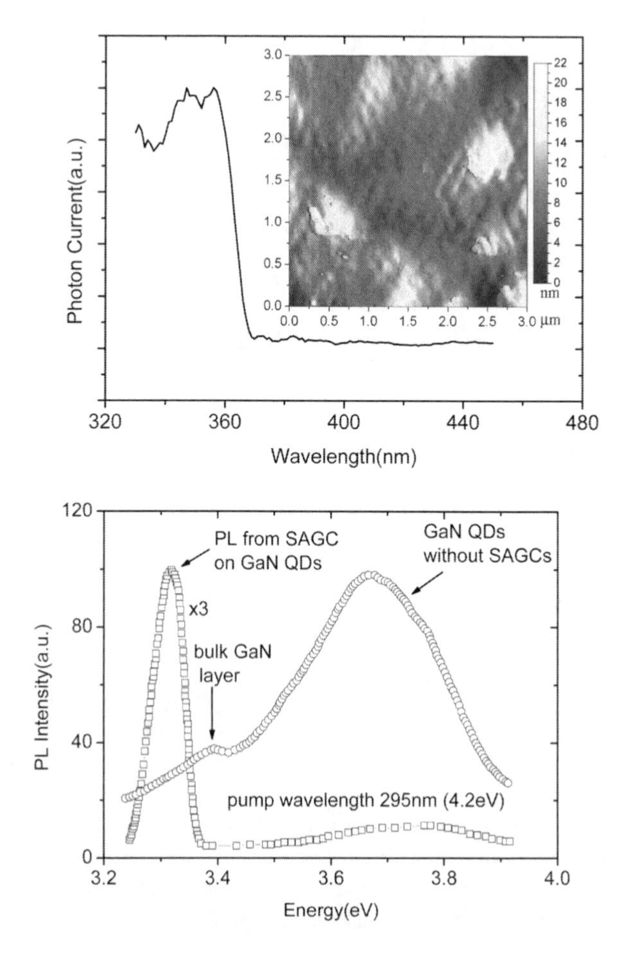

Fig. 5.4 (**a**) Photocurrent spectrum and AFM image of ribbon-like SAGC showing semiconductor band-edge characteristics (**b**) Enhancement of photoluminescence in self-assembled guanosine molecules conjugated to GaN quantum dots due to radiative energy transfer [54]

surface plasmon interaction. We have recently synthesized SAGCs confined to <100 nm spatial pits and measured the refractive index of SAGC thin film by variable angle spectroscopic ellipsometry [53]. The dielectric constant was estimated to be very low ($\sim$1.19) compared to 2.4 for GaN, thereby offering a large refractive index contrast exceeding 2.

SAGCs thus offer a unique combination of properties, which can be used for conjugation with GaN-based nanostructures. These molecules will avoid aggregation, in addition to providing a means for bioconjugation. Most importantly, they will provide stability to the biophotonic system, by means of covalent attachment of the bioanalytes to the surface functionality present on the SAGC. Our group has synthesized SAGC, which has the lowest oxidation potential among DNA bases and has a strong dipole moment of the order of 7 Debye. The strong dipole moment provides polarity to each self-assembled guanosine conjugated supramolecular structures. We have recently demonstrated resonant energy transfer from GaN quantum dots to enhance the photoluminescence emission from the conjugated self-assembled guanosine when confined within 50–100 nm spaces as shown in Fig. 5.4b [54]. This result opens up the opportunity for the development of guanine-based novel biophotonic devices. We use this promising material system for the development of UV–visible photonic band gap structures by employing the self-assembly properties of SAGCs on GaN nanostructures which has been achieved in our group [54, 60].

The lattice constants of photonic crystal based on GaN are in the order of 100 or 200 nm in UV to blue region. The modified deoxyguanosine molecules can self-assemble into crystal structure inside holes of GaN-based photonic crystals if the deoxyguanosine solution is infiltrated into the holes. By tuning lattice constant, hole radius, and slab thickness, the density of states of light can be tuned in the photonic crystals. So the coupling efficiency of deoxyguanosine to light source can be improved when the deoxyguanosine molecules are detected or work as functional linker to sense or label other biomolecules. Similar method was used in polymer hydrogel photonic crystals to sense glucose and lead [61–63]. The hybrid photonic crystal of GaN and SAGC with slab structure and no defects are simulated in this chapter.

4 Refractive Index Measurement of SAGC by Ellipsometer

We used variable angle spectroscopic ellipsometry (VASE) to measure it. VASE is a very accurate method to measure refractive index of materials. As light is reflected from a surface, s- and p-polarizations change differently from Fresnel formulas which Fresnel derived in 1823 [64].

$$
\begin{aligned}
r_{\mathrm{s}} &= \left(\frac{E_{\mathrm{r}}}{E_{\mathrm{i}}}\right)_{\mathrm{s}} = \frac{n_{\mathrm{i}}\cos(\theta_{\mathrm{i}}) - n_{\mathrm{t}}\cos(\theta_{\mathrm{t}})}{n_{\mathrm{i}}\cos(\theta_{\mathrm{i}}) + n_{\mathrm{t}}\cos(\theta_{\mathrm{t}})} \\
r_{\mathrm{p}} &= \left(\frac{E_{\mathrm{r}}}{E_{\mathrm{i}}}\right)_{\mathrm{p}} = \frac{n_{\mathrm{t}}\cos(\theta_{\mathrm{i}}) - n_{\mathrm{i}}\cos(\theta_{\mathrm{t}})}{n_{\mathrm{t}}\cos(\theta_{\mathrm{i}}) + n_{\mathrm{i}}\cos(\theta_{\mathrm{t}})}
\end{aligned}
\qquad (5.1)
$$

where r_s and r_p are the reflection coefficients of s- and p-polarizations, respectively; E_r and E_i are electric fields of the reflected and incident light, respectively; n_i and n_t are the refractive indexes of medium that incident and refracted lights travel in; and θ_i and θ_t are the incident and refracted angles relative to the normal of the two medium interface. From Snell's law

$$n_i \sin(\theta_i) = n_t \sin(\theta_t) \tag{5.2}$$

If

$$\theta_i + \theta_t = 90^\circ$$

Then,

$$n_i \cos(\theta_t) = n_t \cos(\theta_i) \tag{5.3}$$

Substituting the above equation into Eq. (5.1) for p-polarization, we can see that $r_p = 0$. Under the condition of Eq. (5.3), the incident angle $\theta_B = \theta_i$ is

$$\theta_B = \tan^{-1}\left(\frac{n_t}{n_t}\right)$$

So, the reflection of p-polarization light just vanishes at the incident angle θ_B. This angle is called Brewster's angle. Define

$$\rho = \frac{r_p}{r_s} = \tan(\Psi)\, e^{i\Delta} \tag{5.4}$$

where r_p is the complex reflection coefficient of p-polarization light, r_s is the complex reflection coefficient of s-polarization, tan (ψ) is the amplitude ratio of p- to s-polarization of reflection coefficient, and Δ is the phase change. Ellipsometry results can be very accurate because it measures the relative change of p- and s-polarization light rather than the absolute value. This theory is used for reflection mode of ellipsometer and there are ellipsometers that depend on transmission model. For thin films, ellipsometer can obtain the refractive index and the thickness simultaneously by changing the incident angle. When the incident angle is close to Brewster's angle, the reflection is weak for p-polarization and strong for s-polarization. The change of amplitude and phase is large. So the variable angle spectroscopic ellipsometry takes the measurement at several different incident angles, which are around Brewster's angle. Then the measured data of ψ and Δ are fitted with models depending on the samples structure to obtain refractive index and probably thin film thickness. To measure the refractive index of SAGC, we deposit deoxyguanosine solution on GaN substrate and a thin film is formed. Then the samples were measured with VASE at incident angles of 60°, 65°, and 70° (as shown in Fig. 5.5). The ψ and Δ were fitted with Cauchy relation for refractive index. The refractive index of SAGC was found to be about 1.19 at around 400 nm.

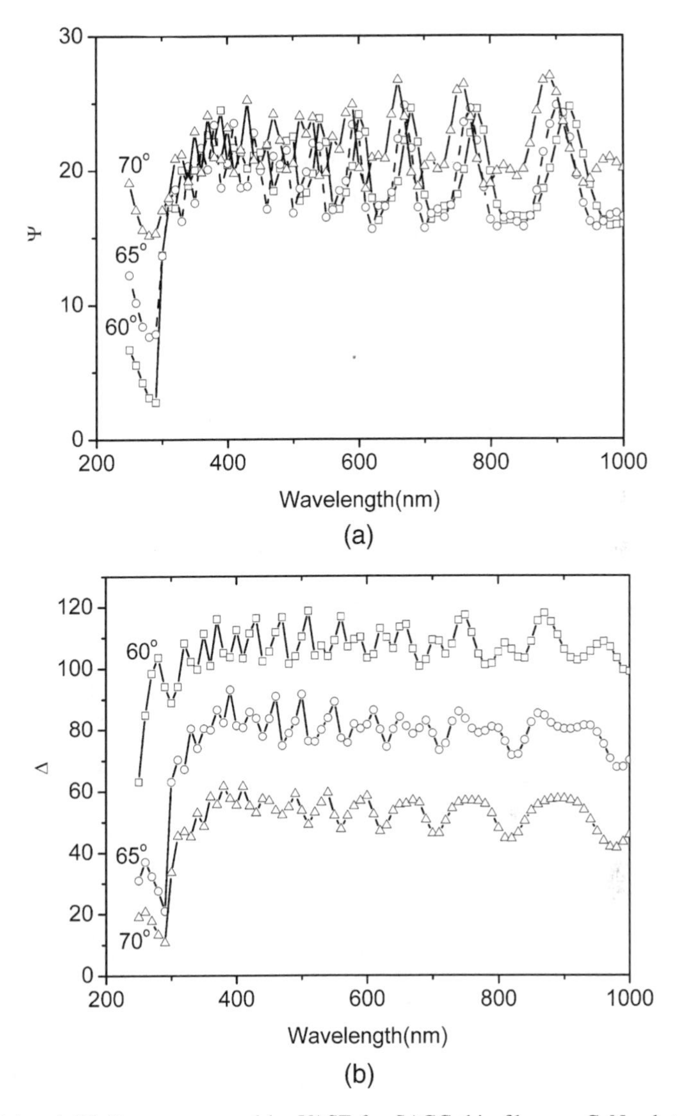

Fig. 5.5 (**a**) and (**b**) Curves measured by VASE for SAGC thin films on GaN substrate at the incident angles of 60°, 65°, and 70°

5 Modeling of Photonic Crystal

5.1 Design of Photonic Crystal with Software MPB

The substrate of photonic crystal is a freestanding GaN slab and the air is above and beneath the slab. There are hole arrays in the slab to form photonic crystals. The PC slab can also be considered as periodic in the direction perpendicular to the slab if the slab is sandwiched between two air layers and the air layers are thick enough [65]. So the guided modes in neighboring slab do not interfere. The guided modes can be separated to TE- and TM-like modes because the slab is mirror symmetric about the center plane [65]. We only design the triangular structure because triangular has larger band gap in TE-like mode. The refractive index of GaN is about 2.53 around 400 nm [66]. The band gap is small because the refractive index contrast is small between SAGC/air and GaN.

Because the calculation of band structure is much slower in three dimension than that in two dimension, first we find the hole radius in two dimension so that the band gap is large. From Fig. 5.6, we can see that the maximum gap size is reached when the hole radius is about $0.4a$, where a is the lattice constant. Here, we use unit based on lattice constant because Maxwell's equations are scale-invariant. After the infiltrating deoxyguanosine inside holes, the band gap reduces because of the refractive index change. Because the band gap of PC slab depends on the thickness of the slab [65], we are trying to find the trend of band gap changing with slab thickness, h, at certain hole radius, r, and concentrate around the regime with hole radius $r = 0.4a$.

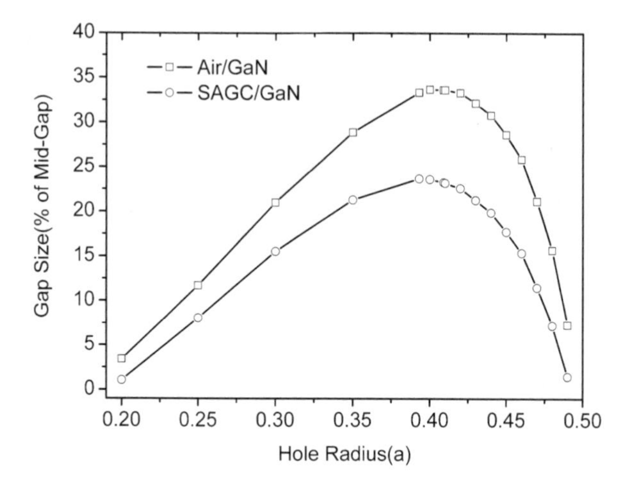

Fig. 5.6 The gap size of 2D triangular lattice vs. hole radius for air/SAGC and GaN photonic crystal

5.2 Photonic Crystal Slab with $r = 0.4a$

From Fig. 5.7, we can see that the lowest two bands of SAGC/GaN have lower energy than that of air/GaN with the same lattice parameters. The difference between the conduction bands under two conditions is even larger compared with the difference of valence bands. The band gap size of SAGC/GaN is smaller because of the lower refractive index contrast. The band gap size under two situations has maximum values, but at different slab thickness. The maximum band gap is at

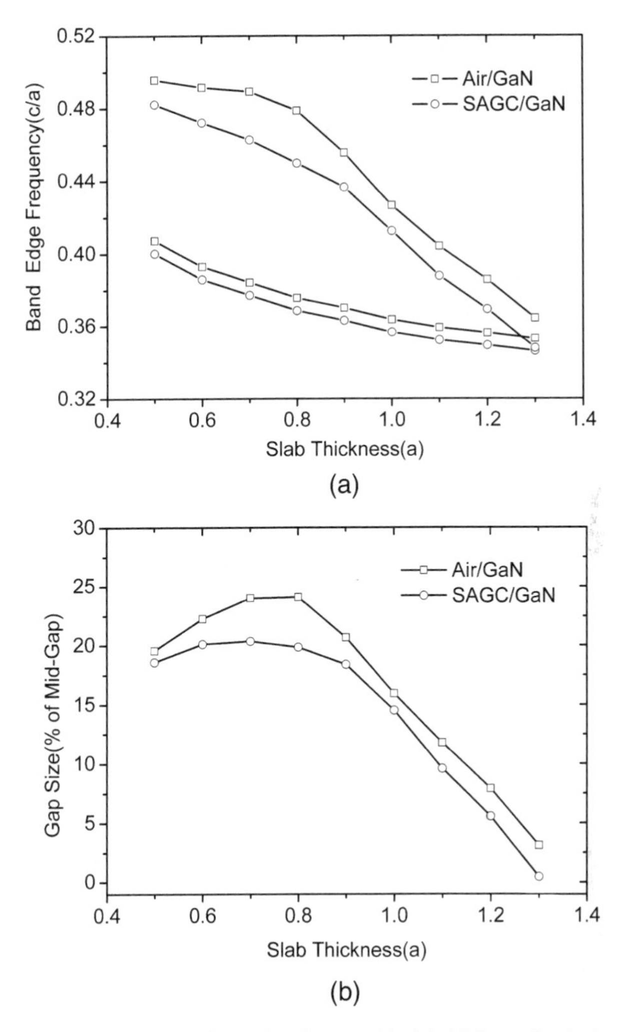

Fig. 5.7 (**a**) Band edges and (**b**) band gap size change with slab thickness for the lowest two bands of photonic crystal slab at hole radius $r = 0.4a$

$h = 0.8a$ and $h = 0.7a$ for air/GaN and SAGC/GaN, respectively. The band gap size increases slowly with the increasing of slab thickness, then reduces rapidly after the maximum. Finally, the band gap disappears. The trend of band gap changing with slab thickness just likes what was discussed by Johnson et al. [65]. If the slab thickness is too large, the energy difference between two neighboring bands is too small and probably cannot be differentiated. New modal plane will be formed in vertical direction with very little energy. On the other hand, the mode inside the slab can be weakly guided if the slab thickness is too thin. The slab is only a weak perturbation to the background. So, there exists a optimum thickness with maximum band gap.

5.3 Photonic Crystal Slab with $r = 0.44a$

Figure 5.8 shows the band edge of the lowest two bands and the band gap information changing with slab thickness for the hole radius $r = 0.44a$. The curves show similar characteristics to that of the structure with $r = 0.4a$. But the differences of conduction and valence bands between air/GaN and SAGC/GaN are larger than that of the structure with $r = 0.4a$. The band edges are higher and the band gap is

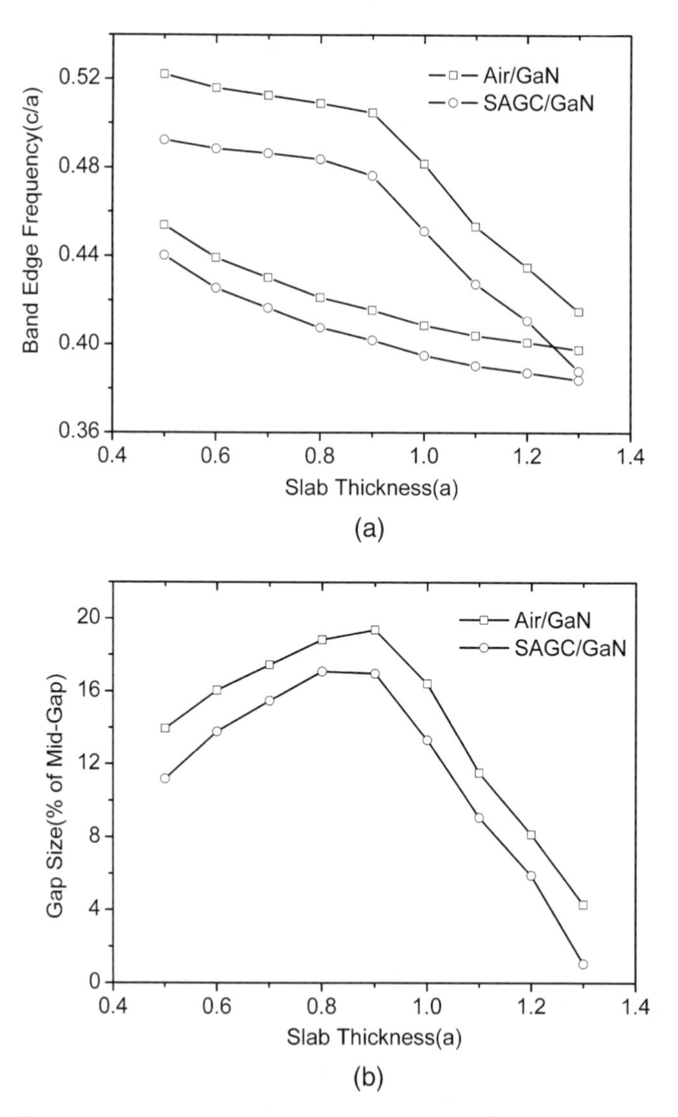

Fig. 5.8 (**a**) Band edges and (**b**) band gap size change with slab thickness for the lowest two bands of photonic crystal slab at hole radius $r = 0.44a$

smaller when $r = 0.44a$. The band gaps reach the maximum values at $h = 0.9a$ and $h = 0.8a$ for air/GaN and SAGC/GaN systems, respectively.

5.4 Photonic Crystal Slab with $r = 0.35a$

Figure 5.9 shows the band edge of the two lowest bands and the band gap change with slab thickness for the hole radius $r = 0.35a$. The band edges have a similar trend to the other two hole radii and the differences of the conduction and the valence bands between air/GaN and SAGC/GaN PC slab are smaller than that of the structure for $r = 0.4a$ and $r = 0.44a$. But the band edges are at lower energy. The band gaps are small and reach the maximum value at $h = 0.7a$ and $h = 0.6a$ for air/GaN and SAGC/GaN PC slab, respectively.

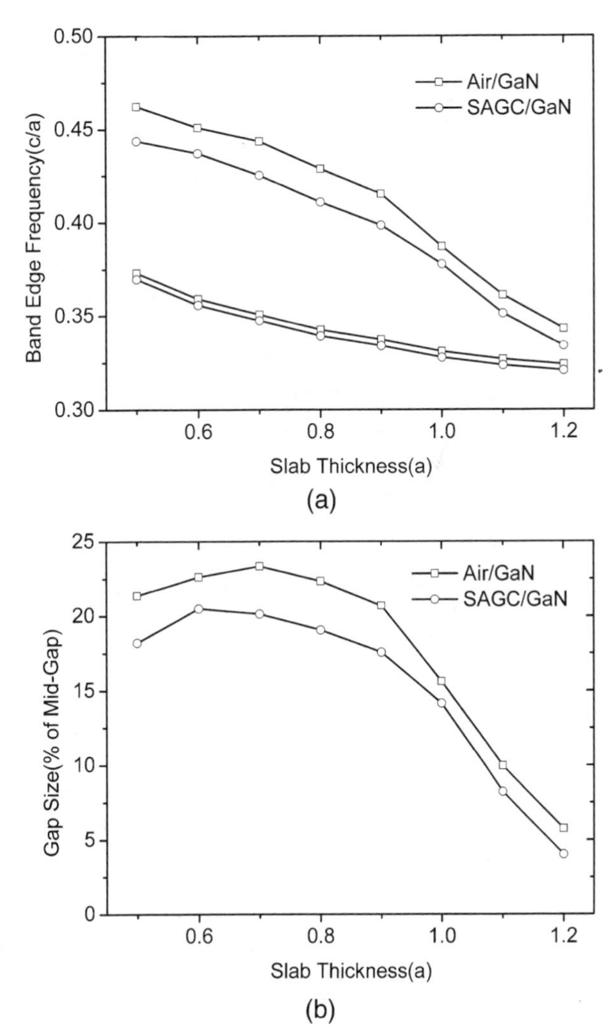

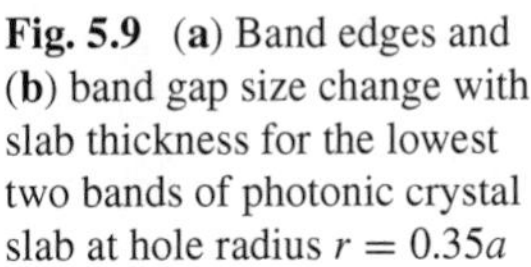

Fig. 5.9 (**a**) Band edges and (**b**) band gap size change with slab thickness for the lowest two bands of photonic crystal slab at hole radius $r = 0.35a$

5.5 Verification of the Photonic Crystal Designs by EMPLab^{TM}

According to the above analysis, the structures in Table 5.1 are simulated by the software EMPLab^{TM}. The center of the band gap is at 3.1 eV (400 nm), in the blue–UV region. Because electromagnetic modes in band gap cannot exist in photonic crystal, transmission simulation is used to get the band gap. In the simulation, light source is a modulated Gaussian pulse that does not have direct current component. The frequency spectrum is a Gaussian peak with center shifted from zero frequency. The perfect-matched-layer absorbing boundary has eight layers.

Table 5.1 Lattice parameters of the photonic crystal slab

r/a	h/a	a (nm)	r (nm)	h (nm)	AG VB (eV)	AG CB (eV)	SG VB (eV)	SG CB (eV)
0.40	0.7	175	70	122	2.73	3.47	2.68	3.28
0.44	0.9	191	84	172	2.69	3.51	2.60	3.14

Figures 5.10 and 5.11show the simulation results. The left part of each figure is the band diagram that is solved by the software MPB. The black curve is the edge of the light cone. In the light cone, electromagnetic mode can exist in both air and PC slab. So, the light that travels inside PC slab can leak into the air and the transmission gets lower. The right part of the figures is the transmission spectrum which is obtained from the simulation results by the software EMPLab^{TM}. The transmission figures are rotated so that it can align with the band diagrams for easy comparing. We simulated two light propagation directions, ΓM and ΓK, as shown in the figures. The light gray strip in the figures shows the band gap of the structure. The valence band edge in band diagram matches with the drop of transmission. At the conduction band edge, the transmission changes relatively slow and has a small discrepancy probably due to the leakage of light from PC slab to the air. The transmission spectrum shows that the band gap shrinks as deoxyguanosine molecules self-assemble inside holes. And the energy of the valence band edge is also decreased, just as shown by MPB.

6 Discussion

The periodic structure based on GaN and SAGC can have photonic band gaps. The band gap size and the band edges can be tuned by tuning lattice parameters. Light propagation and emission can be tuned by photonic crystals. So, the hybrid photonic crystal can be potentially used to detect deoxyguanosine molecules. If deoxyguanosine molecules are used as functional linker to other biomolecules which usually absorb or emit light in the blue to UV region, the hybrid photonic crystal can also be used to tune the coupling of light source to deoxyguanosine molecules, and then to other biomolecules.

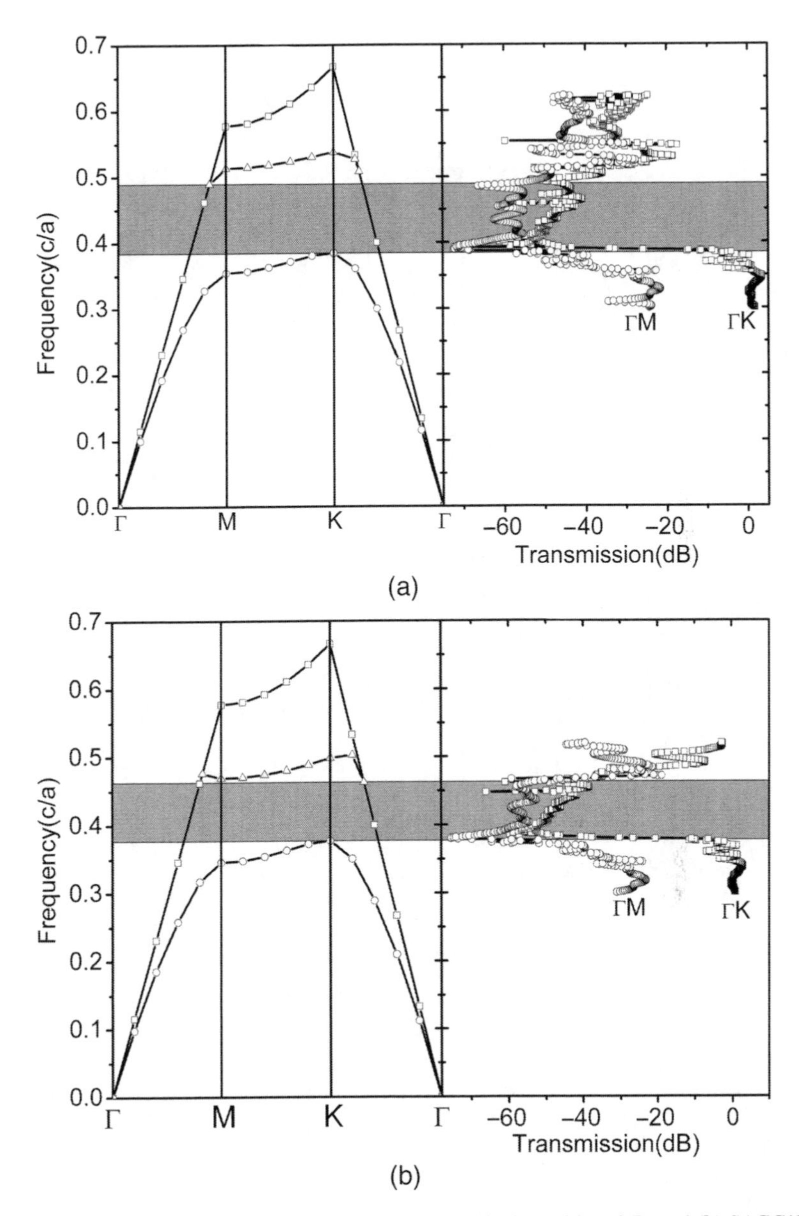

Fig. 5.10 Simulation results of (**a**) air/GaN PC with $r = 0.40a$ and $h = 0.7a$ and (**b**) SAGC/GaN PC with $r = 0.40a$ and $h = 0.7a$. In each figure, the *left* part is the band diagram and the *right* part is the transmission result obtained by EMPLab^{TM} in ΓK and ΓM direction. The light gray strip is the forbidden band of the structure

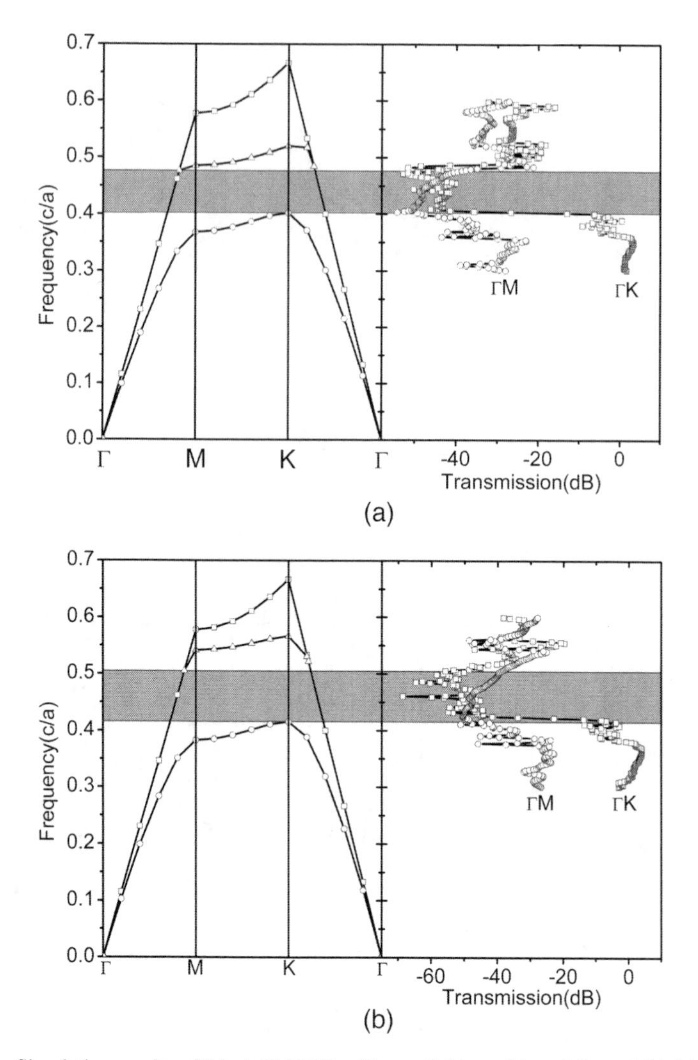

Fig. 5.11 Simulation results of (**a**) air/GaN PC with $r = 0.44a$ and $h = 0.9a$ and (**b**) SAGC/GaN PC with $r = 0.44a$ and $h = 0.9a$. In each figure, the *left* part is the band diagram and the *right* part is the transmission result obtained by EMPLab^{TM} in ΓK and ΓM direction. The light gray strip is the forbidden band of the structure

References

1. Moore, G.E.: Cramming more components onto integrated circuits. Electronics **38**, 114 (1965)
2. Chutinan, A., John, S.: Light localization for broadband integrated optics in three dimensions. Phys. Rev. B **72**, 161316(R) (2005)
3. Peercy, P.S.: The drive to miniaturization. Nature **406**, 1023 (2000)
4. Kingon, A.I., Maria, J.-P., Streiffer, S.K.: Alternative dielectrics to silicon dioxide for memory and logic devices. Nature **406**, 1032 (2000)

5. Mitzi, D.B., Chondroudis, K., Kagan, C.R.: Organic-inorganic electronics. IBM J. Res. Dev. **45**, 29 (2001)
6. Mitzi, D.B.: Synthesis, structure, and properties of organic-inorganic perovskites and related materials. In: Karlin, K.D. (ed.) Progress in Inorganic Chemistry. 58, 1. Wiley, New York, NY (1999)
7. Lakhtakia, A., Mackay, T.G.: Meet the metamaterials. Optics Photonics News **18**, 32 (2007)
8. Braun, E., Eichen, Y., Sivan, U., Ben-Yoseph, G.: DNA-templated assembly and electrode attachment of a conducting silver wire. Nature **391**, 775 (1998)
9. Javey, A., Kim, H., Brink, M., Wang, Q., Ural, A., Guo, J., McIntyre, P., McEuen, P., Lundstrom, M., Dai, H.: High-κ dielectrics for advanced carbon-nanotube transistors and logic gates. Nat. Mater. **1**, 241 (2002)
10. Keren, K., Berman, R.S., Buchstab, E., Sivan, U., Braun, E.: DNA-templated carbon nanotube field-effect transistor. Science **302**, 1380 (2003)
11. Maruccio, G., Visconti, P., Arima, V., D'Amico, S., Biasco, A., D'Amone, E., Cingolani, R., Rinaldi, R., Masiero, S., Giorgi, T., Gottarelli, G.: Field effect transistor based on a modified DNA base. Nano Lett. **3**, 479 (2003)
12. Tans, S.J., Verschueren, A.R.M., Dekker, C.: Room-temperature transistor based on a single carbon nanotube. Nature **393**, 49 (1998)
13. Yablonovitch, E.: Inhibited spontaneous emission in solid-state physics and electronics. Phys. Rev. Lett. **58**, 2059 (1987)
14. John, S.: Strong localization of photons in certain disordered dielectric superlattices. Phys. Rev. Lett. **58**, 2486 (1987)
15. Yablonovitch, E., Gmitter, T.J.: Donor and acceptor modes in photonic band structure. Phys. Rev. Lett. **67**, 3380 (1991)
16. Noda, S., Tomoda, K., Yamamoto, N., Chutinan, A.: Full three-dimensional photonic bandgap crystals at near-infrared wavelengths. Science **289**, 604 (2000)
17. Benisty, H., Weisbuch, C., Labilloy, D., Rattier, M., Smith, C.J.M., Krauss, T.F., de la Rue, R.M., Houdre, R., Oesterle, U., Jouanin, C., Cassagne, D.: Optical and confinement properties of two-dimensional photonic crystals. J. Lightwave Techn. **17**, 2063 (1999)
18. Painter, O., Lee, R.K., Scherer, A., Yariv, A., O'Brien, J.D., Dapkus, P.D., Kim, I.: Two-dimensional photonic band-gap defect mode laser. Science **284**, 1819 (1999)
19. Johnson, S.G., Joannopoulos, J.D.: Three-dimensionally periodic dielectric layered structure with omnidirectional photonic band gap. Appl. Phys. Lett. **77**, 3490 (2000)
20. Qi, M., Lidorikis, E., Rakich, P.T., Johnson, S.G., Joannopoulos, J.D., Ippen, E.P., Smith, H.I.: A three-dimensional optical photonic crystal with designed point defects. Nature **429**, 538 (2004)
21. Ogawa, S., Imada, M., Yoshimoto, S., Okano, M., Noda, S.: Control of light emission by 3D photonic crystals. Science **305**, 227 (2004)
22. Pisignano, D., Persano, L., Gigli, G., Visconti, P., Stomeo, T., De Vittorio, M., Barbarella, G., Favaretto, L., Cingolani, R.: Planar organic photonic crystals fabricated by soft lithography. Nanotechnology **15**, 766 (2004)
23. Cao, W., Muñoz, A., Palffy-Muhoray, P., Taheri, B.: Lasing in a three-dimensional photonic crystal of the liquid crystal blue phase II. Nature Materials **1**, 111 (2002)
24. Saravanamuttu, K., Blanford, C.F., Sharp, D.N., Dedman, E.R., Turberfield, A.J., Denning, R.G.: Sol-gel organic-inorganic composites for the 3-D holographic lithography of photonic crystals. Chem. Mater. **15**, 2301 (2003)
25. Winfree, E., Liu, F., Wenzler, L.A., Seeman, N.: Design and self-assembly of two-dimensional DNA crystals. Nature **394**, 539 (1998)
26. Saravanamuttu, K., Andrews, M.P.: Spontaneously ordered sol-gel composites with sub-micron periodicity. Chem. Mater. **15**, 14 (2003)
27. Coquillat, D., Torres, J., Peyrade, D., Legros, R., Lascaray, J.P., Le, M., Vassor d'Yerville, E. Centeno, Cassagne, D., Albert, J.P., Chen, Y., De La Rue, R.M.: Equifrequency surfaces in a two-dimensional GaN-based photonic crystal. Opt. Express **12**, 1097 (2004)

28. Chen, Y., Faini, G., Launois, H., Etrillard, J.: Fabrication of two-dimensional photonic lattices in GaAs: the regular graphite structures. Superlattices and Microstructures **22**, 109 (1997)
29. Barra, A., Cassagne, D., Jouanin, C.: Visible light control by GaN photonic band gaps. Phys. Stat. Sol. **176**, 747 (1999)
30. Chowdhury, A., Ng, H.M., Bhardwaj, M., Weimann, N.G.: Second-harmonic generation in periodically poled GaN. Appl. Phys. Lett. **83**, 1077 (2003)
31. Shoji, I., Kondo, T., Ito, R.: Second-order nonlinear susceptibilities of various dielectric and semiconductor material. Opt. Quantum Electron. **34**, 797 (2002)
32. Neogi, A., Everitt, H., Morkoç, T., Kuroda, T., Tackeuchi, A.: Size dependence of radiative efficiency in GaN/AlN quantum dots. IEEE Trans. Nanotechnology **4**, 723 (2004)
33. Kasic, A., Schubert, M., Einfeldt, S., Hommel, D., Tiwald, T.E.: Free-carrier and phonon properties of n- and p-type hexagonal GaN films measured by infrared ellipsometry. Phys. Rev. B **62**, 7365 (2000)
34. Oder, T.N., Shakya, J., Lin, J.Y., Jian, H.X.: III-Nitride photonic crystals. Appl. Phys. Lett. **83**, 1231 (2004)
35. Shakya, J., Kim, K.H., Lin, J.Y., Jian, H.X.: Enhanced light extraction in III-nitride ultraviolet photonic crystal light-emitting diodes. Appl. Phy. Lett. **85**, 142 (2004)
36. Braun, E., Eichen, Y., Sivan, U., Ben-Yoseph, G.: DNA-templated assembly and electrode attachment of a conducting silver wire. Nature **391**, 775 (1998)
37. Okahata, Y., Kobayashi, T., Tanaka, K., Shimomura, M.: Anisotropic electric conductivity in an aligned DNA cast film. J. Am. Chem. Soc. **120**, 6165 (1998)
38. Porath, D., Bezryadin, A.: Direct measurement of electrical transport through DNA molecules. Nature **403**, 635 (2000)
39. Uchihasi, T., Okada, T., Sugawara, Y., Yokoyama, K., Morita, S.: Self-assembled monolayer of adenine base on graphite studied by noncontact atomic force microscopy. Phys. Rev. B **60**, 8309 (1999)
40. Sowerby, S.J., Edelwirth, M., Heckl, W.M.: Self-assembly at the prebiotic solid liquid interface: structures of self-assembled monolayers of adenine and guanine bases formed on inorganic surfaces. J. Phys. Chem. B **102**, 5914 (1998)
41. Bumm, L.A., Arnold, J.J., Cygan, M.T., Dunbar, T.D., Burgin, T.P., Jones II, L., Allara, D.L., Tour, J.M., Weiss, P.S.: Are single molecular wires conducting? Science **271**, 1705 (1996)
42. Friend, R.H., Gymer, R.W., Holmes, A.B., Burroughes, J.H., Marks, R.N., Taliani, C., Bradley, D.D.C., Dos Santos, D.A., Brédas, J.L., Lögdlund, M., Salaneck, W.R.: Electroluminescence in conjugated polymers. Nature **397**, 121 (1999)
43. Vollmer, M.S., Effenberg, F., Stecher, R., Gompf, B., Eisenmenger, W.: Steroid-A bridged thiophenes: synthesis and self-organization at the solid/liquid interface. Chem. Eur. J. **5**, 96 (1999)
44. Samori, P., Francke, V., Mullen, K., Rabe, J.P.: Self-assembly of a conjugated polymer: from molecular rods to a nanoribbon architecture with molecular dimensions. Chem. Eur. J. **5**, 2312 (1999)
45. Rinaldi, R., Branca, E., Cingolani, R., Masiero, S., Spada, G.P., Gottarelli, G.: Photodetectors fabricated from a self-assembly of a deoxyguanosine derivative. Appl. Phys. Lett. **78**, 3541 (2001)
46. Cingolani, R., Rinaldi, R., Maruccio, G., Biasco, A.: Nanotechnology approaches to self-organized bio-molecular devices. Physica E **13**, 1229 (2002)
47. Williamson, J.M.: Guanine quartets. Curr. Opin. Struct. Biol. **3**, 357 (1993)
48. Gottarelli, G., Masiero, S., Mezzina, E., Pieraccini, S., Rabe, J.P., Samorí, P., Spada, G.P.: The self-assembly of lipophilic guanosine derivatives in solution and on solid surfaces. Chem. Eur. J. **6**, 3242 (2000)
49. Gottarelli, G., Masiero, S., Mezzina, E., Spada, G.P., Mariani, P., Recantini, M.: The self assembly of a lipophilic deoxyguanosine derivative and the formation of a liquid-crystalline phase in hydrocarbons solvents. Helv. Chim. Acta **81**, 2078 (1998)

50. Forman, S.L., Fettinger, J.C., Pieraccini, S., Gottarelli, G., Davis, J.T.: Toward artificial ion channels: a lipophilic G-quadruplex. J. Am. Chem. Soc **122**, 4060 (2000)
51. Marsh, T.C., Vesenka, J., Henderson, E.: A new DNA nanostructure, the G-wire, imaged by scanning probe microscopy. Nucl. Acids Res. **23**, 696 (1995)
52. Sessler, J.L., Sathiosatham, M., Doerr, K., Lynch, V., Abboud, K.A.: A G-quartet formed in the absence of a templating metal cation: a new 8-(*N*,*N*-dimethylaniline) guanosine derivative. Angew. Chem. Int. Ed. **39**, 1300 (2000)
53. Neogi, A., Sarkar, A., Morkoç, H., Neogi, P.B.: Self-assembled DNA molecules conjugated to GaN semiconductor nano-structures for radiative decay engineering. Conference on Lasers and Electrooptics, San Francisco, CThF7 (2005)
54. Neogi, A., Lee, C.W., Everitt, H., Yablonovitch, E.: Enhancement of spontaneous emission rate due to resonant surface plasmon coupling. Phys. Rev. BI **66**, 153305 (2002)
55. Neogi, A., Li, J., Sarkar, A., Neogi, P.B., Morkoç, H.: Self-assembled modified deoxyguanosines conjugated to GaN quantum dots for biophotonic applications. Electron. Lett. **40**, 1605 (2004)
56. Morkoc, H.: Nitride Semiconductors and Devices. Springer Series in Materials Science, vol. 32. Springer, Heidelberg (1999)
57. Huang, D., Reshchikov, M.A., Morkoç, H.: Growth, structures, and optical properties of III-nitride quantum dots. Int. J. High Speed Electron. Syst. **12**, 79 (2002)
58. Martin, B.R., Dermody, D.J., Reiss, B.D., Fang, M., Lyon, L.A., Natan, M.J., Mallouk, T.E.: Orthogonal self-assembly on colloidal gold-platinum nanorods. Adv. Mater. **11**, 1021 (1999)
59. Neogi, A., Morkoç, H.: Resonant surface plasmon-induced modification of photoluminescence from GaN/AlN quantum dots. Nanotechnology **15**, 1252 (2004)
60. Neogi, A., Everitt, H., Morkoç, H., Kuroda, T., Tackeuchi, A.: Enhanced radiative efficiency in GaN quantum dots grown by molecular beam epitaxy. IEEE Transactions on Nanotechnology **2**, 10 (2003)
61. Asher, S.A., Alexeev, V.L., Goponenko, A.V., Sharma, A.C., Lednev, I.K., Wilcox, C.S., Finegold, D.N.: Photonic crystal carbohydrate sensors: low ionic strength sugar sensing. J. Am. Chem. Soc. **125**, 3322 (2003)
62. Asher, S.A., Peteu, S.F., Reese, C.E., Lin, M., Finegold, D.: Polymerized crystalline colloidal array chemical-sensing materials for detection of lead in body fluids. Anal. Bioanal. Chem. **373**, 632 (2002)
63. Reese, C.E., Asher, S.A.: Photonic crystal optrode sensor for detection of Pb^{2+} in high ionic strength environments. Anal. Chem. **75**, 3915 (2003)
64. Born, M., Wolf, E.: Principles of Optics: Electromagnetic Theory of Propagation. Interference and Diffraction of Light, 7th ed. Cambridge University Press, Cambridge (1999)
65. Johnson, S.G., Fan, S., Villeneuve, P.R., Joannopoulos, J.D., Kolodziejski, L.A.: Guided modes in photonic crystal slabs. Phys. Rev. B **60**, 5751 (1999)
66. Yu, G., Wang, G., Ishikawa, H., Umeno, M., Soga, T., Egawa, T., Watanabe, J., Jimbo, T.: Optical properties of wurtzite structure GaN on sapphire around fundamental absorption edge (0.78–4.77 eV) by spectroscopic ellipsometry and the optical transmission method. Appl. Phys. Lett. **70**, 3209 (1997)

Chapter 6
Carbon Nanotubes for Optical Power Limiting Applications

Shamim Mirza, Salma Rahman, Abhijit Sarkar, and George Rayfield

Abstract Optical limiters are nonlinear materials that exhibit a drop in transmittance as the energy of incident laser pulses increases, usually above a certain threshold. They have the potential for protecting optical sensors, and possibly even human eyes, from laser pulse damage. The problem of optical power limiting has been a subject of increasing interest for more than two decades now. The interest is due to the increasingly large number of applications based on lasers that are currently available. Several research groups have been attempting to develop novel OPL materials based on nonlinear optical (NLO) chromophores. As a result, there are a large number of publications and patents on this subject. Some of the best-performing optical limiters are materials containing carbon nanotubes (CNTs); however, such materials are difficult to prepare and have problems with stability. In this chapter, the origin of OPL as well as the mechanism of OPL has been discussed. Ways to modify CNTs and to make them suitable for OPL applications have also been discussed.

Keywords Carbon nanotubes · Optical power limiting (OPL) · OPL chromophores · OPL mechanisms · Nonlinear optics · Photonics · OPL chromophores · Hyperbranched polymers

1 Introduction

Laser light is a high-intensity monochromatic radiation having extremely high coherence. An increasingly large number of applications based on lasers are currently available. Most applications in the consumer sector incorporate low-intensity lasers such as compact discs, DVDs, and other optical devices while high-intensity lasers largely remain in the research, medical, defense, industrial, nuclear, and astronomy sectors. Low-energy lasers are commonly used in law enforcement and warfare for target illumination. The affordable cost of producing lasers and easy

A. Sarkar (✉)
Michigan Molecular Institute, 1910 West St. Andrews Rd., Midland, MI 48640, USA
e-mail: sarkar@mmi.org

Z.M. Wang, A. Neogi (eds.), *Nanoscale Photonics and Optoelectronics*,
Lecture Notes in Nanoscale Science and Technology 9,
DOI 10.1007/978-1-4419-7587-4_6,

availability have led to their being used as weapons causing harmful radiation damage to human eyes and to the multitude of light-sensitive receptors that are employed by the military and in civilian applications. Lasers directed at aircraft pilots, optical sensors, or through the sights of tank gunners can do serious damage to the operator and/or the equipment. Keeping transmitted light below a certain maximum intensity is useful in protecting such light-sensitive receptors [1]. The main requirement, especially in the case of high-sensitivity devices, is high transparency under low intensity and ambient conditions combined with opacity under high-intensity radiation. Typically, there are two ways to protect optically sensitive materials against damage caused by overexposure to high-intensity light: active and passive [2]. The active approach is based on "smart structures" comprising electronic circuitry that reacts when a harmful intensity of light is detected and activates mechanical barriers between the light source and the optically sensitive material. The passive approach relies on the inherent property of the "smart material" to form a light barrier and prevent transmission of light when its intensity surpasses a threshold value. Optical power limiters (OPLs) are "smart materials" that follow passive approaches to provide laser protection.

Optical power limiters (OPLs) are materials and devices designed to allow normal transmission of light at low intensities and limited transmission of light of higher intensities (Fig. 6.1) [3]. A barrier is formed as a direct response of such a material to excessive intensity of light. There are various important considerations that go into the design of an OPL device. The speed at which light travels dictates that an OPL device must be able to react almost instantaneously to changing light intensity. Fast response time favors a material-based device over a mechanical one. The material must be able to bear the brunt of prolonged exposure to high-intensity light, as well as allow for continuous transparency in regions outside the path of the high-intensity light. All these factors favor a device incorporating a liquid or solid film of molecules, which acts as a stand-alone optical power limiter.

For over two decades, several research groups have been attempting to develop novel OPL materials based on nonlinear optical (NLO) chromophores [3–8]. As a result, there are a large number of publications and patents on this subject. In order to be used for practical applications, an OPL material must fulfill the following requirements:

1. It must have a fast response time.
2. It should operate over a broad wavelength range.
3. The on–off cycle must be extremely fast and ideally it should follow the speed of the cycle of the laser pulse it is responding to.

OPL devices rely on one or more nonlinear optical mechanisms that include:

- Reverse saturable absorption (RSA)
- Multiphoton absorption (MPA)
- Induced scattering
- Photo-refraction

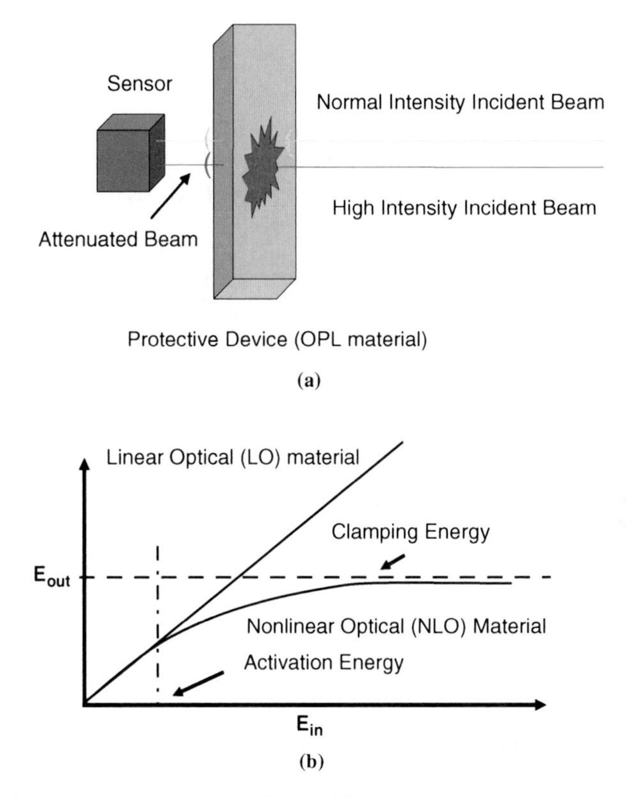

Fig. 6.1 (**a**) Concept of optical power limiters (OPLs). (**b**) Output energy (E_{out}) vs. input energy (E_{in}) for an ideal optical limiter

The above-mentioned nonlinear optical processes for various materials have been extensively studied for OPL applications [9–14]. To date, however, there is not a single OPL material available which, *taken individually*, can provide ideal and smooth attenuation of an output beam. Therefore, the design and development of radically novel types of materials for OPL is urgently required. In this regard, some attempts were made with combinations of nonlinear optical materials in cascading geometries, such as multiplate or tandem cells [15] and use of two intermediate focal planes of a sighting system [16]. These efforts have shown promising results but much more effort is still needed.

The human eye is a very sensitive optical sensor with a very low damage threshold for the retina ($\sim$1 μJ). This imposes stringent demands on materials for laser protection. Existing nonlinear optical materials can respond to such low energies only when the light is tightly focused – this is achieved most easily in an optical system which provides focal planes at which the nonlinear material can be positioned. Protection applications demand materials with the following characteristics:

- high linear transmission across the response band of the sensor,
- sensitive nonlinear response to pulses of duration 1 ns and below,

- resistance to permanent optical damage, and
- stability in the working environment.

The ideal OPL material is expected to have (1) a fast response time, (2) high transparency at normal illumination conditions, and (3) an increased broadband spectral response for protection of eyes and sensors from laser beams.

2 Mechanisms of Optical Power Limiting

For eye protection against laser threat, mainly three classes of nonlinear optical (NLO) materials are promising: multiphoton absorbers (MPA), reverse saturable absorbers (RSA), and nonlinear scattering (NLS) materials. Multiphoton absorbers generally exhibit high linear transmittance with a more favorable colorimetry (from yellowish to colorless) in solution, in a solid matrix, or in organic crystals [17]. Reverse saturable absorbing molecules exhibit very efficient optical limiting properties both in solution [18, 19] and in a solid matrix [20, 21]. Nonlinear scattering materials occur principally in suspensions of absorbing nanoparticles in organic solvents. When a laser beam impinges on such medium, heating of the particles causes evaporation of the surrounding solvent and sublimation of the particles themselves, leading to fast growth of strongly scattering gaseous cavities.

2.1 Nonlinear Absorption

2.1.1 Reverse Saturable Absorption (RSA)

A more effective method by which a molecule could function as an OPL centers on the idea of sequential two-photon absorption (STPA) [22]. It is sometimes referred as excited-state absorption (ESA) or reverse saturable absorption (RSA). In general, reverse saturable absorption (RSA) may occur whenever the excited-state absorption cross section of a molecule is larger than its ground-state absorption cross section at the input wavelength. To understand the process, we can consider the electronic energy levels represented in Fig. 6.2, where $\sigma(S_{0-1})$ and $\sigma(T_{0-1})$ symbols designate absorption cross section of the singlet, $S_0 \rightarrow S_1$, and the triplet, $T_0 \rightarrow T_1$, optical transitions, respectively. Following an initial $S_0 \rightarrow S_1$ absorption event, the excited S_1 state can either relax to the ground state (via $S_1 \rightarrow S_0$ emission) or undergo a phonon-mediated S_1 to intersystem conversion (isc). If the latter process is faster than the former one, a significant buildup of the T_0 state population would occur, given that the $T_0 \rightarrow S_0$ transition is spin forbidden. $S_1 \rightarrow S_n$ transitions are equally important as they too contribute in populating the triplet states. Once the T_0 state is populated, the new absorption path ($T_0 \rightarrow T_1$) opens up for the incoming radiation. Finally, if $\cdot(S_{0-1}) < \sigma(S\backslash T_{0-1})$, all necessary requirements for RSA are met and the molecule behaves as an optical limiter. Under appropriate conditions, the population accumulates in the lowest triplet state during the exposure to a strong light source and it provides an increasing concentration to the total absorption of the system. The

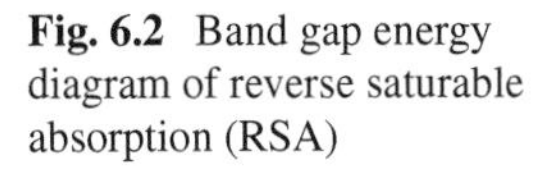

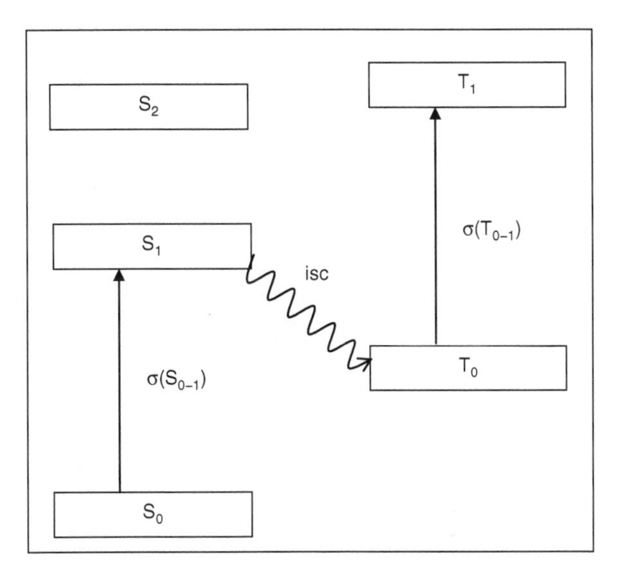

Fig. 6.2 Band gap energy diagram of reverse saturable absorption (RSA)

significant parameters for the effectiveness of this mechanism are the excited state lifetimes, the ISC time, and the values of the excited-state absorption cross sections.

2.1.2 Multiphoton Absorption (MPA)

Multiphoton absorption, which also includes two-photon absorption, is a particularly efficient technique for optically limiting short (sub-nanosecond to picosecond) pulses of high-intensity radiation. Unlike other OPL phenomena requiring a fluid state to enable free molecular motion, MPA chromophores have the advantage of being able to perform even in the solid state.

Two-photon absorption (TPA) which has been known for a long time [23] has experienced a renewed surge of interest in recent years. This can be largely attributed to the availability of materials with enhanced two-photon absorption cross section [24] that are particularly well suited for effective optical limiting. In the case of TPA, the electron is promoted from the ground state to an excited state through a virtual intermediate state, by simultaneously absorbing two photons. A possible band gap energy model for the relevant molecular photonic absorption processes is depicted in Fig. 6.3. The electronic transition to a high-energy state may take place either by (1) absorption of two photons from the same optical field of frequency ω such that the transition takes place to an excited state resonance at 2ω or by (2) a two-beam two-photon absorption process involving simultaneous absorption of two photons of different frequencies. One of the beams is called the pump and the other represents the probe beam. Usually the intensities of the two beams are different, i.e., $I_{\text{probe}} << I_{\text{pump}}$. TPA offers the advantage of high transmission at low incident intensity for light with a frequency well below the band gap frequency. It is a fast third-order nonlinear process and the mechanism is efficient for short pulses.

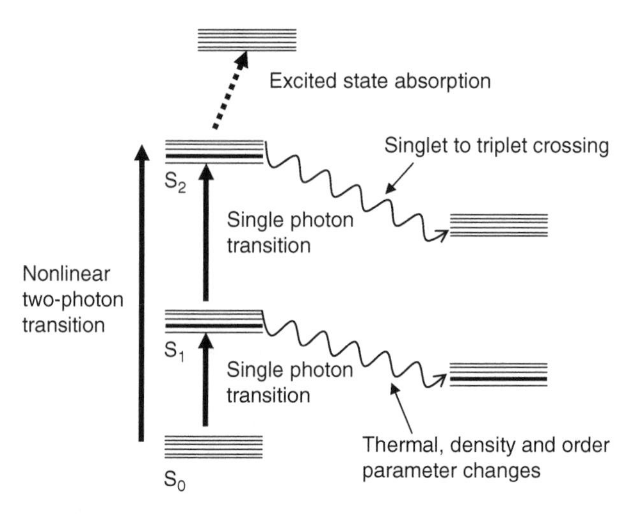

Fig. 6.3 Band gap energy model for the relevant molecular photonic absorption processes including two-photon absorption (TPA). S_0 is ground electronic states, S_1 is manifold of ground electronic states (virtual state), and S_2 is excited electronic states

A three-photon absorption (or multiphoton absorption) mechanism works on the same principle as TPA, differing only in the number of photons involved in the excitation [24]. For example, in three-photon absorption, three photons are absorbed by a three-beam three-photon absorption process. This mechanism is suited for achieving OPL effects in the 400–800 nm wavelength region. Three-photon absorption can be enhanced if delocalization of π-electrons occurs in a conjugated molecule. Fluorophores for MPA should be subjected to the same scrutiny as those intended for single-photon investigations. The probes should have large absorption cross sections at convenient wavelengths, high quantum yields, a low photobleaching rate, and the lowest possible degree of chemical and photochemical toxicity. The fluorophores should also be able to withstand high-intensity illumination from the laser source without significant degradation. Thus, a suitable conjugated molecule can be selected or designed which will show significant higher-order optical nonlinearity.

2.2 Nonlinear Refraction

When a nonlinear optical (NLO) material is subjected to light of high intensity it may demonstrate a change in refractive index. This change can be seen as either positive or negative and causes the material to either defocus or focus the light [25].

Self-focusing and defocusing are optical nonlinear effects which can be used in optical limiting applications. This nonlinear refraction effect in a material can manifest itself as beam broadening or narrowing in the far field, i.e., causing the focusing or defocusing which is dependent on the irradiance input. Figure 6.4 illustrates these

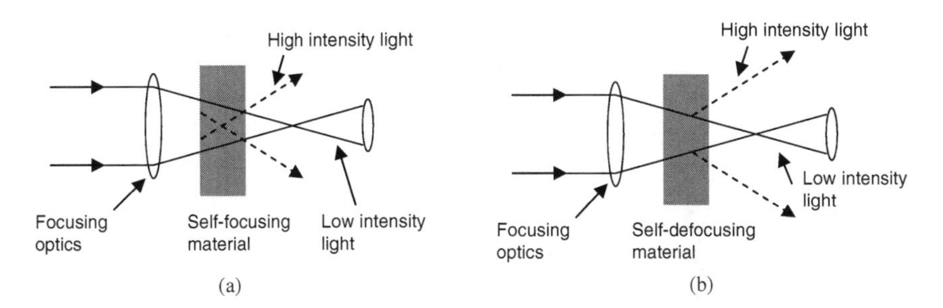

Fig. 6.4 (**a**) Effect of a self-focusing device and (**b**) effect of a self-defocusing device

concepts. Assuming that both bound and free carriers cause the change in the index of refraction, we get the relation [26]:

$$\Delta n = \gamma I + \sigma_r N \tag{6.1}$$

where $\gamma(\mathrm{m}^2/\mathrm{W})$ is the nonlinear index that is due to bound electrons and σ_r is the change in the index of refraction per unit photo-excited charge-carrier density N. γ is related to the real part of the third-order susceptibility, $\chi^{(3)}$, the speed of light, c, the refractive index in the absence of free carriers, n_0, and the free-space permittivity, ε_0, in such a way that [27]

$$\gamma = \frac{3}{4n_0^2 c\varepsilon_0} R_e(\chi^{(3)}) \tag{6.2}$$

The term in Eq. (6.1) related to the nonlinear refraction due to bound electrons (γI) will be dominant at relatively low irradiance levels, whereas the free-carrier refraction ($\sigma_\mathrm{r} N$) will dominate at high irradiance levels. Since the carrier nonlinearity ($\sigma_\mathrm{r} N$ in Eq. (6.1)) is proportional to a temporal integral of I^2 [27], this will be an effective fifth-order nonlinearity. In the case of two-photon absorption, this fifth-order nonlinearity is a sequential $I_\mathrm{m}\chi^{(3)}$ process (i.e., two-photon absorption) followed by an $R_\mathrm{e}\chi^{(1)}$ process (i.e., a linear index change from the carriers). Since the electronic Kerr effect is a third-order effect, the nonlinearity due to carriers generated by two-photon absorption will dominate above a certain irradiance level.

Experiment and calculations have shown that the maximum of γ will occur close to the resonance for two-photon absorption and that it will change sign from positive to negative when the photon energy is greater than about 70% of band gap energy (E_g). Since γhas been shown to have a E_g^{-4} dependence [28], a strong nonlinear refraction is expected for semiconductors with a small band gap.

A problem with self-focusing is, however, that it may cause damage to the nonlinear material due to the high energy density in the focal spot. Self-defocusing should for this reason be preferable.

Another nonlinear refraction related mechanism is molecular reorientation. A strong electric field applied to a system which contains anisotropic molecules tends to align the induced dipole moments of the molecules along the direction of the field, changing the refractive index. Therefore, an intense incident optical field will orient the molecules along the direction of polarization and will thus encounter a different index of refraction than will a weak field.

2.3 Induced Scattering

Two types of nonlinear optical scattering (NOS) may occur due to the refractive index of the materials. The first possible NOS is a fundamental manifestation of the interaction between matter and radiation, resulting from inhomogeneities in the refractive index, which decrease transmission. Photoinduced resonant scattering or nonresonant scattering is especially attractive for sensor protection from laser light by optical limiting. One of the best known and perhaps most effective example of a resonant device is the carbon black suspension limiter [29], in which the diffusion centers are created following absorption of light and subsequent heating of carbon particles. This implies an appropriate choice of the frequency range. An example of the second type of NOS is the Kerr effect that need not involve any resonances. In this case, the index of refraction n is intensity dependent and follows a nonlinear relation

$$n = n_0 + n_2 I \tag{6.3}$$

where n_0 is the linear index of refraction, n_2 is the nonlinear (or Kerr) index of refraction, and I is the light intensity. A two-component medium, having good index matching, acts as a transparent filter at low intensity. At high intensity of light, the refractive index of the components gets perturbed, making the medium highly scattering [30].

2.4 Photo-refraction

The photo-refractive effect is another mechanism related to the intensity-dependent refractive index. The photo-refractive material acts as a volume grating. It describes a modulation in the refractive index due to spatial variation of the intensity [27, 31] formed by the interference of two coherent incident beams. So the photo-refractive effect can be used for sensor protection utilizing materials acting as nonlinear optical limiters. Due to high-intensity laser irradiation, free charges are liberated in photo-refractive materials. The change in the space-charged field because of the radiation causes alterations in the index of refraction. Laser damage thresholds and the time response of the photo-refractive effect are important issues related to the use of photo-refractive materials in optical limiting applications.

One effect of the photo-refractive mechanism that is used for optical limiting is beam fanning [27, 32] which is a phenomenon that has been discussed in relation to optical power limiting. The beam fanning effect is described as a beam entering

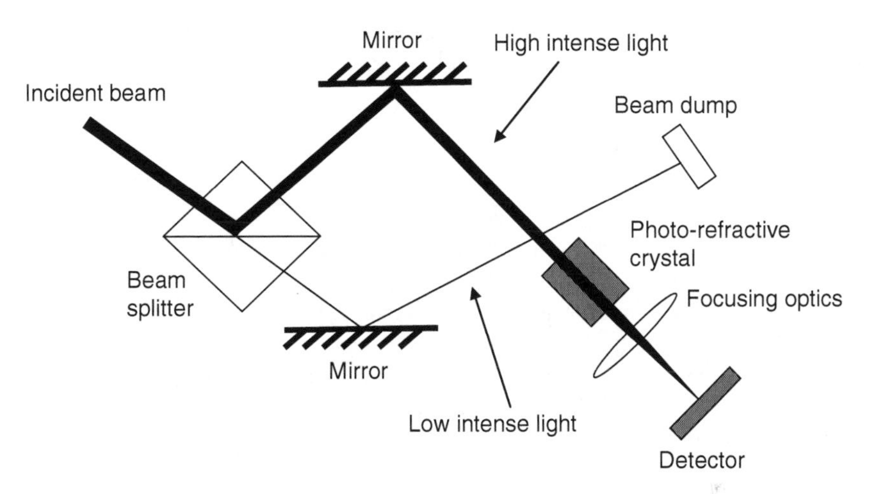

Fig. 6.5 Principle of photo-refractive optical limiting using the beam-fanning method

a photo-refractive material and causing a set of asymmetrical beams to be formed after passing through the photo-refractive crystal.

The method of beam fanning can be improved by splitting the incident radiation into weak- and high-intensity beams before entering the photo-refractive crystal. The two beams are coupled in the crystal and the high-intensity radiation is transferred to the weak beam path and thereafter dumped (Fig. 6.5). The process is dependent on highly coherent radiation, i.e., broadband irradiation is not coupled to the beam dump. The optical limiter is activated by high-intensity coherent radiation. A disadvantage of optical limiters using the photo-refractive effect is the rather slow response time of the material since diffusion processes are the governing mechanism.

3 Optical Power Limiting (OPL) Chromophores

3.1 Organics and Organometallics

Over the years a large variety of organic structures, such as porphyrins, fullerene, and carbocyanines, have been investigated for NLO properties and optical power limiting. Some other classes of chromophores exhibit dramatically changed OPL performance when incorporating a heavy metal atom (e.g., Pb, Pt, Pd, Au, Ag). Porphyrins (Pf) have been much studied with respect to electron and energy transfer processes. Metal ion complexation often increases the third-order nonlinear optical susceptibility. Metallo-Pf typically have a Soret band (π–π^* transition to second excited state) in the region of 400–450 nm and a less intense Q band (to first excited state) in the region of 500–600 nm. Additional charge transfer bands are sometimes present and arise from Pf–metal interactions. The Q band is also affected by the presence of a metal. The photophysics of this class of compounds is well known and the Pfs generally show good thermal stability [33]. Several Pf systems have been studied with respect to optical limiting and have shown good characteristics.

3.2 Multiphoton Absorbers

An appropriate choice of MPA chromophores is important for the efficiency of the OPL material. Two MPA chromophores have promising attributes for organic photonics applications, namely, stilbene-3 and POPOP (Fig. 6.6). These organic chromophores have large excitation cross sections and are transparent in the visible region. Additionally, these chromophores can be blended with the polymer matrix via solubilization in common organic solvents. Silver nanoparticles can be used in conjunction with the MPA chromophores. Silver nanoparticles have been reported to enhance multiphoton absorption of organic molecules by five orders of magnitude [34].

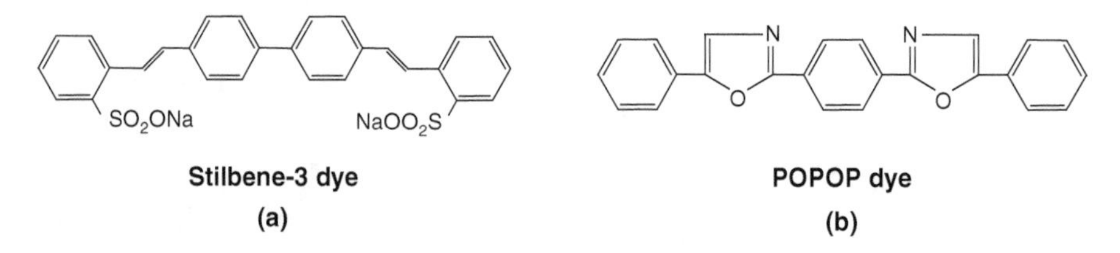

Fig. 6.6 Chemical structure of MPA chromophores: (**a**) stilbene-3 and (**b**) POPOP

3.3 Reverse Saturable Absorbers

Two RSA chromophores are among the better ones for organic photonics materials, namely carbocyanine chromophore, 1,1′,3,3,3′,3′-hexamethylindotricarbocyanine iodide (HITCI), and fullerene (C_{60}) (Fig. 6.7).

The C_{60} is a good organic chromophore because of its excellent visible broadband transmission under normal illumination and equally fast OPL response. HITCI,

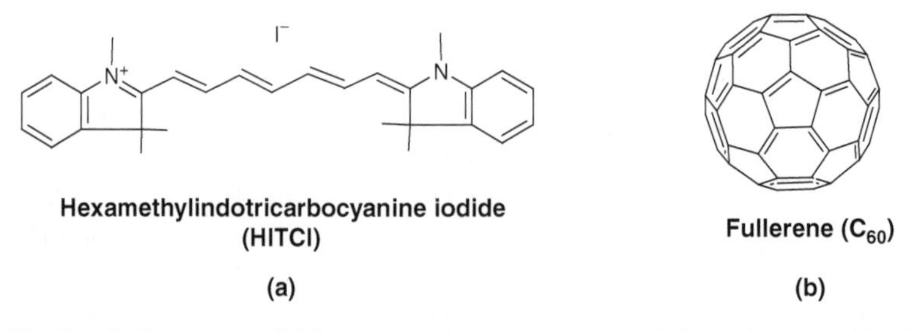

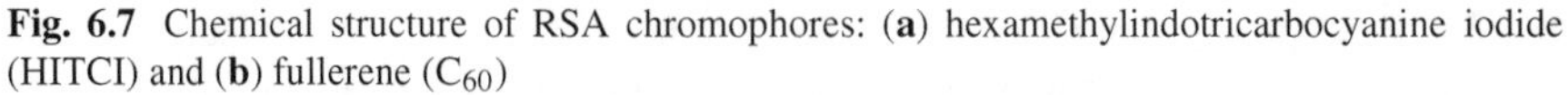

Fig. 6.7 Chemical structure of RSA chromophores: (**a**) hexamethylindotricarbocyanine iodide (HITCI) and (**b**) fullerene (C_{60})

on the other hand, is a good organic chromophore because of its high efficiency. The RSA properties of this chromophore have been well studied [35]. The combination of an RSA dye (HITCI) and CNT in an ethanol solvent has been reported to have the best OPL response so far reported [36]. It has fast (ps) response, broadband transmission, low threshold response (10 mJ/cm^2), and high fluence clamping. These chromophores can be blended with the polymer matrix using a sonicator and a solid-state polymer film with uniform nano-sized chromophore particle distribution can be achieved [37, 38].

3.4 Azo Dyes

The origin of OPL properties of azobenzene films is in their photoinduced anisotropy, which includes photoinduced birefringence as well as photoinduced dichroism, and varies with the intensity of incident light [39]. The transition moment of the azobenzene molecule lies along the molecular axis and only those molecules with their axis oriented parallel to the electric field vector absorb light and isomerize from the *trans* to *cis* form. The *cis* form is thermodynamically unstable and eventually reverts to the *trans* form where it may again be isomerized to the *cis* form if its transition moment is not perpendicular to the electric field direction of light. Figure 6.8 shows the *cis–trans* isomerization process in 4-nitroazobenzene. Repeated *trans–cis–trans* isomerization results in the alignment of azobenzene molecules in a direction that is perpendicular to the polarization of the excitation beam. This then causes photoinduced birefringence that makes the output polarization elliptical, and photoinduced dichroism that makes the semimajor axis of the ellipse shift through a rotation angle from the origin. Both mechanisms contribute to the OPL property, although, in the case of azobenzene, photoinduced birefringence is expected to be the major contributor.

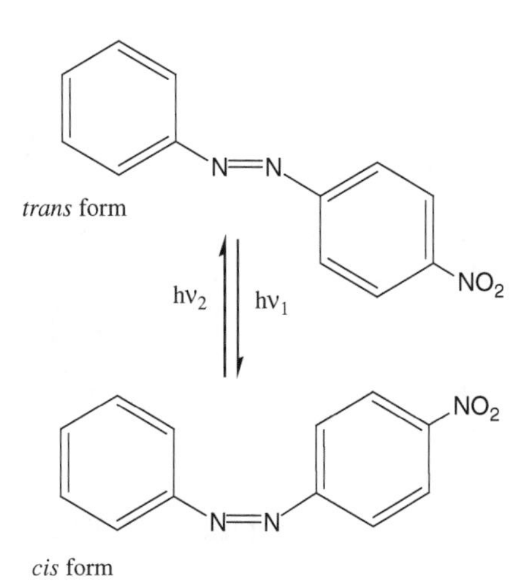

Fig. 6.8 Isomerization process in 4-nitroazobenzene

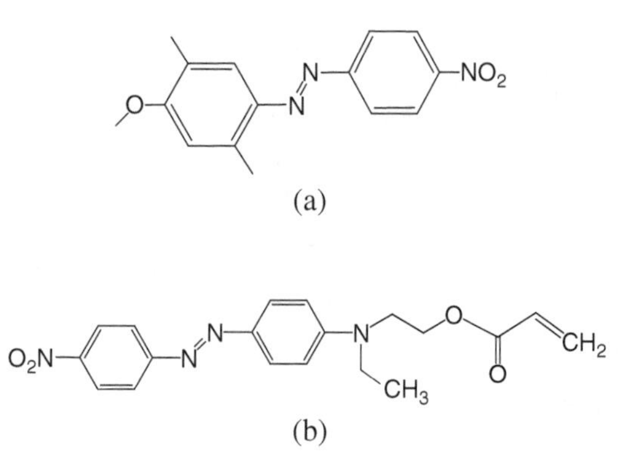

Fig. 6.9 Structure of the NIR azo dyes: (**a**) DMNPAA and (**b**) disperse red 1 acrylate

For example, a NIR azo dye, namely 2,5-dimethyl-(4-*p*-nitrophenylazo)anisole (DMNPAA), is a promising molecule (Fig. 6.9). This dye should work particularly well for OPL in the 800–1,400 nm region.

Another azo chromophore that is of interest for use in solid OPL material is disperse red 1 acrylate (Fig. 6.9). This dye can interact with the polymer matrix due to the acrylate functionality, thereby keeping it uniformly distributed in the OPL filter. One of the goals of OPL material development is to decrease optical limiting threshold energy. Azo dyes are known to undergo *cis–trans* transition at low input energy.

4 Carbon-Based Materials for Optical Power Limiting

4.1 Fullerene and Carbon Black Suspension (CBS)

Carbon exists in four different stable allotropes: graphite, diamond, fullerene, and carbon nanotubes. After the discovery of fullerene in 1985 by Kroto et al., numerous articles on the synthesis, physics, chemistry, and functionalization of fullerene (especially C_{60}) have been published [40–46]. Fullerene and its derivatives in solution, similar to porphyrin and phthalocyanine systems, show nonlinear OPL behavior via an RSA mechanism and subsequent nonlinear refraction and scattering [41, 42, 47, 48]. These materials exhibit a positive nonlinear absorption coefficient.

Carbon black suspensions (CBS) act as nonlinear optical limiters, especially in the case of high-intensity laser pulses. CBS undergo dramatic changes in transmittance due to laser irradiation as well as thermally induced nonlinear scattering which can be attributed to the formation and rapid expansion (within the pulse width) of microplasmas initiated by rapid heating and thermo-ionization of the carbon particles [29, 49–51].

4.2 Carbon Nanotubes (CNTs)

Carbon nanotubes (CNTs), first discovered by Iijima in 1991, are a family of new materials with unique structure and excellent mechanical, electrical, thermal, and

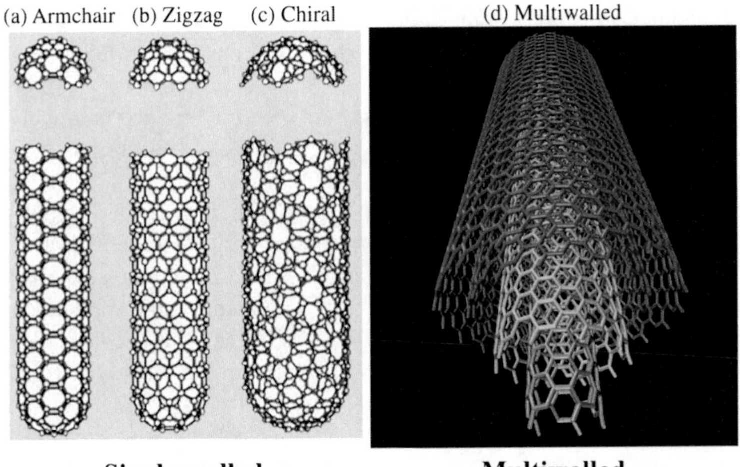

Fig. 6.10 Different types of carbon nanotubes: (**a**) armchair, (**b**) zigzag, (**c**) chiral nanotubes [57], and (**d**) multiwalled carbon nanotube (reprinted with permission from Alain Rochefort, Assistant Professor, Engineering Physics Department, Nanostructure Group, Center for Research on Computation and its Applications (CERCA))

optical properties [52, 53–56]. CNTs have fullerene-related structures and consist of graphene cylinders closed at both end with caps containing pentagonal rings. There are three categories of CNT structures: "armchair," "zigzag," and "chiral" based on the arrangement of hexagons around the circumference (Fig. 6.10) [57]. An electronic structure study showed that the CNTs have a lower work function, steeper Fermi edge, lower electron-binding energy, and stronger plasma excitation in comparison with graphite [58].

CNTs are generally classified as single-walled carbon nanotubes (SWNTs) and multiwalled carbon nanotubes (MWNTs). Typical images of these two types of CNTs are shown in Fig. 6.11. Single-walled carbon nanotubes (SWNTs) can be considered to be formed by rolling of a single layer of graphite into a seamless

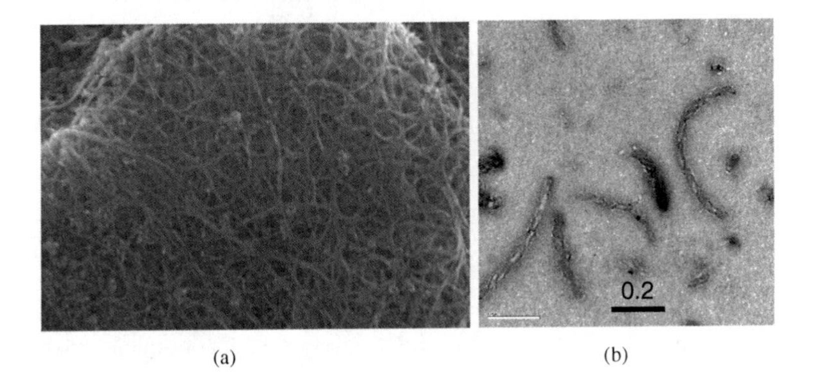

Fig. 6.11 (**a**) FESEM image of SWNT (95% purity, visible particles might be catalysts and amorphous carbon) and (**b**) TEM image of short-MWNT in polymer matrix

cylinder (diameter of SWNTs on the order of a nanometer) whereas multiwalled carbon nanotubes (MWNTs) contain more than one concentric cylindrical shell of graphene sheets coaxially arranged around a hollow core (diameter of MWNTs on the order of tens of nanometers). Highly symmetric structure and extremely small size allow CNTs for remarkable quantum effects, magnetic, electronic, and lattice properties [59].

A lot of work has been performed to exploit the properties of CNTs. High tensile strength and elasticity of CNTs make them suitable for aerospace and fiber industries [60]. On the other hand, electronic conductance and unique semiconducting characteristics are ideal for nanoelectronics and semiconductor applications [60–62]. Hydrogen adsorption (storage) capacity of CNTs can be utilized for application in hydrogen-based fuel cells [63, 64], as well as electronic sensitivity in different chemical environments, allowing CNTs to be useful for novel environmental sensors [65–68]. The broadband absorption property of CNTs has also been utilized for various optical applications, e.g., optical power attenuation, optical filter. As a result of these versatile properties of CNTs, a wide range of potential applications, such as field emission devices [54, 69, 70], electrochemical devices [71–73], and nanotube-based composites or thin films, for protection against laser damage are now within reach [74–76].

CNTs possess one of the most extensive delocalized π-electron systems and it has been observed to be a broadband optical limiting material which is outside the range of many optical limiting materials [47, 49, 77]. The general onset of optical power limiting was observed at $\sim$0.1–0.5 J cm^{-2} at 532 nm [78]. The optical limiting performance of CNT depends on several factors such as the laser pulse duration, aspect ratio of CNT, surrounding host liquids [36]. It was observed that longer laser pulse durations (nanosecond or longer) improved the OPL efficiency which is attributed to more effective solvent bubble growth and sublimation of CNT. This phenomenon is discussed in the next section. CNTs with higher aspect ratio (longer tubes) possess better OPL efficiency [79]. The effect of solvents is also discussed in detail in the next section.

Applications of carbon nanotubes have been impeded by difficulties associated with CNT processing and manipulation. Extremely poor solubility of CNTs in most of solvents due to substantial van der Waals attractions between CNTs along the length axis has made this task a difficult one [80, 81]. For the same reason the nanotubes aggregate easily. Considerable effort has been devoted toward the solubilization of carbon nanotubes both in aqueous media and in organic solvent. Earlier reports have shown that aggregation of CNTs can be prevented by applying a noncovalent surface coating onto them by amphiphilic molecules such as low molecular weight surfactants and polymeric amphiphiles [82–86]. The other strategy that has been followed to prevent undesirable intertube aggregation in solvents is covalent functionalization of CNT surfaces with solvophilic molecules [87–89]. Incorporation of CNTs in various polymer matrices such as polystyrene [90], poly(methyl methacrylate) (PMMA) [91], and epoxy has also been reported [92, 93]. However, most of these methods involve aqueous media or organic solvents.

4.2.1 Solubilized and Suspended Carbon Nanotubes

Both SWNTs and MWNTs in suspended and solubilized form exhibit stronger broadband optical limiting responses. Nonlinear scattering and refraction are probably the dominant mechanisms for optical power limiting. However, the suspended and solubilized carbon nanotubes act differently in optical power limiting. In suspension form, the OPL property is mainly due to nonlinear scattering that originates from microbubbles formed due to absorption-induced heating (Fig. 6.12).

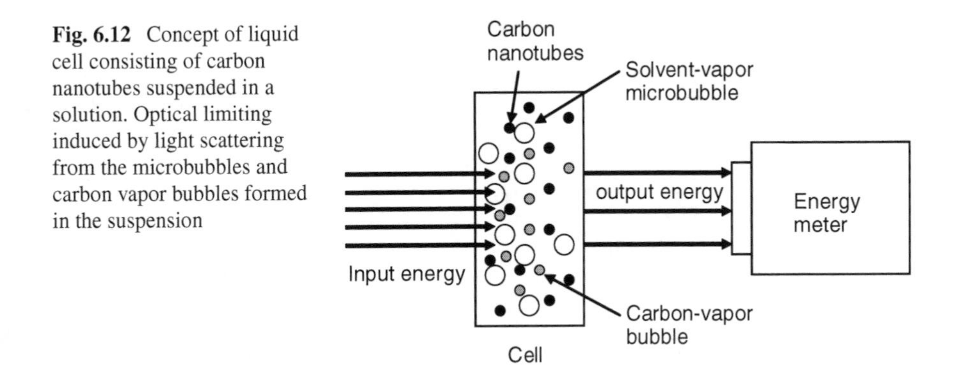

Fig. 6.12 Concept of liquid cell consisting of carbon nanotubes suspended in a solution. Optical limiting induced by light scattering from the microbubbles and carbon vapor bubbles formed in the suspension

Thus, the nonlinear scattering process can be divided into two steps: nucleation of scattering centers followed by their growth. In this process, heating due to laser pulses leads to vaporization and ionization of CNTs which ultimately forms rapidly expanding microplasmas [58]. The microplasma strongly scatters light from the transmitted laser beam direction which leads to a nonlinear decrease in the energy of transmitted light. In solubilized form, the OPL property is observed due to a nonlinear absorption mechanism. The CNTs exhibit a strong optical limiting property, superior to both C_{60} and carbon black (Fig. 6.13) [58]. In case of CNTs, due to their high aspect ratio, coupling to the radiation field is significantly more efficient [36]. Optical limiting properties of carbon nanotube suspensions and solution are presented in Table 6.1. The CNTs show weaker and concentration-dependent OPL responses in solutions than in suspensions.

The electronic structure and optical limiting behavior of carbon nanotubes of different diameters have been investigated in various solvents such as ethanol, water [58, 94]. A lower optical limiting threshold was observed in solution than in a solid host which is not unexpected because expansion of a microplasma is easier in solution. Improvement of OPL property of suspended CNTs can be achieved by an appropriate choice of solvent to obtain better dispersion and selection of the diameter and symmetry of the nanotubes. Better optical limiting efficiency can be achieved from CNT suspension in a solvent with lower boiling point, surface tension, and viscosity [59].

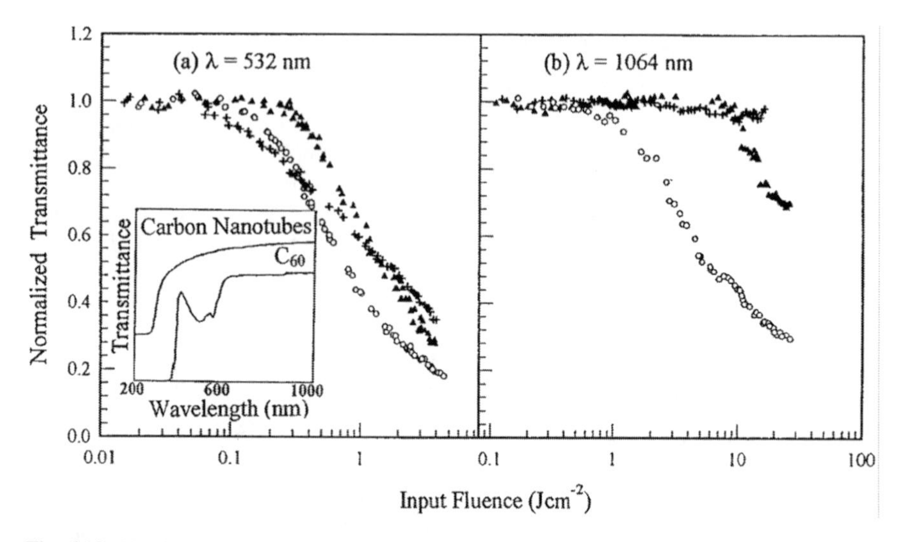

Fig. 6.13 Nonlinear transmission of the carbon nanotubes in ethanol (○); C_{60} in toluene (+); and carbon black in distilled water (▲). The nonlinear transmission was measured with 7 ns laser pulses at (**a**) 532 nm and (**b**) 1,064 nm wavelength. The linear transmittance of the three systems has all been normalized to unity. The inset of (**a**) shows the optical transmission spectra recorded in the wavelengths between 200 and 1,200 nm for the carbon nanotubes suspended in ethanol (*top curve*) and C_{60} dissolved in toluene (*bottom curve*). The transmission spectrum of the nanotubes has been shifted vertically for clear presentation. The spectrum of the carbon black suspension is identical to that (*top curve*) of the carbon nanotubes (reprinted with permission from Ref. [58])

Table 6.1 Optical limiting properties of the nanotube suspensions and solutions of 70% linear transmittance at 532 nm (reproduced with permission from Ref. [87])

Compounds	Medium	Saturated I_{out}(J cm^{-2})[a]	Limiting threshold (J cm^{-2})[b]
C_{60}	Toluene	0.10	0.21
Carbon black suspension	Water	0.12	0.31
SWNT suspension	Water[c]	0.11	0.31
S-SWNT suspension	Water[c]	0.16	0.39
S-SWNT-PPEI-EI	Chloroform	0.35	0.98
S-MWNT suspension	Water[c]	0.14	0.38
S-MWNT-PPEI-EI	Chloroform	0.26	0.64
S-MWNT-octadecylamine	Chloroform	0.29	0.71

[a]for the value of I_{out} at I_{in} = 1 J cm^{-2} when there is no optical limiting plateau.
[b]Defined as the value of I_{in} when I_{out} is 50% of what is expected on the basis of the linear transmittance (0.35 I_{in}).
[c]Also contains 5% Triton X-100.
[d]PPEI-EI = poly(propionylethylenimine-co-ethylenimine)

Single-Walled Carbon Nanotubes

Nonlinear scattering is the primary optical limiting phenomenon for SWNT suspensions due to thermally induced solvent bubble formation at low fluences and long pulses, and due to bubbles from nanotube sublimation at high fluences and

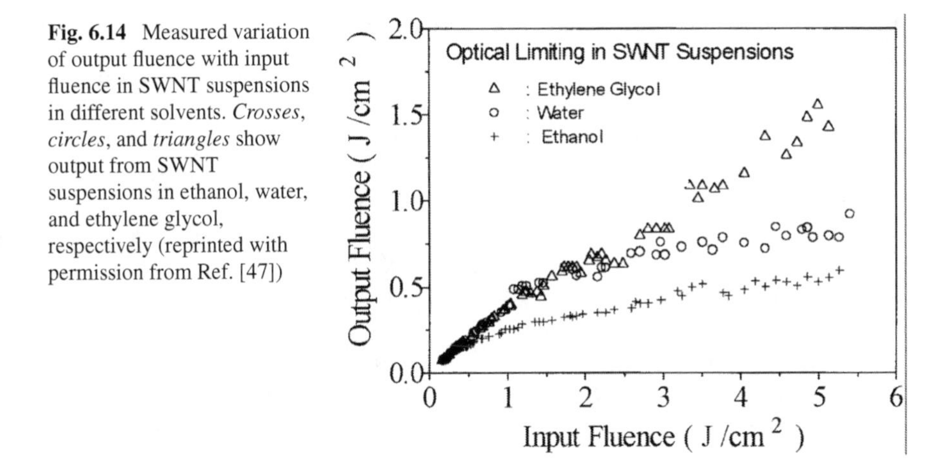

Fig. 6.14 Measured variation of output fluence with input fluence in SWNT suspensions in different solvents. *Crosses*, *circles*, and *triangles* show output from SWNT suspensions in ethanol, water, and ethylene glycol, respectively (reprinted with permission from Ref. [47])

short pulses [95–97]. The time evolution of the radius and concentration of the scattering centers were determined by Vivien et al. using a model based on Mie theory [78]. SWNT is considered as a broadband optical limiter (visible and near-infrared region) due to the large optical densities and small nonlinear threshold. The effects of pulse duration and wavelength on the OPL property of SWNT in solvents over a broadband wavelength region were investigated [97]. The OPL behavior of a suspension of SWNT in different solvents such as water, ethanol, chloroform, and ethylene glycol has been reported (Fig. 6.14) [36, 47, 97]. According to this report, optical limiting behavior is strongly solvent-dependent which rules out plasma formation as being an important mechanism. The optical limiting response of SWNT in ethanol suspension was found to be the strongest. The OPL behavior in glycol suspension was observed to be weaker than that in water. This phenomenon suggests that absorption-induced refractive index inhomogeneities did not have much contribution to nonlinear scattering [59].

Multiwalled Carbon Nanotubes

OPL behavior of MWNT in different solvents such as water, chloroform, dichlorobenzene, methanol and a host matrix such as polymethyl methacrylate (PMMA) was also studied at 532 nm as well as in the broadband (visible–NIR) region with nanosecond laser pulses [49, 98–100]. Liu et al. found better optical limiting performance of MWNT in chloroform than in dichlorobenzene. On the other hand, Pratap et al. observed better optical limiting behavior at 532 nm (limiting threshold measured = 2 J cm^{-2}) than in the near-infrared region at 1,064 nm (limiting threshold measured = 20 J cm^{-2}). In the study of size-dependent optical limiting performance, the MWNTs with large aspect ratio were observed more stable to nanosecond pulsed laser and also showed stronger limiting properties which are more prominent at low input fluence [79]. OPL efficiency of MWNT suspensions in various solvents was studied in the temperature range from room temperature to the boiling point of the solvents. As the temperature was increased, the OPL efficiencies decreased [101].

4.2.2 Combination of CNTs and Other OPL Components

The CNT system alone is not enough to fulfill all the specifications of broadband optical power limiting. A major limitation of CNTs is their inability to show optical limiting behavior in the picosecond time regime; the CNT suspensions are only effective for nanosecond laser pulses [98]. The other disadvantage of CNT is the high limiting threshold.

Various systems need to be coupled to extend the spectral and temporal ranges of effective limiting. These include modifying CNT by polymer coating/wrapping, substitutional doping, blending with various optically active/absorbing dyes, and covalent or noncovalent sidewall functionalization of CNTs. Considerable effort has also been made by using two intermediate focal planes [102, 103]. It is expected that CNT along with OPL-active components in a host matrix will exhibit OPL properties with faster response time. At the same time, a processible material will provide enough flexibility (as thin films or coatings) so as to allow OPL material incorporation into optical systems at intermediate focal planes and also applicable to wide field-of-view (FOV) and fast f-number optical systems.

Polymer/Carbon Nanotube Composites

Carbon nanotube composites have been extensively studied for their thermal, mechanical, electronic, and optical properties [104]. SWNTs were used to augment the thermal properties of epoxy [105]. As for the use of CNTs in practical OPL and other potential applications, efforts have been made in the research and development of films and coatings from the suspension of polymer/carbon nanotube composite materials [106]. One of the approaches to make the composites involves mixing dispersed CNTs in an appropriate solvent with polymer solutions and then evaporating the solvents [59]. Another approach is to incorporate CNTs in thermoplastic polymers at temperature above the polymer's melting point [107, 108]. CNTs can also be introduced to suitable monomers, such as aniline, styrene, via in situ polymerization [109]. Incorporation of CNTs into various other matrices, e.g., metals [110] and ceramics [111] to form composite materials, has also been reported.

MWNTs were sometimes covalently bound to polymers to increase OPL efficiency. Optical limiting efficiency of two soluble polymer-bound MWNTs (poly(*N*-vinylcarbazole)-MWNTs (PVK-MWNTs) and polybutadiene-MWNTs (PB-MWNTs)) and fullerene (C_{60}) was reported by Li et al. [112]. PVK-MWNTs exhibit better OPL response (limiting threshold is 0.24 J cm^{-2}) than that of PB-MWNTs (limiting threshold is 0.40 J cm^{-2}) and C_{60} (limiting threshold is 0.70 J cm^{-2}) which is due to the stronger electron-donating competence of PVK than PB and C_{60}. Poly[3-octyl-thiophene-2,5-diyl]-[*p*-aminobenzylidene-quinomethane]-bonded MWNTs (POTABQ-MWNTs) showed large third-order nonlinear optical responses mainly due to the formation of intermolecular photoinduced charge transfer system of polymers and carbon nanotubes [113]. Strong ground-state interaction between MWNTs and POTABQ was observed.

The polymer/CNT forms a stable solution in many organic solvents which improves the solubility of suspended CNTs in liquids. A simple and novel technique

(a) (b)

Fig. 6.15 (**a**) Neat hyperbranched polymer and (**b**) carbon nanotubes loaded hyperbranched polymer (>2 years)

to obtain extremely stable and solventless suspensions of CNTs in hyperbranched polymers, e.g., hyperbranched polycarbosiloxane (HB-PCS), is reported (Fig. 6.15) [37]. These suspensions were easily processed and fabricated into freestanding thin films or fabricated onto various substrates (e.g., glass and quartz). Thus, solid-state CNT optical power limiting materials can be prepared.

A special polymer (poly(*p*-phenylenevinylene-*co*-2,5-dioctyloxy-*m*-phenylenevinylene)) (PmPV) can "trap" the CNTs to form a temporally stable polymer–CNT dispersion [114]. MWNT–PmPV composites exhibit OPL property (corresponds to reduction in normalized transmission) at high incident pulse energy densities ($\sim$17–18 J cm^{-2}). Improvement in OPL performance was observed with the increase of MWNT content [59]. OPL behavior of SWNT/MWNT-containing polymers, such as polyimide (SWNT-polyimide) [115], poly(ethylene)glycol (MWNT-PEG), poly(2-vinylpyridine) (MWNT-PVP) [116], poly(phenylacetylenes) (MWNT-PPAs) [117], and double-fullerence-end-capped poly(ethyleneoxide) (MWNT-FPEOF) [118], has been reported. Enhanced optical limiting behavior was observed in the case of composites than with CNT itself in some of the cases; the reason for this is not clear yet. A uniform dispersion of CNT can be achieved by tethering the nanotubes to polymeric chains. For example, it has been shown that carbon nanotubes can be combined with polyurethane–urea polymers (via chemical grafting) to provide composite films which are suitable as an optical limiting material [119].

Coated Carbon Nanotubes

Thin film coatings on CNT using various coating materials, such as polycrystalline gold (Au), polycrystalline silver (Ag), silicon carbide (SiC), and silicon nitride (Si_xN_y), have also been demonstrated [120, 121]. Plasma-enhanced chemical vapor deposition (PECVD) and electron beam evaporation technique have been utilized for coatings. It was observed that Au and Ag coatings enhanced the overall OPL effect at 532 nm while not much enhancement was observed at 1,064 nm [112]. The enhancement was mainly attributed to a surface plasmon absorption process in the metal films as surface plasmon resonant peaks at 520 and 420 nm were observed for the metal-coated CNTs. Au and Ag also exhibit OPL performances for picosecond lasers [122].

Carbon Nanotubes Doped with Small Molecules for Optical Power Limiting

Doped CNTs exhibit extraordinary electronic and mechanical properties compared to undoped CNTs [123, 124]. Doping of SWNTs with an electron acceptor such as halogens (e.g., B, I) and an electron donor such as alkali metals (e.g., Li, K, Cs) did change the absorption spectra [125]. Slight enhancement of OPL performance has also been observed in the case of boron-doped MWNTs (Fig. 6.16) [126]. The modification resulted in high metallic electronic character in comparison to undoped CNTs. Xu et al. concluded that the primary OPL property originates due to dissipative absorption which results in "microbubble" or "microplasma" generation through heat transfer to the solvent and ultimately leading to enhanced scattering. However, study of photonic devices based on doped CNTs remains a field with many unclear issues. A deeper understanding of the physical mechanism and optical properties due to chemical doping is needed [125].

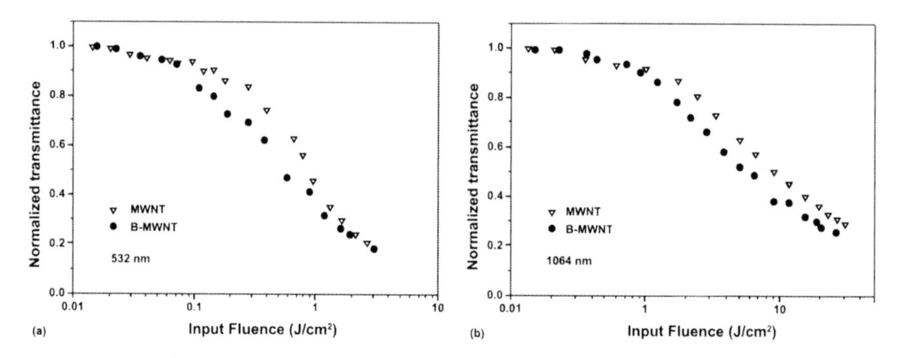

Fig. 6.16 Comparison of the optical limiting behavior of pure and boron-doped MWNT (B-MWNT) suspension at (**a**) 532 nm and (**b**) 1,064 nm (reprinted with permission from Ref. [126])

Blending of Optically Active Dyes with Carbon Nanotubes

As mentioned earlier, several optical limiting mechanisms as well as OPL materials need to be combined synergistically into one system to obtain an ideal OPL limiter or to extend the performance of optical limiters due to the cumulative effects. These materials include: (a) nonlinear scattering materials (e.g., CNTs), (b) two- or multiphoton absorption dyes (e.g., stilbene-3, POPOP, and platinum acetylide), and (c) reverse saturable absorption dyes (e.g., HITCI, Zn-TPP, and lead phthalocyanine). Izard et al. reported blending of MPA dye (stilbene-3) with nonlinear scattering CNTs [119]. The blend gives slightly better OPL efficiency than that of pure CNTs in the case of nanosecond laser pulses. MPA dye is also effective for sub-nanosecond pulses as strong multiphoton absorption cross sections are responsible for OPL.

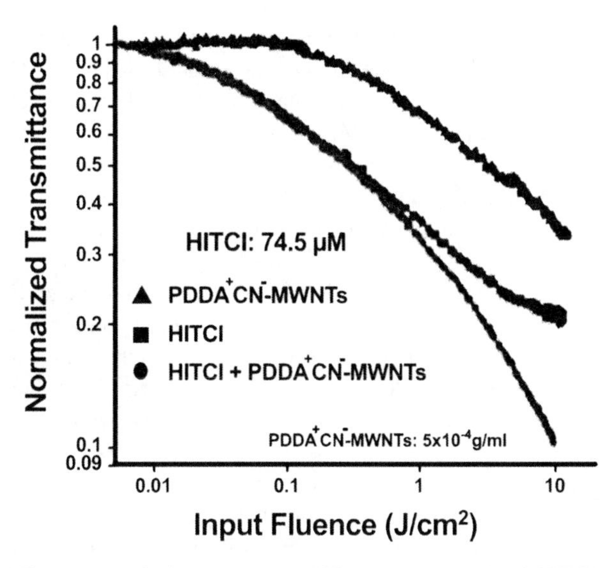

Fig. 6.17 Nonlinear transmission spectra at 532 nm wavelength of HITCI, $PDDA^+CN_x^-$-MWNTs, and blended HITCHI and $PDDA^+CN_x^-$-MWNTs at molar concentration of HITCI of 74.5μM (reprinted with permission from Ref. [35])

Webster et al. reported fabrication of a new CNT-based OPL material by combining CNTs (specifically, nitrogen-doped MWNTs) with RSA dye HITCI (Fig. 6.17) [35].

The nitrogen-doped MWNTs were functionalized for compatibilization in solvent. The blended suspension exhibited superior OPL performance to the individual constituent materials.

Carbon-rich molecules are known to liberate more carbon as a decomposition product when irradiated with intense laser beams. Therefore, incorporation of carbon-rich organic molecules (such as anthracene) in addition to CNTs will provide an additional barrier for protection of optical materials against high-intensity lasers. The mechanism in this case is via instant blackening due to high energy-mediated decomposition of the carbon-rich molecule into carbon.

Although OPL has been observed in the past using carbon nanotubes in a suitable solvent, it occurs in these solvents due to solvent bubble formation which scatters light at high fluence. The carbon nanotubes absorb incident light and transfer energy to the solvent to form bubbles.

Siloxane-based polymers can be processed into thin films and coatings that are mechanically robust and optically clear. Siloxane-based polymers can also be formed into three-dimensional networks via a sol–gel process resulting in a homogeneous material [127–129].

Hyperbranched (HB) polycarbosiloxane (PCS)–CNT hybrid networks can also be synthesized by sol–gel processing and lead to stable and durable materials. The novelty of using a dye covalently linked to the HB-PCS matrix leads to the reduction of CNT aggregation. The homogeneous and transparent character of the HB-PCS/CNT composite films makes them appropriate for use as an OPL material. HB-PCS-based films have the added advantage of immunity to temperature fluctuations. A series of other RSA dyes, MPA dyes, and azo dyes can also be used, in combination, to fine-tune the laser-blocking material.

A proprietary HBP family, namely, HB-PCS that forms excellent optical quality films and coatings has been used as the host materials for this invention. The salient feature of this system is the combination of three different OPL mechanisms in an additive way to provide efficient protection from laser beam damage. For a polymer composite coating to work as a laser-blocking material, an appropriate choice and formulation of the active components is necessary. Toward this end, a model system has been formulated using HB-PCS as the polymer matrix, fullerene as the RSA dye, stilbene-3 as the MPA dye, disperse red 1 (DR1) as the azo dye, and sMWNT.

The experimental results show that when sMWNT are dispersed in a HB-PCS host, they absorb incident light and convert the polymer into a state that scatters light. OPL onset occurs at about 10 μJ of input energy, very similar to that observed in the liquid systems. With azo dyes containing samples, the OPL onset occurs at much lower threshold energy (∼5 μJ) while the output energy level where clamping occurs is significantly reduced ∼10 μJ. However, the most promising results were obtained for the multilayer film configuration (Fig. 6.18). For this, the OPL onset occurs at a much lower input threshold energy (∼1 μJ) while the output energy level where clamping occurs is even further reduced to ∼3 μJ. Although it is still not understood what happens to the HB-PCS when energy is transferred from the

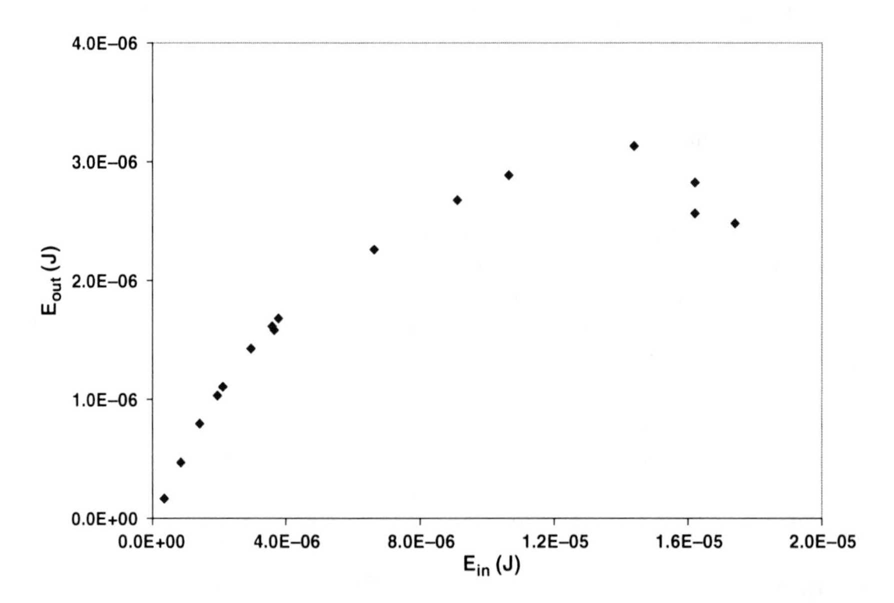

Fig. 6.18 OPL property of multicomponent including azo dye. The laser wavelength was 532 nm

carbon nanotubes to the polymer matrix, it is expected that either a chemical or structural change occurs.

The present work provides a similar system in a solid polymer host. Additionally, the matrix needs to contain RSA and MPA dyes that function in a synergistic fashion to provide better protection against laser irradiation. This provides a much more useful laser-blocking material suitable for incorporation into eye protection devices than presently used dye filters or a similar system.

Thus, the results have provided both positive and promising evidence that this approach will lead to a suitable OPL material that can be coated on a variety of surfaces of different curvatures.

5 Summary and Future Outlook

Carbon nanotubes, if appropriately combined with other organic and inorganic chromophores, possess a huge potential for optical power limiting materials for protection of sensors, including human eyes. The combination of the OPL components makes the simultaneous but synergistic application of the various OPL mechanisms possible. As a result, the OPL materials can have effectively extended application ranges, covering a broad spectral and temporal range. Multifunctional OPLs are expected to have a broader application range for various laser sources. The plasticity and flexibility of various host materials should allow one to design and fabricate a range of optimized structures to meet different requirements.

References

1. Hollins, R.C.: Materials for optical limiters. Curr. Opin. Solid State Mater. Sci. **4**, 189 (1999)
2. Prasad, P.N., Williams, D.J.: Introduction to nonlinear optical effects in molecules and polymers. Wiley, New York, NY (1991)
3. Spangler, C.W., He, M.: The design of optical limiters based on the formation of bipolaron-like dications. Proc. MRS Symp. **479**, 59 (1997)
4. Albota, M., Beljonne, D., Bredas, J.L., Ehrlich, J.E., Fu, J.Y., Heikal, A.A., Hess, S.E., Kogej, T., Levin, M.D., Marder, S.R., McCord-Maughon ,D., Perry, J.W., Rockel, H., Rumi, M., Subramaniam, C., Webb, W.W., Wu, X.L., Xu, C.: Design of organic molecules with large two-photon absorption cross sections. Science **281**, 653 (1998)
5. Hernandez, F.E., Yang, S., Van Stryland, E.W., Hagan, D.J.: High dynamic range cascaded-focus optical limiter. Opt. Lett. **25**, 1180 (2000)
6. Drobizhev, M., Karotki, A., Rebane, A., Spangler, C.W.: Dendrimer molecules with record large two-photon absorption cross section. Opt. Lett. **26**, 1081 (2001)
7. He, G.S., Bhawalkar, J.D., Zhao, C.F., Prasad, P.N.: Optical limiting effect in a two-photon absorption dye doped solid matrix. Appl. Phys. Lett. **67**, 2433 (1995)
8. Perry, J.W., Mansour, K., Marder, S.R., Perry, K.J., Alvarez, D. Jr., Choong, T.: Enhanced reverse saturable absorption and optical limiting in heavy-atom-substituted phthalocyanines. Opt. Lett. **19**, 625 (1994)
9. Van Stryland, E.W., Hagan, D.J., Xia, T., Said, A.A.: Applications of Nonlinear Optics to Passive Optical Limiting. In: Nalwa, H.S., Miyata, S. (eds) Nonlinear optics of organic molecules and polymers. CRC, New York, NY (1997)

10. Crane, R., Lewis, K., Van Stryland, E.W., Khoshnevisan, M. (eds.): Proceedings of the MRS Symposium, Pittsburgh, PA, 374 (1995)
11. Lawson, C.M. (ed.): nonlinear optical liquids and power limiters. Proceedings of SPIE, Bellingham, WA, 3146 (1997)
12. Xia, T., Hagan, D.J., Dogariu Sail, A.A., Van Stryland, E.W.: Optimization of optical limiting devices based on excited-state absorption. Appl. Opt. **36**, 4110 (1997)
13. Kamanina, N.V., Plekhanov, A.I.: Mechanisms of optical limiting in fullerene-doped -conjugated organic structures demonstrated with polyimide and COANP molecules. Opt. Spectrosc. **93**, 403 (2002)
14. Zhou, G., Wang, X., Wang, Z., Shao, Z., Jieng, M.: Upconversion fluorescence and optical power limiting effects based on the two- and three-photon absorption process of a new organic dye BPAS. Appl. Opt. **41**, 1120 (2002)
15. Miles, P.A.: Bottleneck optical limiters: the optimal use of excited-state absorbers. Appl. Opt. **33**, 6965 (1994)
16. Hernandez, F.E., Yang, S.S., Dubikovsky, V., Shensky, W., Van Stryland, E.W., Hagan, D.J.: Dual focal plane visible optical limiter. J. Nonlinear Opt and Mater. **9**, 423 (2000)
17. Feneyrou, P., Doclot, O., Block, D., Baldeck, P.L., Zyss, J.: Two-photon absorption and optical-limiting properties in dimethylaminocyanobiphenyl crystal. J Nonlin Opt and Mater **5**, 767 (1996)
18. Bentivegna, F., Canva, M., Georges, P., Brun, A., Chaput, F., Malier, L., Boilot, J.-P.: Reverse saturable absorption in slid xerogel matrices. Appl. Phys. Lett. **62**, 1721 (1992)
19. Maggini, M., Scorrano, G., Parto, M., Brusatin, G., Innocenzi, P., Guglielmi, M., Renier, A., Signorini, R., Meneghetti, M., Bozio, R.: C_{60} derivatives embedded in sol-gel silica films. Adv. Mater. **7**, 404 (1995)
20. Cha, M., Sariciftci, N.S., Heeger, A.J., Hummelen, J.C., Wudl, F.: Enhanced nonlinear absorption and optical limiting in semiconducting polymer methanofullerene charge transfer films. Appl. Phys. Lett. **67**, 3850 (1995)
21. Kojima, Y., Matsuoka, T., Takahashi, H., Kurauchi, T.: Optical limiting property of fullerene-containing polystyrene. J. Mater. Sci. Lett. **16**, 2029 (1997)
22. Hagan, D.J., Xia, T., Said, A.A., Wei, T.T., Van Stryland, E.W.: High Dynamic Range Passive Optical Limiters. Int. J. Nonlinear Opt. Phys. **2**, 483 (1993)
23. Van Stryland, E.W., Vanherzeele, H., Woodall, M.A., Soileau, M.J., Smirl, A.L., Guha, S., Boggess, T.F.: 2 photon-absorption, nonlinear refraction, and optical limiting in semiconductors. Opt. Eng. **24**, 613 (1985)
24. He, G.S., Gvishi, R., Prasad, P.N., Reinhardt, B.A.: Two-photon absorption based optical limiting and stabilization in organic molecule-doped solid materials. Opt Commun. **117**, 133 (1995)
25. Tutt, L.W., Boggess, T.F.: A review of optical limiting mechanisms and devices using organics, fullerines, semiconductors and other materials. Prog. Quantum Electron. **17**, 299 (1993)
26. Wu, S.-T.: Infrared markers for determining the order parameters of uniaxial liquid crystals. Opt. Eng. **26**, 120 (1987)
27. Sutherland, R.L.: Handbook of nonlinear optics. Marcel Dekker, New York, NY (1996)
28. DeSalvo, R., Said, A.A., Hagan, D.J., Van Stryland, E.W., Sheik-Bahae, M.: Infrared to Ultraviolet Measurements of 2-Photon Absorption and n, in Wide Bandgap Solids. IEEE J. Quantum Electron. **32**, 1324 (1996)
29. Mansour, K., Soileau, M.J., Van Stryland, E.W.: Nonlinear optical properties of carbon-black suspensions (ink). J. Opt. Soc. Am. B **9**, 1100 (1992)
30. Justus, B.L., Campillo, A.J., Huston, A.L.: Thermal-defocusing/scattering optical limiter. Opt. Lett. **19**, 673 (1994)
31 Giuliano, C.R., Hess, L.D.: Nonlinear absorption of light: Optical saturation of electronic transitions in organic molecules with high intensity laser radiation. IEEE Journal of Quantum Electronics QE-3, 358 (1967)

32. Shultz, J.L., Salamo, G.J., Sharp, E.J., Wood, G.L., Anderson, R.J., Neurgaonkar, R.R: Enhancing the response time for photorefractive beam fanning. Proc. SPIE **1692**, 78 (1992)
33 Harriman, A., Hosie, R.J.: Luminescence of porphyrins and metalloporphyrins. Part 4. Fluorescence of substituted tetraphenylporins. J Chem Soc, Faraday Transactions 2 **77**, 1695 (1981)
34. Wenseleers, W., Stellacci, F., Meyer-Friedrichsen, T., Mangel, T., Bauer, C.A., Pond, S.J.K., Marder, S.R., Perry, J.W.: Five orders-of-magnitude enhancement of two-photon absorption for dyes on silver nanoparticle fractal clusters. J. Phys. Chem. B **106**: 6853 (2002)
35. Webster, S., Reyes-Reyes, M., Pedron, X., Lopez-Sandoval, R., Terrones, M., Carroll, D.: Enhanced nonlinear transmittance by complementary nonlinear mechanisms: A reverse-saturable absorbing dye blended with nonlinear-scattering carbon nanotubes. Adv. Mater. **17**, 1239 (2005)
36. Chin, K.C., Gohel, A., Elim, H.I., Chen, W., Ji, W., Chong, G.L., Sow, C.H., Wee, A.T.S.: Modified carbon nanotubes as broadband optical limiting nanomaterials. J. Mater. Res. **21**, 2758 (2006)
37. Rahman, S.R., Mirza, S., Dvornic, P.R., Sarkar, A.: Unpublished results. Chem. Comm. (Submitted) (2009)
38. Sarkar A, Rahman SR, Mirza, S, Rayfield, GR, (2009). US Patent appln., 61/210, 066
39. Wu, P., Rao, D.V.G.L.N., Kimball, B.R., Nakashima, M., DeCristofano, B.S.: Transient optical modulation with disperse red-1 doped polymer film. Appl. Opt. **39**, 814 (2000)
40. Kroto, H.W., Heath, J.R., O'Brien, S.C., Curl, R.F., Smalley, R.E.: C_{60}: Buckminsterfullerene. Nature **318**, 162 (1985)
41. Sun, Y.P., Riggs, J.E.: Organic and inorganic optical limiting materials. From fullerenes to nanoparticles. Int. Rev. Phys. Chem. **18**, 43 (1999)
42. Brustain, G., Signorini, R.: Linear and nonlinear optical properties of fullerenes in solid state materials. J. Mater. Chem. **12**, 1964 (2002)
43. Geckeler, K.E., Samal, S.: Syntheses and properties of macromolecular fullerenes, a review. Polym. Int. **48**, 743 (1999)
44. Dai, L., Mau, A.W.H.: Controlled synthesis and modification of nanostructures for advanced polymeric composite materials. Adv Mater **13,** 899 (2001)
45. Wang, C., Guo, Z.X., Fu, S., Wu, W., Zhu. D.: Polymers containing fullerene or carbon nanotube structures. Prog. Polym. Sci. **29**, 1079 (2004)
46. Kadish, K.M., Ruoff, R.S. (eds.): Fullerenes chemistry, physics, and technology. Wiley, New York, NY (2000)
47. Mishra, S.R., Rawat, H.S., Mehendale, S.C., Rustagi, K.C., Sood, A.K., Bandyopadhyay, R., Govindaraj, A., Rao, C.N.R.: Optical limiting in single-walled carbon nanotube suspensions. Chem. Phys. Lett. **317**, 510 (2000)
48. Lefrant, S., Anglaret, E., Vivien, L., Riehl, D.: Utilisation des fullerènes et des nanotubes de carbone pour la réalisation de limiteurs optiques. Vide-Sci. Technol. Appl. **56**, 288 (2001)
49. Sun, X., Yu, R.Q., Xu, G.Q., Hor, T.S.A., Ji, W.: Broadband optical limiting with multiwalled carbon nanotubes. Appl. Phys. Lett. **73**, 3632 (1998)
50. Durand, O., Grolier-Mazza, V., Frey, R.: Picosecond-resolution study of nonlinear scattering in carbon black suspensions in water and ethanol. Opt. Lett. **23**, 1471 (1998)
51. Tiwari, S.K., Joshi, M.P., Laghate, M., Mehendale, S.C.: Role of host liquid in optical limiting in ink suspensions. Opt. Laser Techn. **34**, 487 (2002)
52. Iijima, S.: Helical microtubules of graphitic carbon. Nature **354**, 56 (1991)
53. Iijima, S., Ichihashi, T.: Synthesis of single-wall nanotubes [Single-shell carbon nanotubes] of 1-nm diameter. Nature **363**, 603 (1993)
54. Baughman, R.H., Zakhidov, A.A., de Heer, W.A.: Carbon Nanotubes–the Route Toward Applications. Science **297**, 787 (2002)
55. Dreher, K.L.: Health and environmental impact of nanotechnology: toxicological assessment of manufactured nanoparticles. Toxicol. Sci. **77**, 3 (2004)

56. Ajayan, P.M., Charlier, J.C., Rinzler, A.G.: Carbon nanotubes: From macromolecules to nanotechnology. Proc. Natl. Acad. Sci. USA 96, 14199 (1999)
57. Saito, R., Dresselhaus, G., Dresselhaus, M.S.: Physical properties of carbon nanotubes. Imperial College Press, London (2005)
58. Chen, P., Wu, X., Sun, X., Lin, J., Ji, W., Tan, K.L.: Electronic structure and optical limiting behavior of carbon nanotubes. Phys. Rev. Lett. **82**, 2548 (1999)
59. Chen, Y., Lin, Y., Liu, Y., Doyle, J., He, N., Zhuang, X., Bai, J., Blau, W.J.: Carbon nanotube-based functional materials for optical limiting. J. Nanosci. Nanotechnol. **7**, 1268 (2007)
60 Ajayan, P.M., Zhou, O.Z.: Applications of carbon nanotubes," Topics Appl Phys **80**, 391 (2001)
61. Dai, H., Wong, E.W., Liebert, C.M.: Probing electrical transport in nanomaterials: conductivity of individual carbon nanotubes. Science **272**, 523 (1996)
62. Rueckes, T., Kim, K., Joselevich, E., Tseng, G.Y., Cheung, C.-L., Lieber, C.: Carbon Nanotube-Based Nonvolatile Random Access Memory for Molecular Computing. Science **289**, 94 (2000)
63. Dillon, A.C., Jones, K.M., Bekkedahl, T.A., Kiang, C.H., Bethune, D.S., Heben, M.J.: Storage of hydrogen in single-walled carbon nanotubes. Nature **386**, 377 (1997)
64. Cheng, H.M., Yang, Q.H., Liu, C.: Hydrogen storage in **carbon** nanotubes. Carbon **39**, 1447 (2001)
65. Kong, J., Franklin, N.R., Zhou, C., Chapline, M.G., Peng, S., Cho, K., Dai, H.: Nanotubes Molecular Wires as Chemical Sensors. Science **287**, 622 (2000)
66. Collins, P.G., Bradley, K., Ishigami, M., Zettl, A.: Extreme oxygen sensitivity of electronic properties of carbon nanotubes. Science **287**, 1801 (2000)
67. Chopra, S., McGuire, K., Gothard, N., Rao, A.M., Pham, A.: Selective gas detection using a carbon nanotube sensor. Appl. Phys. Lett. **83**, 2280 (2003)
68. Chopra, S., Pham, A., Gaillard, J., Parker, A., Rao, A.M.: Carbon-nanotube-based resonant-circuit sensor for ammonia. Appl. Phys. Lett. **80**, 4632 (2002)
69. Nguyen, C.V., So, C., Stevens, R.M., Li, Y., Delziet, L., Sarrazin, P.: High Lateral Resolution Imaging with Sharpened Tip of Multi-Walled Carbon Nanotube Scanning Probe. J. Phys. Chem. B **108**, 2816 (2004)
70. Kim, Y.C., Sohn, K.H., Cho, Y.M., Yoo, E.H.: Vertical alignment of printed carbon nanotubes by multiple field emission cycles. Appl. Phys. Lett. **84**, 5350 (2004)
71. Wang, J., Deo, R.P., Poulin, P., Mangey, M.: Carbon Nanotube Fiber Microelectrodes. J. Am. Chem. Soc. **125**, 14706 (2003)
72. Wang, J., Musameh, M.: Carbon nanotube/teflon composite electrochemical sensors and biosensors. Anal. Chem. **75**, 2075 (2003)
73. Zhang, M., Gorski, W.: Electrochemical sensing based on redox mediation at carbon nanotubes. Anal. Chem. **77**, 3960 (2005)
74. Zengin, H., Zhou, W., Jin, J., Czerw, R., Smith, D.W., Echegoyen, L., Carroll, D.L., Foulger, S.H., Ballato, J.: Carbon nanotube doped polyaniline. Adv Mater **14**, 1480 (2002)
75. Saran, N., Parikh, K., Suh, D.S., Munoz, E., Kolla, H., Manohar, S.K.: Fabrication and Characterization of Thin Films of Single-Walled Carbon Nanotube Bundles on Flexible Plastic Substrates. J. Am. Chem. Soc. **126**, 4462 (2004)
76. Regev, O., El Kati, P.N.B., Loos, J., Koning, C.E.: Preparation of conductive nanotube-polymer composites using latex technology. Adv Mater **16**, 248 (2004)
77. Xie, R.H., Rao, Q. Chem. Third-order optical nonlinearities of chiral graphene tubules. Phys. Lett. **313**, 211 (1999)
78. Vivien, L., Riehl, D., Delouis, J.F., Delaire, J.A., Hache, F., Anglaret, E.: Picosecond and nanosecond polychromatic pump-probe studies of bubble growth in carbon-nanotube suspensions. J. Opt. Soc. Am. B **19**, 208 (2002)
79. Jin, Z.X., Huang, L., Goh, S.H., Xu, G., Ji, W.: Size-dependent optical limiting behavior of multi-walled carbon nanotubes. Chem. Phys. Lett. **352**, 328 (2002)

80. Wang, H., Hobbie, E.K.: Amphiphobic Carbon Nanotubes as Macroemulsion Surfactant. Langmuir **19**, 3091 (2003)
81. Girifalco, L.A., Hodak, M. Lee, R.S.: Carbon nanotubes, buckyballs, ropes and a universal graphitic potential. Phys. Rev. B **62**, 13104 (2000)
82. Bandyopadhyaya, R., Nativ-Roth, E., Regev, O., Yerushalmi-Rozen, R.: Stabilization of Individual Carbon Nanotubes in Aqueous Solutions. Nano Lett. **2**, 25 (2002)
83. Moore, V.C., Strano, M.S., Haroz, E.H., Hauge, R.H., Smalley, R.E.: Individually suspended single-walled carbon nanotubes in various surfactants. Nano. Lett. **3**, 1379 (2003)
84. Islam, M.F., Rojas, E., Bergey, D.M., Johnson, A.T., Yodh, A.G.: High weight fraction surfactant solubilization of single-wall carbon nanotubes in water. Nano Lett. **3**, 269 (2003)
85. Wang, D., Ji, W.X., Li, Z.C., Chen, L.: A biomimetic "polysoap" for single-walled carbon nanotube dispersion. J. Am. Chem. Soc. **128**, 6556 (2006)
86. Chatterjee, T., Yurekli, K., Hadjiev, V.G., Krishnamoorti, R.: Single-Walled Carbon Nanotube Dispersions in Poly(ethylene oxide). Adv. Funct. Mater. **15**, 1832 (2005)
87. Riggs, J.E., Guo, Z., Carroll, D.L., Sun, Y.P.: Strong Luminescence of Solubilized Carbon Nanotubes. J. Am. Chem. Soc. **122**, 5879 (2000)
88. Huang, W., Fernando, S., Allard, L.F., Sun, Y.P.: Solubilization of Single-Walled Carbon Nanotubes with Diamine-Terminated Oligomeric Poly(ethylene Glycol) in Different Functionalization Reactions. Nano Lett. **3**, 565 (2003)
89. Fernando, S., Lin, Y., Sun, Y.P.: High aqueous solubility of functionalized single-walled carbon nanotubes. Langmuir **20**, 4777 (2004)
90. Yang, Y.L., Gupta, M.C., Dudley, K.L., Lawrence, R.W.: Conductive carbon nanoriber-polymer foam structures. Nano Lett. **5**, 2131 (2005)
91 Kim, H.M., Kim, K., Lee, C.Y., Joo, J., Cho, S.J., Yoon, H.S., Pejakovic, D.A., Yoo, J.W., Epstein, A.: Electrical conductivity and electromagnetic interference shielding of multiwalled carbon nanotube composites containing Fe catalyst. Appl Phys Lett **84**, 589 (2004)
92. Bryning, M.B., Islam, M.F., Kikkawa, J.M., Yodh, A.G.: Carbon Nanotube Aerogels. Adv. Mater. **17**, 1186 (2005)
93. Li, N., Huang, Y., Du, F., He, X., Lin, X., Gao, H., Ma, Y., Li, F., Chen, Y., Eklund, P.C.: Electromagnetic interference (EMI) shielding of single-walled carbon nanotube epoxy composites. Nano Lett. **6**, 1141 (2006)
94. Vivien, L., Anglaret, E., Riehl, D., Bacou, F., Journet, C., Goze, C., Andrieux, M., Brunet, M., Lafonta, F., Bernier, P., Hache, F.: Single-wall carbon nanotubes for optical limiting. Chem. Phys. Lett. **307**, 317 (1999)
95 Vivien, L., Lancon, D., Riehl, D., Hache, F., Anglaret, E.: Carbon nanotubes for optical limiting. Carbon **40**, 1789 (2002)
96. Vivien, L., Riehl, D., Hache, F., Anglaret, E.: Nonlinear scattering origin in carbon nanotube suspensions. J. Nonlinear Opt. Phys. Mater. **9**, 297 (2000)
97. Vivien, L., Riehl, D., Lancon, P., Hache, F., Anglaret, E.: Pulse duration and wavelength effects on optical limiting behavior of carbon nanotube suspensions. Opt. Lett. **26**, 223 (2001)
98. Sun, X., Xiong, Y.N., Chen, P., Lin, J.Y., Ji, W., Lim, J.H., Yang, S.S., Hagan, D.J., Van Stryland, E.W.: Investigation of an optical limiting mechanism in multiwalled carbon nanotubes. Appl. Opt. **39**, 1998 (2000)
99. Liu, L.Q., Zhang, S., Qin, Y.J., Guo, Z.X., Ye, C., Zhu, D.B.: Solvent effects of optical limiting properties of carbon nanotubes. Synth. Met. **135**, 853 (2003)
100. Pratap, A., Shan, A.L., Singh, A.R., Pal, S., Tyagi, R.K., Dawar, A.L., Chaturvedi, P., Lamba, S., Harsh, M.B.: Linear and non-linear optical transmission from multi-walled carbon nanotubes. J. Mater. Sci. **40**, 4185 (2005)
101. Yu, H.J., Kim, S.W.: Temperature Effects in an Optical Limiter Using Carbon Nanotube Suspensions. J. Korean Phys. Soc. **47**, 610 (2005)

102. Miles, P.A.: Bottleneck optical limiters: the optimal use of excited-state absorbers. Appl. Opt. **33**, 6965 (1994)
103 Hernandez, F.E., Yang, S.S., Van Stryland, E.W., Hagan, D.J.: High dynamic range cascaded-focus optical limiter. Opt Lett **25**, 1180 (2000)
104. Zhang, W., Sakalkar, V., Koratkar, N.: In situ health monitoring and repair in composites using carbon nanotube additives. Appl. Phys. Lett. **91**, 133102 (2007)
105 Biercuk, M.J., Llaguno, M.C., Radosavljevic, M., Hyun, J.K., Fischer, J.E., Johnson, A.T.: Carbon nanotube composites for thermal management. Appl Phys Lett **80**, 2767 (2002)
106. Xie, X.L., Mai, Y.W., Zhou, X.P.: Dispersion and alignment of carbon nanotubes in polymer matrix: A review. Mater. Sci. Eng. R **49**, 89 (2005)
107. Sennettt, M., Welsh, E., Wright, J.B., Li, W.Z., Wen, J.G., Ren, Z.F.: Dispersion and alignment of carbon nanotubes in polycarbonate. Appl. Phys. A **76**, 111 (2003)
108. Puglia, D., Valentini, L., Kenny, J.M.: Analysis of the Cure Reaction of Carbon Nanotubes/Epoxy Resin Composites Through Thermal Analysis and Raman Spectroscopy. J. Appl. Polym. Sci. **88**, 452 (2003)
109. Hughes, M., Chen, G.Z., Shaffer, M.S.P., Fray, D.J., Windle, A.H.: Electrochemical Capacitance of a Nanoporous Composite of Carbon Nanotubes and Polypyrrole. Chem. Mater. **14**, 1610 (2002)
110. Chen, W.X., Tu, J.P., Wang, L.Y., Gan, H.Y., Xu, Z.D., Zhang, X.B.: Tribological application of carbon nanotubes in a metal-based composite coating and composites. Carbon **41**, 215 (2003)
111. Zhan, Z.D., Kuntz, J.D., Wan, J.L., Mukherjee, A.K.: Single-wall carbon nanotubes as attractive toughening agents in alumina-based nanocomposites. Nat. Mater. **2**, 38 (2003)
112. Li, C., Liu, C., Li, F., Gong, Q.: Optical limiting performance of two soluble multi-walled carbon nanotubes. Chem. Phys. Lett. **380**, 201 (2003)
113. Yi, W.H., Feng, W., Xu, Y.L., Wu, H.C.: Synthesis and Third-Order Optical Nonlinearities of Conjugated Polymer-Bonded Carbon Nanotubes. Jpn. Appl. Phys. Part 1 **44**, 3022 (2005)
114 Murphy, R., Coleman, J.N., Cadek, M., McCarthy, B., Bent, M., Drury, A., Barklie, R.C., Blau, W.J.: High Yield, Non-destructive Purification and Quantification Method for Carbon Nanotubes. J Phys Chem B **106**, 3087 (2002)
115. Chen, Y.C., Raravikar, N.R., Schadler, L.S., Ajayan, P.M., Zhao, Y.P., Lu, T.M., Wang, G.C., Zhang, X.C.: Ultrafast optical switching properties of single-wall carbon nanotube polymer composites at 1.55 μm. Appl. Phys. Lett. **81**, 975 (2002)
116. Jin, Z.X., Sun, X., Xu, G.Q., Goh, S.H., Ji, W.: Nonlinear optical properties of some polymer/multi-walled carbon nanotube composites. Chem. Phys. Lett. **318**, 505 (2000)
117. Tang, B.Z., Xu, H.: Preparation, Alignment, and Optical Properties of Soluble Poly(phenylacetylene)-Wrapped Carbon Nanotubes. Macromolecules **32**, 2569 (1999)
118. Goh, H.W., Goh, S.H., Xu, G.Q., Lee, K.Y., Yang, G.Y., Lee, Y.W., Zhang, W.D.: Optical Limiting Properties of Double-C_{60}-End-Capped Poly(ethylene oxide), Double-C_{60}-End-Capped Poly(ethylene oxide)/Poly(ethylene oxide) Blend, and Double-C_{60}-End-Capped Poly(ethylene oxide)/Multiwalled Carbon Nanotube Composite. J. Phys. Chem. B **107**, 6056 (2003)
119. Izard, N., Menard, C., Riehl, D., Doris, E., Mioskowski, C., Anglaret, E.: Combination of carbon nanotubes and two-photon absorbers for broadband optical limiting. Chem. Phys. Lett. **391**, 124 (2004)
120. Chin, K.C., Gohel, A., Chen, W.Z., Elim, H.I., Ji, W., Chong, G.L., Sow, C.H., Wee, A.T.S.: Gold and silver coated carbon nanotubes: An improved broad-band optical limiter. Chem. Phys. Lett. **409**, 85 (2005)
121. Chin, K.C., Gohel, A., Elim, H.I., Ji, W., Chong, G.L., Lim, K.Y., Sow, C.H., Wee, A.T.S.: Optical limiting properties of amorphous SixNy and SiC coated carbon nanotubes. Chem. Phys. Lett. **383**, 72 (2004)
122. Philip, R., Kumar, G.R., Sandhayarani, N., Pradeep, T.: Picosecond optical nonlinearity in monolayer-protected gold, silver, and gold-silver alloy nanoclusters. Phys. Rev. B **62**, 13160 (2000)

123. Li, T., Wang, J.N., Zhang, Y.M.: Electrical transport in doped one-dimensional nanostructures. J. Nanosci. Nanotechnol. **5**, 1435 (2005)
124. Huang, J.W., Bai, S.J.: Light emitting diodes of fully conjugated heterocyclic aromatic rigid-rod polymer doped with multi-wall carbon nanotubes. Nanotechnology **16**, 1406 (2005)
125. Zhao, J., Chen, X., Xie, J.R.H.: Optical properties and photonic devices of doped carbon nanotubes. Anal. Chim. Acta **568**, 161 (2006)
126. Xu, J., Xiao, M., Czerw, R., Carroll, D.L.: Optical limiting and Enhanced Optical Nonlinearity in Doped Carbon Nanotubes. Chem. Phys. Lett. **389**, 247 (2004)
127 Dvornic, P.R., Hu, J., Meier, D.J., Nowak, R.M.: US Patent 6,534,600 (2003)
128 Dvornic, P.R., Hu, J., Meier, D.J., Nowak, R.M.: US Patent 6,646,089 (2003)
129 Dvornic, P.R., Hu, J., Meier, D.J., Nowak, R.M.: Silicon-Containing Hyperbranched Polymers Via Bimolecular Polymerization. Polym Prepr **45**, 585 (2004)

Chapter 7
Field Emission Properties of ZnO, ZnS, and GaN Nanostructures

Y. Mo, J.J. Schwartz, M.H. Lynch, P.A. Ecton, Arup Neogi, J.M. Perez, Y. Fujita, H.W. Seo, Q.Y. Chen, L.W. Tu, and N.J. Ho

Abstract We review the growth and field emission (FE) properties of ZnO, ZnS, and GaN nanostructures. For ZnO nanostructures, we discuss in detail solution-based growth techniques and the effects of residual gas exposure on the FE properties. We present new results showing that O_2 and CO_2 exposures do not have a significant effect on the FE properties of ZnO nanorods, but N_2 exposure significantly degrades them. We also present new results showing that Cs deposition significantly improves the FE properties of GaN nanorods.

Keywords Field emission · ZnO · ZnS · GaN · Nanostructures

1 Introduction

ZnO, ZnS, and GaN nanostructures have been recently investigated as field emission (FE) electron sources for potential applications in flat panel displays and other vacuum microelectronic applications. In particular, ZnO nanostructures have received considerable interest [1–15]. The motivation is to discover new FE materials with improved properties over carbon nanotubes and Spindt-type metal microtips such as Mo microtips. Carbon nanotubes and metal microtips have been extensively studied for FE applications due to their low turn-on voltages of 1–10 V/μm [16–26]. However, the FE properties of carbon nanotubes and metal microtips degrade significantly with exposure to O_2 and oxygen-containing gases typically found in vacuum containers of flat panel displays [17, 18, 23–26]. This leads to operating lifetimes that are too short for commercial applications. The degradation is thought to be due to oxidation of carbon and metals under the high-electric-field conditions present during FE. It is thought that oxide-based materials such as ZnO may be less susceptible to oxidation since it is difficult to further oxidize ZnO [27]. In addition, wide bandgap semiconductors such as ZnO (3.4 eV), ZnS (3.6 eV), and GaN (3.5 eV)

J.M. Perez (✉)
Department of Physics, University of North Texas, Denton, TX 76203, USA
e-mail: jperez@unt.edu

Z.M. Wang, A. Neogi (eds.), *Nanoscale Photonics and Optoelectronics*,
Lecture Notes in Nanoscale Science and Technology 9,
DOI 10.1007/978-1-4419-7587-4_7,

may exhibit band bending at the surface such as negative electron affinity (NEA) that may reduce the barrier for FE [28].

In this review, we discuss the growth and FE properties of ZnO, ZnS, and GaN nanostructures, with an emphasis on ZnO nanorods. We discuss in detail solution-based techniques for the growth of ZnO nanorods that are well suited for commercial applications such as flat panel displays. The effects of residual gases such as O_2, CO_2, N_2, H_2, Ar, and other gases on the FE properties of ZnO and GaN nanostructures are discussed in detail. We present new results showing that the FE properties of ZnO nanorods grown using the DC arc discharge technique are not significantly degraded by exposure to O_2 and CO_2, but are significantly degraded by exposure to N_2. This is thought to be due to the higher reactivity of nitrogen with ZnO. In addition, we present new results on the effects of Cs deposition on the FE properties of GaN nanorods. Cs deposition is observed to significantly reduce the turn-on voltage by approximately 50%. We propose that this is due to a reduced barrier or NEA surface induced by the Cs. It would be interesting to investigate if Cs deposition reduces the turn-on voltage of ZnO, ZnS, and other wide bandgap nanostructures.

2 ZnO Nanostructures

A wide range of techniques have been successfully employed in the production of ZnO nanostructures. Such techniques can generally be categorized as vapor or solution-based techniques. The ZnO nanostructures produced using these methods may exist as one of a variety of different possible morphologies such as ZnO nanorods, nanoneedles, nanoribbons, nanodisks, and nanorings. Vapor-based techniques were the first used to grow ZnO nanostructures and include vapor–liquid–solid (VLS) [29–32], chemical vapor deposition (CVD) [33], plasma-enhanced CVD [34], metal–organic CVD (MOCVD) [35–39], metal–organic vapor phase epitaxy (MOVPE) [40, 41], molecular beam epitaxy [42, 43], and template-assisted growth processes [44]. Vapor-based techniques can be used to grow a variety of ZnO nanorods including vertically aligned arrays of ZnO nanorods, nanorod heterostructures [45], and alloyed and doped nanorods [46–48]. The FE properties of vapor-grown ZnO nanostructures have been extensively studied and found to be comparable to those of carbon nanotubes [1–15]. Recently, solution-based or hydrothermal techniques using zinc salts in an aqueous solution were developed in which ZnO nanostructures could be grown at low temperatures of about 100°C [49–59]. Nanorods could be grown as suspended particles in solution or attached to a substrate in a random or vertically aligned configuration. Solution-based growth techniques can be more easily scaled to large deposition areas since they are less expensive and use significantly lower temperatures than vapor-based techniques. The FE properties of vapor and solution-grown ZnO nanostructures have been reported to be the same [28, 60–62]. Thus, solution-based techniques are well suited for mass production of flat panel displays and other large-area FE applications.

In VLS techniques, catalyst nanoparticles such as Au, Ni, Co, Cu, or Sn are used and remain embedded in the tips of the ZnO nanostructures as they grow [29–32]. The growth mechanism involves the dissolution of Zn vapor into the catalyst nanoparticles at high temperatures of about 900°C. The Zn precipitates out upon saturation and oxidizes to form ZnO. This technique can be used to grow nanorod heterostructures having different compositions along the axial direction. However, the interfaces in the heterostructures are usually not sufficiently abrupt to produce quantum confinement effects due to alloying caused by the embedded nanoparticles [63]. The embedded nanoparticles also produce defect sites that may lead to oxidation and degradation of FE properties. Another disadvantage of VLS techniques is that they involve the use of vacuum that makes the process costly.

Recently, MOCVD and MOVPE growth techniques have been developed that do not require the use of catalyst nanoparticles [35–41]. These techniques can be used to grow nanorods in a random, vertically aligned, or mosaic pattern depending on the substrate, growth temperature, and Zn/O precursor ratio [39]. Using MOVPE, it is possible to grow high-quality nanorod heterostructures having abrupt interfaces exhibiting quantum effects [45]. Other vapor-based techniques include physical vapor deposition such as thermal evaporation of powders of Zn [64, 65], ZnO [66, 67], ZnO, and SnO_2 or In_2O_3 [68, 69], or ZnO and graphite [70].

In addition, there are pulsed laser deposition (PLD) [71] and DC arc discharge [72] techniques. PLD has, until recently, received only limited attention as a means to prepare ZnO nanorods compared to the previous methods discussed, despite its widespread use in the manufacture of thin films of ZnO and other materials. ZnO nanorods only form within a certain range of temperatures and pressures by PLD, outside of which a thin film is deposited instead. This restriction on growth conditions, as well as the expense of the equipment necessary for the process, may contribute to PLD's lower popularity in ZnO nanostructure preparation compared to other methods. The PLD method does, however, possess the advantages of one's ability to dynamically and easily alter the partial pressures of the gases present during deposition, as well as the close correspondence of the chemical makeup of the deposited ZnO to that of the source material, allowing one to easily control the level of doping. The DC arc discharge technique involves a DC arc discharge in air. Figure 7.1 shows a scanning electron microscopy (SEM) image of ZnO nanorods grown using the DC arc discharge technique with a Zn target, air at a pressure of 610 Torr, and arc current of 30 A [72]. The nanorods measure approximately 100–200 nm in width and 300–600 nm in length. They are n-type due to donor formation by oxygen vacancies and impurities.

Solution-based or hydrothermal techniques for the growth of ZnO nanostructures use an aqueous solution of a zinc salt that is thermally decomposed at low temperatures of about 100°C to produce ZnO [49–59]. The growth solution is typically made of an equimolar mixture of a zinc salt, such as $Zn(NO_3)_2{\cdot}6H_2O$, and hexamethylenetetramine, also known more simply as hexamine or methenamine, with concentrations on the order of 10 mM [49]. Additionally, an amine complexing agent, such as ammonia, is regularly added to the growth solution in order to promote heterogeneous epitaxial growth of the ZnO during nanorod formation. A

Fig. 7.1 Scanning electron microscope image of ZnO nanorods grown using the arc discharge technique. Scale denotes 600 nm

variety of substrates can be used such as glass, single crystal sapphire, Si/SiO_2 wafers, ZnO thin films, Zn foils, and other metals. The solution is heated to a temperature ranging between 70 and 100°C for several hours producing random or vertically aligned ZnO nanorods depending on the substrate material, temperature, Zn^{2+} concentration, and pH of the solution [49].

Figures 7.2, 7.3, 7.4, and 7.5 show SEM images of various ZnO nanostructures grown using solution-based techniques. Uniformly distributed and well-aligned ZnO nanorods were grown perpendicular to the substrate surface as single crystals with an easily distinguishable hexagonal-rod structure. The growth solution for the nanostructures in Figs. 7.2, 7.3, 7.4, and 7.5 was prepared using equimolar mixtures of a zinc salt and hexamine. The pH was adjusted with a dilute solution of ammonium hydroxide or an acid containing the corresponding anion of the zinc salt used. The substrate was mounted vertically inside an autoclavable glass vial containing the prepared growth solution. The sealed vial was immersed in a water bath at the desired temperature maintained by a programmable hot plate. The zinc salts used to prepare the growth solution include zinc nitrate (NO_3^-), sulfate (SO_4^{2-}), chloride (Cl^-), and acetate (CH_3COO^-), and ranged in concentration from 0.1 to 5.0 mM. The ZnO nanostructures were allowed to grow for 2–5 h at temperatures between 60 and 80°C. A variety of substrates were used including silver, copper, nickel, silicon, and glass.

3 Field Emission Properties

FE is the tunneling of electrons from the conduction band to the vacuum level under a high electric field. Its explanation by Fowler and Nordheim in 1928 was one of the first successful applications of quantum mechanics [73]. In a typical experiment, the

Fig. 7.2 High-resolution scanning electron microscope images of ZnO nanorods prepared on a silver-coated Si wafer. Equimolar mixtures of 0.05 M $Zn(NO_3)_2$ and hexamine were used as the growth solution. The nanorods were grown at 60°C for 3 h. (**a**) Tilted view (58°) of the edge of the sample showing a highly aligned structure of nanorods with uniform height. Also visible is a ZnO nanorod cluster that is embedded into the layer grown on the surface. (**b**), (**c**), (**d**) Increasingly magnified images of a disordered section on the surface that allows an excellent assessment of the individual nanorod morphology and size. Nanorods are shown to grow with uniform size distribution with clearly defined crystal facets and hexagonal cross sections

sample is placed in a vacuum chamber and an anode is positioned a distance from the sample. A voltage is applied between the anode and sample, and the resulting current measured. The FE current, I, versus anode-sample voltage, V, is described by the Fowler–Nordheim (FN) equation:

$$I = AV^2 exp\left(\frac{-b\varphi^{3/2}}{\beta V}\right) \tag{7.1}$$

where I is the current, A and b are positive constants, ϕ is the work function, and β the geometric field enhancement factor [73]. ZnO nanorods can have β as high as 2×10^6 [3] due to their nanometer-scale features, although values of β for ZnO nanorods and carbon nanotubes are typically on the order of 1,000. The large value of β results in significant FE at low applied fields on the order of volts per micron.

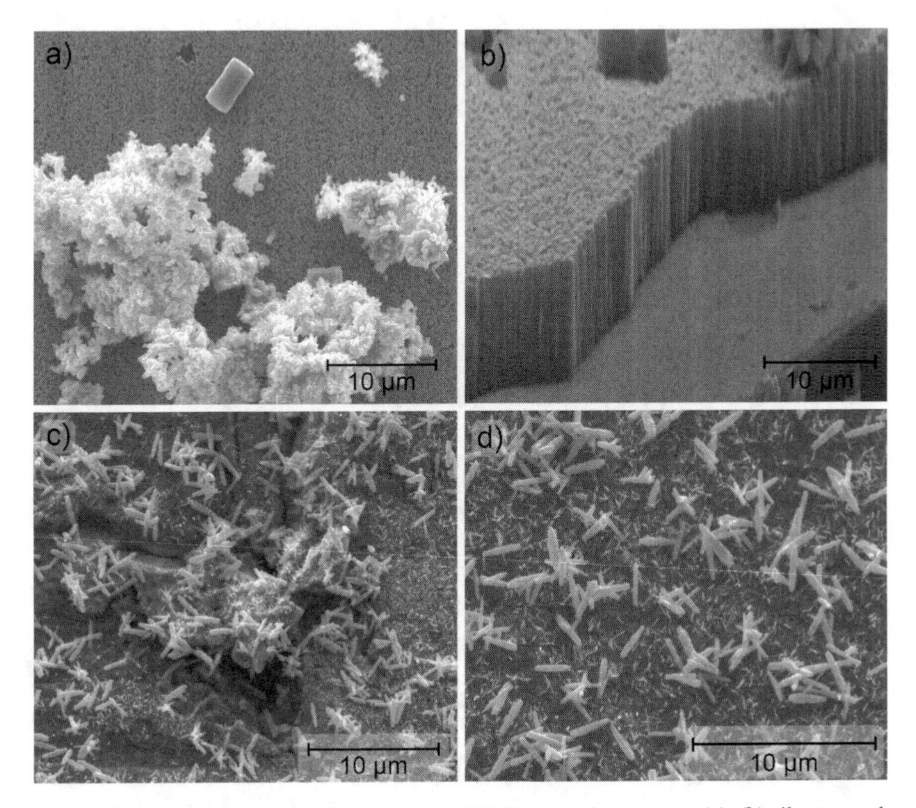

Fig. 7.3 Scanning electron microscope images of ZnO nanorods grown on (**a**), (**b**) silver-coated Si wafer and (**c**), (**d**) brass-coated Cu surface. Equimolar solutions of (**a**), (**b**) 0.05 M and (**c**), (**d**) 5.0 mM $Zn(NO_3)_2$ and hexamine were used as the growth solution. Nanorods grown at (**a**), (**b**) 60°C for 3 h and (**c**), (**d**) 70°C for 2 h. (**a**) Top view of sample surface showing large clusters of ZnO nanorods on top of a well-aligned ZnO nanorod layer. (**b**) Tilted view of the edge of the sample surface showing a densely packed layer made up of tall, thin, highly aligned ZnO nanorods. (**c**), (**d**) Randomly oriented ZnO nanorods grown on a rough brass substrate. Two different sizes are seen: large ZnO nanorods appearing to lie on top with shorter, thinner, nanorods that may have grown directly on the substrate

The turn-on voltage and turn-on field for FE are defined as those values that produce a certain threshold current or current density. In addition to ϕ and β, the turn-on voltage and turn-on field depend on anode geometry, anode–sample distance, and location on the sample [74]. I versus $V(I - V)$ data are typically plotted $\ln(I/V^2)$ versus $1/V$ to allow comparison with the straight-line behavior predicted for FE by the FN equation. We refer to such plots as FN curves. The slope of an FN curve is proportional to $\varphi^{3/2}/\beta$. Thus, changes in slope of an FN curve after prolonged operation or exposure to gases are due to changes in ϕ, β, or both.

FE from ZnO nanostructures was first reported by Lee et al. [1] who used vertically well-aligned ZnO nanorods grown using the VLS technique with Co nanoparticles at 550°C. The FE measurements were carried out in a vacuum chamber at 2×10^{-7} Torr using a flat anode having an area of approximately 30 mm^2

Fig. 7.4 Scanning electron microscope images of ZnO nanorods grown on (**a**), (**b**) Si wafer and (**c**), (**d**) nickel-coated Si wafer. Equimolar solutions of (**a**), (**b**) 0.5 mM and (**c**), (**d**) 0.05 M $Zn(NO_3)_2$ and hexamine were used as the growth solution with (**a**), (**b**) 80 mM NH4OH and (**c**), (**d**) ammonia. Nanorods grown at (**a**), (**b**) 75°C for 5 h and (**c**), (**d**) 60°C for 3 h. (**a**) Nanoneedles grown in clusters on a Si wafer. (**b**) Magnified view of (**a**) showing a "nanoflower" of ZnO nanoneedles. (**c**), (**d**) ZnO nanorods on (**c**) Ni and (**d**) Si substrate showing a lower density coverage than on other samples

and placed approximately 250 μm from the sample. A stable FE current with a turn-on field of about 6 V/μm at a current density of 0.1 $\mu A/cm^2$ was measured. Increasing the field to 11 V/μm increased the current density to 1 mA/cm^2. These values were not as good as those of carbon nanotubes that had turn-on fields of around 1 V/μm at a current density of 90 $\mu A/cm^2$. However, subsequent investigations on ZnO nanorods reported that by reducing the rod diameter and tip radius, improving the vertical alignment, and selecting the optimum areal density, the FE properties of ZnO nanorods could be improved to equal those of carbon nanotubes [2–11]. These reports investigated ZnO nanowires several tens of micrometers in length grown on W tips [2], ZnO nanoneedles having diameters of only several nanometers [3–5], ZnO nanopins [6], ZnO nanowires having different areal densities on planar substrates [7, 8], tetrapod-like ZnO nanostructures [9], and ZnO nanorod arrays with different morphologies [10, 11].

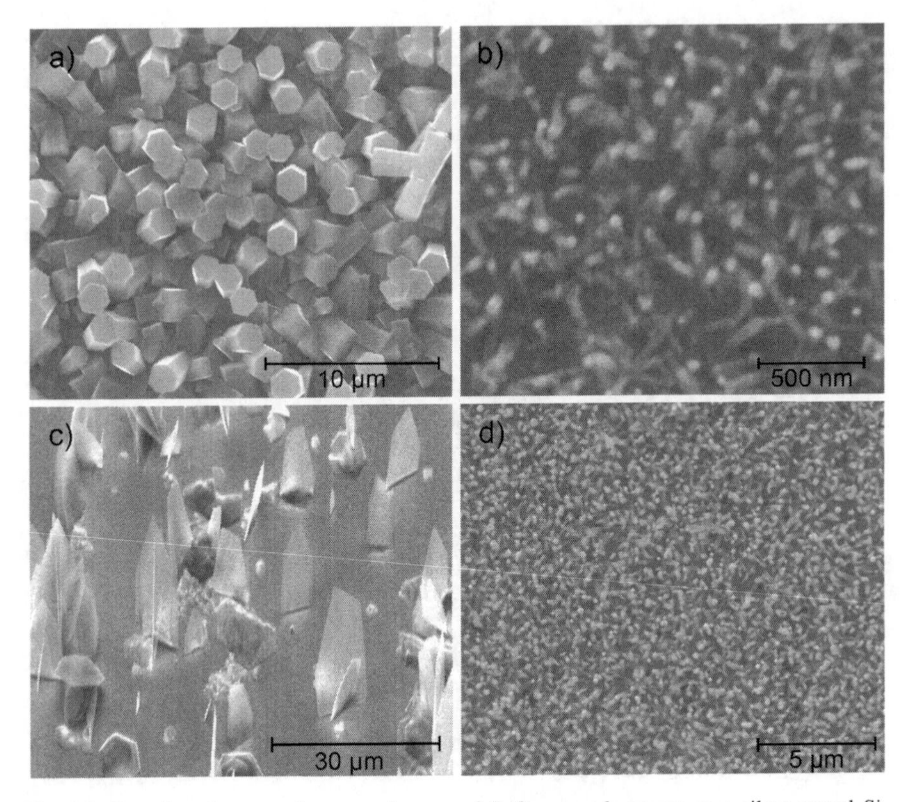

Fig. 7.5 Scanning electron microscope images of ZnO nanorods grown on a silver-coated Si wafer. Equimolar solutions of 0.05 M Zn^{2+} and hexamine were used as the growth solution. Nanorods were grown at 60°C for 3 h. (**a**) Thick (~1.7 μm), densely packed ZnO nanorods grown using $Zn(CH_3COO)_2$. (**b**) Thin (~50 nm) ZnO nanowires grown using $ZnCl_2$. (**c**) Thin nanosheets grown using $ZnSO_4$. (**d**) Large uniform area of well-aligned ZnO nanorods grown using $Zn(NO_3)_2$

The effects of residual gases on the FE properties of ZnO nanostructures have not been extensively studied. Kim et al. [13] reported the effects of O_2, N_2, Ar, air, and H_2 exposure on the FE properties of ZnO nanorods grown using thermal evaporation of ZnO and graphite powders, and Co nanoparticles on a Si substrate. The vacuum chamber was at a base pressure of 2×10^{-5} Torr and gases were introduced to a pressure of 2×10^{-4} Torr. The anode–sample distance was 200 μm, and the anode consisted of a W plate having an unspecified area. They reported that O_2, N_2, Ar, and air exposure degraded the FE properties, but there was full recovery of the FE properties after about 20 min of having evacuated the chamber to base pressure. They also found that H_2 exposure improved the FE properties and the improvement continued after evacuating to base pressure. In Ref. [13], the data included the FE current as a function of time for the period covering the gas exposure and evacuation. However, FN curves before and after each exposure were not presented. Jang et al. [14] also reported enhanced FE from ZnO nanowires by hydrogen gas exposure.

Yeong et al. [15] reported the effects of O_2 and H_2 exposure and UV illumination on the FE properties of individual ZnO nanowires grown on electrochemically sharpened Pt tips using evaporation and oxidation of Zn. The nanowires had lengths of a few microns. The chamber was at a base pressure of ~10^{-9} Torr and gases were introduced to a pressure of 7.5×10^{-7} Torr for 5 min. They found that initially O_2 exposure caused an increase in turn-on voltage of about 20%; however, the turn-on voltage returned to its original value after continued operation at base pressure. Upon additional exposure to O_2, the turn-on voltage permanently increased by 10% and did not return to its original value after operation at base pressure. They found that H_2 exposure caused a permanent reduction in the turn-on voltage of about 15%. They attributed their observations to the ionosorption of oxygen and hydrogen that created a doubly charged layer on the surface whose field extended throughout the nanowire. This changed the carrier concentration and induced band bending. They also found that UV illumination increased the FE current by up to two orders of magnitude.

We have investigated the effects of O_2, CO_2, N_2, H_2, and Ar on the FE properties of ZnO nanorods shown in Fig. 7.1 grown using the arc discharge method [75]. A mixture consisting of 1 mg of ZnO nanorods in 20 ml of ethanol was dispersed using ultrasonication for 1 h. The mixture was deposited onto a silver-coated Si substrate and allowed to dry. The thickness of the resulting ZnO nanorod layer was approximately 1 μm. The experiments were carried out using the system shown schematically in Fig. 7.6 in a chamber at a base pressure $<10^{-9}$ Torr. The FE current was collected from a localized region of the sample using a spherical platinum anode having a diameter of approximately 1.0 mm. A spherical anode was used instead of a planar anode to avoid experimental errors caused by orientation variations between the anode and sample surfaces, and anode edge effects [74]. However,

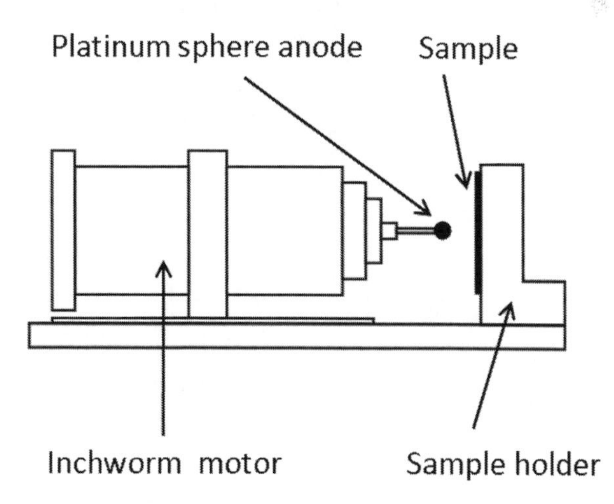

Fig. 7.6 Schematic of the system used to measure the Fowler–Nordheim curves consisting of a spherical platinum anode approximately 1 mm in diameter positioned 25 μm from the sample using an inchworm motor

since it is difficult to calculate the effective emission area using a spherical anode [74], only the total current was reported. The anode was positioned 25 ± 0.25 μm from the sample using a scanning tunneling microscopy system with a piezoelectric inchworm motor made by EXFO Burleigh [76]. The positioning was accomplished by slowly bringing the anode into contact with the substrate and then retracting the anode. A distance of 25 μm was used to simulate typical distances between emitter and anode in triode structures [77]. Low FE currents of approximately 20–2,000 pA were used to avoid sample heating.

Figures 7.7, 7.8, and 7.9 show FN curves obtained before and after exposure in our experiments [75]. In Figs. 7.7, 7.8, and 7.9, the positive slope of the FN curves at low voltages is due to a small leakage current on the order of 0.1 pA across the connectors. At high voltages, the FN curves have a negative slope, in agreement with the FN equation. The turn-on voltage for FE was defined as the smallest voltage at

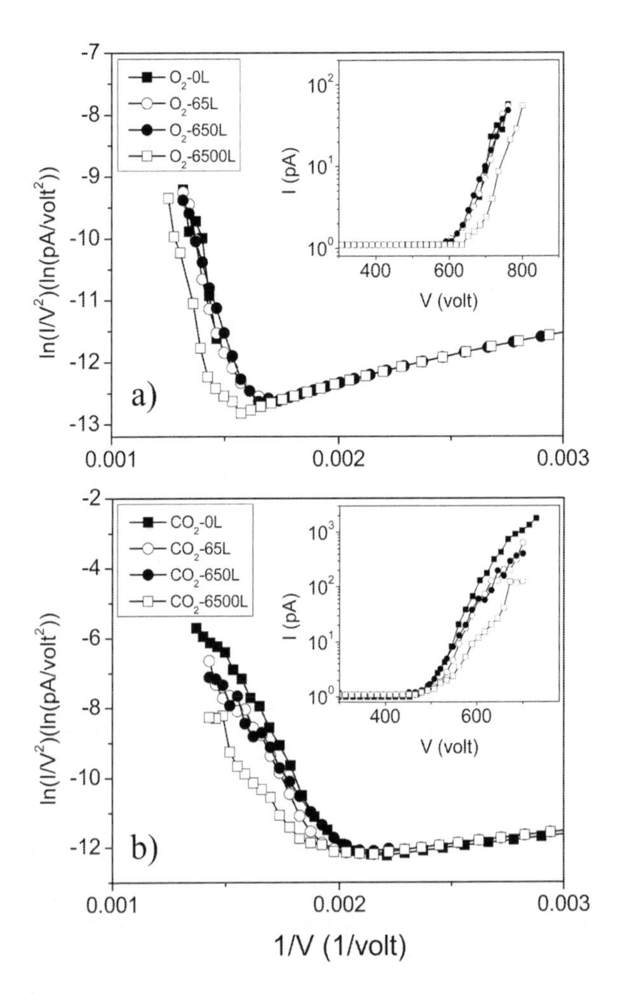

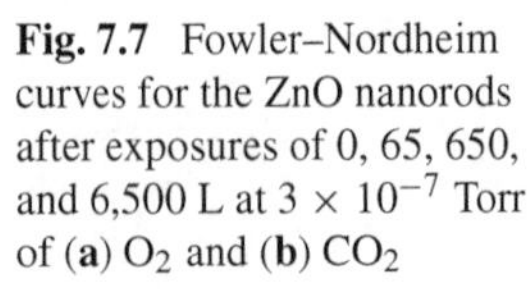

Fig. 7.7 Fowler–Nordheim curves for the ZnO nanorods after exposures of 0, 65, 650, and 6,500 L at 3×10^{-7} Torr of (**a**) O_2 and (**b**) CO_2

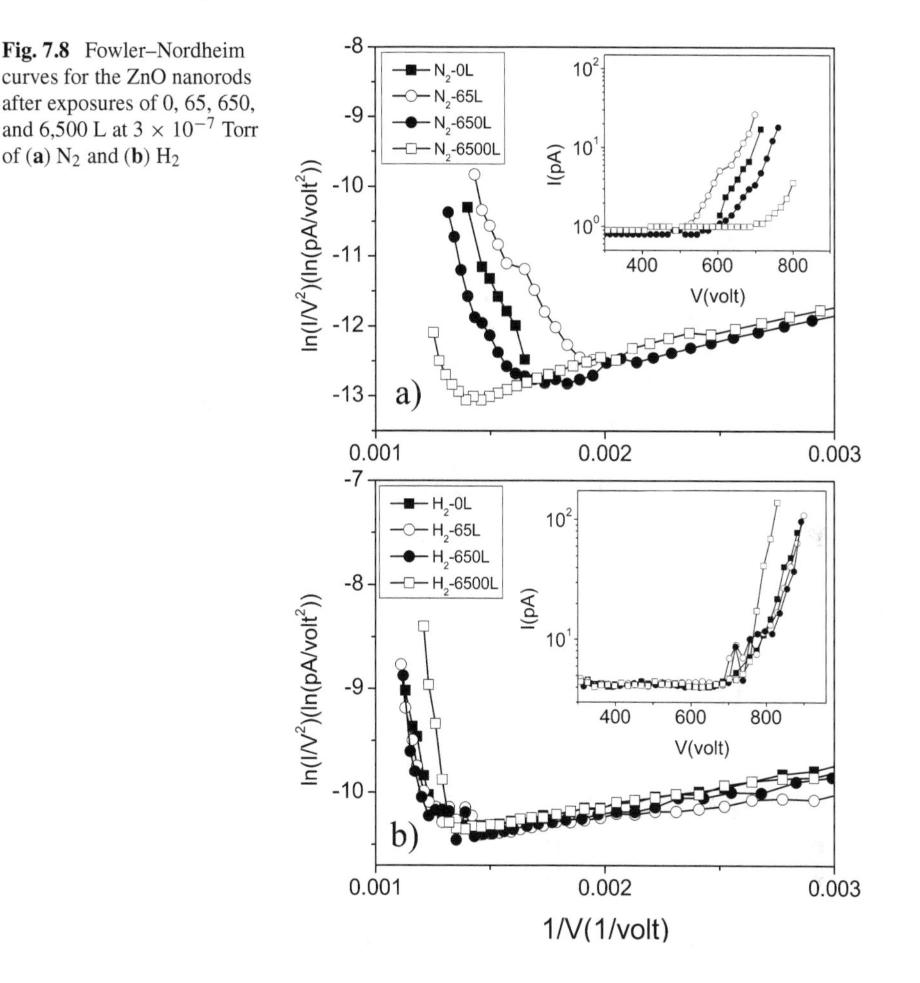

Fig. 7.8 Fowler–Nordheim curves for the ZnO nanorods after exposures of 0, 65, 650, and 6,500 L at 3 × 10^{-7} Torr of (**a**) N_2 and (**b**) H_2

which the FE current was observed above the leakage current. The insets in Figs. 7.7, 7.8, and 7.9 are the corresponding *I–V* curves plotted using a log-linear scale after subtraction of the leakage current.

Before exposure, the *I–V* curves were measured in a vacuum $<10^{-9}$ Torr, as shown by the solid squares in Figs. 7.7, 7.8, and 7.9. To achieve 65 L of exposure (1 L = 10^{-6} Torr s), gas was introduced to a pressure of 3 × 10^{-7} Torr for 216 s while the sample was biased to produce approximately 20 pA of FE current. The bias was then turned off, the vacuum chamber evacuated to $<10^{-9}$ Torr, and the FE *I–V* curves measured, as shown by the open circles in Figs. 7.7, 7.8, and 7.9. Using this procedure, the ZnO nanorods were not exposed to gases during the *I–V* curve measurement. To achieve 650 and 6,500 L of exposure, the procedure described above was repeated for 2,160 and 21,600 s, respectively. For each set of gas exposures, the anode was positioned over a different area of the sample.

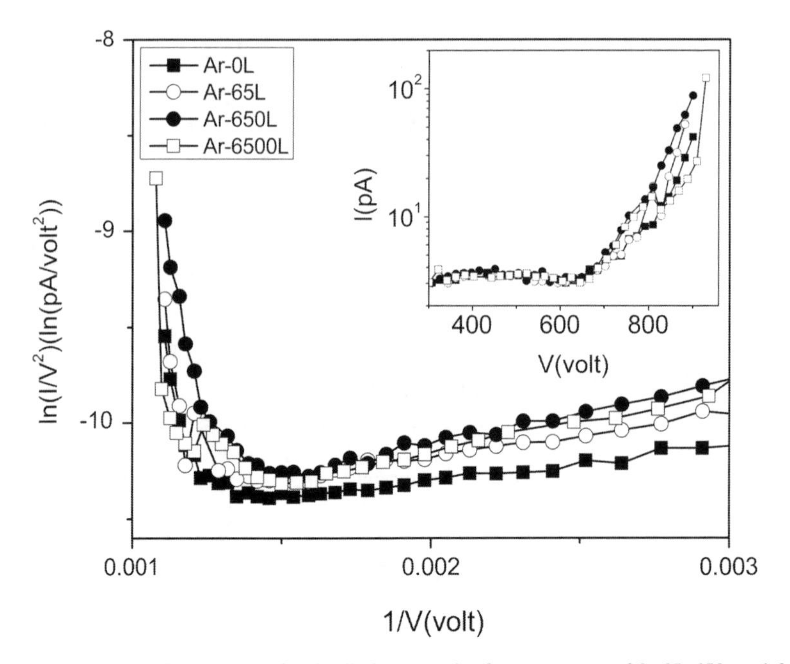

Fig. 7.9 Fowler–Nordheim curves for the ZnO nanorods after exposures of 0, 65, 650, and 6,500 L at 3×10^{-7} Torr of Ar

As shown in Fig. 7.7a, O_2 exposure did not have a significant effect on the FE properties of the ZnO nanorods. After 65 and 650 L of O_2 exposure, the turn-on voltage decreased slightly from approximately 640 to 610 V, while the FE current at higher voltages remained approximately the same. After 6,500 L of O_2 exposure, the turn-on voltage returned approximately to its value before exposure of 640 V, and the FE current decreased slightly from 60 to 20 pA at 760 V. Straight-line fits to the FN curves in Fig. 7.7a showed that all the slopes were within 10% of one another, indicating that ϕ or β did not significantly change with exposure.

As shown in Fig. 7.7b, CO_2 exposure also did not have a significant effect on the FE properties. After 65 and 650 L of CO_2 exposure, the turn-on voltage remained approximately the same, and the FE current decreased slightly from approximately 70 to 50 pA at 690 V. After 6,500 L of exposure, the turn-on voltage increased slightly from approximately 500 to 530 V, and the FE current decreased from approximately 70 to 10 pA at 690 V. Straight-line fits to the FN curves showed that the slopes were the same to within 10%.

In contrast, as shown in Fig. 7.8a, N_2 exposure resulted in a significant degradation of the FE properties. Initially, after 65 L of N_2 exposure, the FE properties improved slightly. The turn-on voltage decreased from approximately 580 to 520 V, and the FE current increased slightly from approximately 7 to 15 pA at 680 V. A slight improvement in FE properties was also observed after 65 and 650 L of O_2 exposure, as shown previously. The initial improvement in FE properties may be due to removal of adsorbates on the ZnO surface by chemical reaction or ion bombardment. Adsorbates have been reported to degrade the FE properties of carbon nanotubes and Mo microtips by increasing the work function, and initial exposure to gases has been shown to improve the FE properties by surface cleaning [17, 18, 23–26]. However, after 650 L of N_2 exposure, the turn-on voltage increased to 590 V, and the FE current decreased from 20 to 4 pA at 700 V. After 6,500 L of N_2 exposure, the turn-on voltage increased significantly to 740 V, and the FE current at 700 V decreased from 20 pA to a value less than the leakage current. In contrast, the FE properties of carbon nanotubes and Mo microtips are not significantly degraded by N_2 exposure, but are significantly degraded by O_2 and CO_2 exposures [17, 18, 23–26].

As shown in Fig. 7.8b, 65 and 650 L of H_2 exposure did not have a significant effect on the turn-on voltage or FE current of ZnO nanorods. However, after 6,500 L of H_2 exposure, the FE current at higher voltages increased significantly from approximately 20 to 200 pA at 830 V. This improvement is consistent with previous reports [13–15]. We note that it has been reported that H_2 exposure significantly improves the FE properties of Mo microtips [17, 18] and diamond-coated microtips [78, 79] due to surface cleaning and formation of a hydrogen layer that reduces the barrier for FE. As shown in Fig. 7.9, Ar exposure did not have a significant effect on the FE properties of ZnO nanorods. After 650 L of Ar exposure, the turn-on voltage decreased slightly from approximately 750 to 705 V, and the FE current increased from 44 to 80 pA at 885 V. After 6,500 L of Ar exposure, the turn-on voltage and FE current were approximately the same as before exposure.

We propose that O_2 and CO_2 exposures do not have a significant effect on the FE properties of ZnO nanorods because O_2 and oxygen-containing species present under high-field conditions do not significantly react with ZnO due to the difficulty of further oxidizing ZnO. It has been reported that the conductivity and electronic properties of ZnO films improve after exposure to an oxygen plasma, due to the suppression of oxygen vacancies [80]. In contrast, it has been reported that a nitrogen plasma significantly reacts with ZnO films resulting in the incorporation of nitrogen into the ZnO lattice and an increase in resistivity, and the formation of Zn and NO [81, 82]. In our experiments, we propose that nitrogen species present under high-field conditions may increase the resistivity of ZnO nanorods and structurally damage the ZnO lattice, which may decrease β and increase the turn-on voltage. These results indicate that it is important to reduce the amount of residual N_2 in vacuum chambers containing ZnO field emitters.

4 ZnS Nanostructures

Like ZnO, ZnS is a wide bandgap semiconductor that has attracted considerable interest for its luminescent and electrical properties with applications in light-emitting diodes [83], flat panel displays [84, 85], transparent conductive coatings [86], and buffer layers in solar cells [87]. Numerous methods have been used in the preparation of ZnS thin films [87, 88] and nanostructures [89–93], including thermal evaporation [89, 90], RF sputtering [87], and solution-based growth processes [88, 91, 92]. ZnS nanostructures such as nanorods, nanowires, and nanobelts can be grown as single crystals with high aspect ratios and in good electrical contact with conducting metallic substrates making them ideal candidates for FE experiments [89, 91].

Vapor phase deposition processes, such as thermal evaporation, rely on heating ZnS at high temperatures inside a deposition chamber under an inert atmosphere (Ar or N_2) containing small amounts of reducing gasses (H_2 or CO) [89, 90]. The ZnS nanostructures form from the vaporized precursor material through the VLS and VS growth modes with a uniform distribution across the substrate surface. Electron microscopy, however, reveals a randomly oriented arrangement, unlike the well-aligned structure of ZnO nanorods produced through a similar growth process, which is less desirable for FE due to lower geometric field enhancement factors [84, 89]. Although vapor deposition techniques are commonly used to produce high-quality ZnS nanostructures, these methods possess several distinct disadvantages making them undesirable for large-scale adoption. Such deposition methods can involve corrosive and highly toxic hydrogen sulfide gas [84, 87] as well as require vacuum systems and often employ high growth temperatures in excess of 1,000°C [89, 90], which limit the choice of substrates suitable for deposition. Additionally, high-temperature deposition of ZnS requires a thorough exclusion of O_2 gas from the growth environment due to the ease with which ZnS is converted to ZnO with even minor traces of O_2 present [90].

As an alternative to vapor deposition, high-quality ZnS thin films and nanostructures may instead be grown through aqueous solution-based processes [88, 91–93]. Aqueous solution growth of ZnS nanostructures is a straightforward, low-temperature process that uses relatively safe and inexpensive precursor materials and does not require specialized equipment, unlike thermal vapor deposition. Using a water-soluble zinc salt, such as $ZnCl_2$, $ZnSO_4$, or $Zn(CH_3COO)_2$, and a sulfide ion source, such as thiourea ($CS(NH_2)_2$) or Na_2S, ZnS nanostructures may be grown in a variety of forms. Under alkaline conditions, thiourea decomposes in solution to give off bisulfide ions, SH^-, which in turn react to form sulfide ions via the following reactions [88]:

$$\begin{aligned} CS(NH_2)_2 + OH^- &\rightarrow SH^- + CH_2N_2 + H_2O \\ SH^- + OH^- &\leftrightarrow S^{2-} + H_2O \end{aligned} \tag{7.2}$$

The Zn^{2+} and S^{2-} ions present in the solution precipitate out as ZnS. Complexing agents may also be used, particularly in thin-film deposition, in order to increase the amount of ZnS formed and promote a more heterogeneous deposition on the substrate surface, which results in the formation of smoother and more optically transparent film [88]. Ammonia is a common complexing agent for Zn^{2+} that can be used in combination with sodium citrate, which has been shown to increase the amount of ZnS deposited [88]. These complexing agents also have the added benefit of acting to raise the pH of the growth solution, which is necessary for sulfide ion formation.

Lu et al. report the large-scale growth from aqueous solution of well-aligned ZnS nanobelts and their field-emitting properties [91, 94]. The nanobelts were synthesized from an aqueous growth solution prepared using a mixture of $Zn(CH_3COO)_2$, thiourea, and ethylenediamine (EDA), using NaOH to adjust the pH. A zinc foil substrate was immersed in the growth solution and sealed inside an autoclavable vial which was heated to 160°C for 10 h. During this initial growth phase a layer of $ZnS(EDA)_{0.5}$ is reported to be deposited on the zinc foil substrate. Following the growth of the $ZnS(EDA)_{0.5}$ layer, the substrate was washed in DI water and ethanol then dried under vacuum at 70°C for 5 h. The substrate was then heat-treated at 250°C under vacuum for 30 min in order to decompose the $ZnS(EDA)_{0.5}$ into ZnS. Scanning electron microscopy reveals that ZnS prepared in this manner forms well-aligned, vertically oriented nanobelts which can attain dimensions of approximately 30 nm thick, 300–500 nm wide, and several micrometers in length. The ZnS nanobelts exhibit a highly crystalline structure and grow uniformly over large areas. The zinc metal substrate was found to play a crucial role in the formation of nanobelts, acting as a secondary source of Zn^{2+} ions and providing a structurally compatible surface for ZnS growth. When using a silicon substrate instead of zinc, ZnS nanobelts have been shown to form as smaller and more disorganized nanoparticles which are formed in solution and deposited on the substrate surface. Additional factors which play an important role in growing ZnS nanobelts include the concentration of the zinc salt as well as the pH of the solution, with high aspect ratio, vertically aligned nanobelts formed using a Zn^{2+} concentration of about 12.5 mM and a pH of 10. The ZnS nanobelts were found to have a low turn-on field of around 3.8 V/μm which is thought to be due to their geometric characteristics, such as their high aspect ratio and sharp corners, resulting in calculated field enhancement factors of over 1,800. The relative ease with which highly aligned ZnS nanobelts may be grown along with their excellent field-emitting properties makes them a promising candidate for use in field-emitting device applications.

5 GaN Nanorods

Various nanostructures based on GaN can be produced using a number of methods depending on the form of the structure intended to be produced [95]. Such structures can take the form of nanorods, nonowires, nanofilms, hexagonal pyramids, and possibly other forms. Production could involve the use of PLD, CVD, epitaxial growth, and DC sputtering. Characterization of a given structure can involve methods such as X-ray diffraction, electron microscopy, and other methods. GaN, having an electron affinity of 2.7–3.3 eV [96] and an electron effective mass of 0.2 times the mass of a free electron [97], is considered useful in the potential development of FE-based devices. It also has very high physical and chemical stability, which, while making it a very durable medium, also makes the manipulation of the substance all the more difficult [96]. GaN is of technological interest due to its role in the operation of UV, blue, and green light-emitting diodes, and blue lasers [98] as well as the potential development of thin screen monitors with brightness and efficiency attributes that are superior to liquid crystal display technology [99].

Of particular interest are GaN nanorods which have interesting crystal structures and FE properties. PLD has proven to be an effective method of producing GaN nanorods [100]. GaN nanorods produced using this method have a single crystalline hexagonal Wurzite structure, and have a diameter of 5–20 nm, and a length on the order of microns. Experiments have shown such GaN to have a turn-on voltage of 8.4 V/μm and a current density of 0.96 mA/cm^2 under an applied voltage of 10.8 V/μm measured under a pressure of 2×10^{-6} Torr.

Doped nanorods can also be produced using a simple VLS thermal evaporation process involving the use of GaN-based powders (Ga_2O_3:GaN,Ga_2O_3:GaN) [101]. The powders are mixed with a doping element such as phosphor, and applied to a Si (111) substrate with a 5 nm thick gold film. The combination is then placed in a quartz oven and heated to temperatures over 1,000°K. The resulting p-doped GaN nanorods, which can be characterized using X-ray diffraction and scanning electron microscopy, have a single crystal Wurzite structure with a diameter of 10–40 nm and a length on the order of tens of microns. Unlike nondoped GaN nanorods, they tend to have a rough, curved surface structure that is responsible for the enhanced FE properties (reduced turn-on voltage) of the nanorods. The turn-on voltage for such nanorods depends on the amount of doping applied to the rods. Undoped nanorods have a turn-on voltage, under a pressure of 2×10^{-6} Torr, of 6.1–12.0 V/μm giving a current density of 0.01 mA/cm^2, while doped nanorods have a turn-on voltage of 5.1 V/μm with the same current density. In comparison, an alternative GaN structural format, namely, thin film amorphous GaN, has been shown to have a turn-on voltage of 5 V/μm and a maximum current density of 500 mA/cm^2 [102].

Another nanorod format developed by a Japanese research group includes needle-like bicrystalline GaN which can serve as low cost, large area emitters [103]. These structures are manufactured on a 20 nm gold substrate, and have a sharp tip and bicrystalline structural defects, which is hypothesized to be the cause of its FE properties. The diameter of the nanorods is 200 nm up to a certain length at which point the diameter gradually decreases until it cuts off with a diameter of 10 nm giving the rods a pencil-like structure. The FE of the rods is performed using a rod-like tip with a 1 mm^2 cross section as an anode. The tip is separated from the sample at a distance of 125 μm, and the applied voltage has a range of 0–1,000 V set at increments of 20 V at a pressure of 6×10^{-6} Torr. The turn-on voltage for the sample is 7.5 V/μm at a current density of 0.01 mA/cm^2.

In the previously described methods for the growth of GaN nanostructures, the mechanism of nanostructure formation is VLS [104]. For nanostructures to form, very thin metallic layers, which usually exist as tiny grains, were used as a catalyst to promote nucleation and growth of the nanostructure. Metal catalysts reduce melting point due to alloying effect and seed nucleation and as the nanostructure growth commences, the nano-droplets continue to stay atop, acting as the source for the continuing growth of the nanostructures.

Despite its effectiveness in fostering nanostructure formation, in VLS growth, traces of catalysts inevitably contaminate the growing material, hence altering its electronic band structure. Therefore, if one can replace such extrinsic seeding procedures with an intrinsic means, say by use of the same atomic constituents as a seed or through substrate surface morphology engineering, the impurity problems may be solved. Seo et al. [105, 106] introduced a concept of nano-capillary condensation suggesting that nanorods would nucleate on specific sites, and as islands on the substrate begin to impinge one another, Ga atoms would condense in the nano-valleys [105, 106]. Such nano-capillaries thus exclude extrinsic catalysis in the traditional VLS mechanism based on foreign metallic elements. The nano-capillaries enable the nanorods to form in single directions, resulting in more manageable nanostructures for device applications.

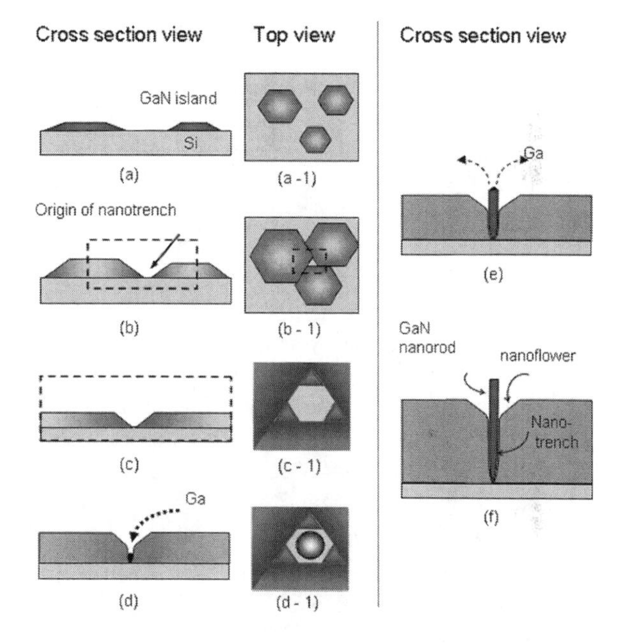

Fig. 7.10 Evolution of nanorod: (**a**) Initial stage of GaN island growth. (**b**) Impinging hexagonal islands and the formation of a triangular void region. (**c**) Corner filling of the triangular void and its evolution into hexagonal shape, precursor to the nanoflower. (**d**) Evolution of the nanoflower and start of the nanotrench formation and capillary condensation of Ga atoms in the trench. (**e**) VLS growth mechanism prevails and the nanorod grows faster, leading to protrusion above the nanoflower. As the protrusion occurs, the condition for capillary condensation diminishes and VS growth mechanism takes over. (**f**) The ultimate structure

Figure 7.10 illustrates the procedure of the nano-capillary condensation and the ensuing nanorod growth. The nucleation and growth process starts with three randomly chosen precursor nuclei on the vertices of an irregular triangle, as shown in Fig. 7.10a. Consequently, a voided region of equilateral triangle would develop as illustrated in Fig. 7.10b when the islands encounter each other. The voided triangular areas are thermodynamically unstable so that a transformation would take place to reduce the surface energy by corner filling via some self-regulated surface diffusion along the edges. This eventually results in a hexagonal nanotrench, as sketched in Fig. 7.10c.

With the hexagonal empty region in place, the islands would continue to grow in the vertical dimension and the six facets surrounding the region would then be elevated, eventually becoming what one observes as a six-leaf nanoflower. The nanotrenches underneath the nanoflower are essentially like an attached capillary tube, as shown in Fig. 7.10d. When the capillary tube is long enough, capillary condensation of Ga atoms occurs as a consequence of the decreases in the equilibrium vapor pressure due to the reduced radius of curvature, r, of a concave surface, as seen from the Thomson (Lord Kelvin) equation: $k_B T \log[P_{Ga}(r)/P_{Ga}(\infty)] \sim -2\gamma^{LV} V_{Ga}/r$.

GaN nanorods grow faster along <0001> via the canonical VLS mechanism [104] by reaction of the nitrogen plasma with Ga liquid clusters while its surrounding GaN islands also grow alongside to form the base material, though at a slower rate. Figures 7.10c–f illustrate this sequence of evolution. Here, as a nanorod outgrows the nanotrench and starts to stick out of the base film surface, the equilibrium vapor pressure would gradually increase, which favors the evaporation of the droplets on top of the nanorods, as shown in Fig. 7.10e. The process eventually reaches a steady state after which the rods grow at the same rate as the surrounding film via the VS growth mechanism.

In addition to the above naturally occurring process, Seo et al. [107] and Chu et al. [108] have also envisaged a method to foster the capillary effects by surface engineering via ion-beam surface modifications. With self-implantation of Si into Si substrates, they were able to control the growth of nanorod arrays. Periodic patterns were realized first by traditional UV lithography followed by Si ion self-implantations. The defects generated as a result, especially the vacancies, coupled with the heating process provide the necessary driving force for nucleation and growth in the implanted area. The density of nanorods in the patterned arrays can be controlled by the energy and dosage of the self-implantation that determines the vacancy concentration near the free surface. Linear arrays of 10 μm in width were patterned on Si (111) substrates by conventional UV lithography such that masked and unmasked regions alternated themselves. The samples were then implanted with 40 kV Si ions to various dosages at room temperature with the Si beam current kept below 100 nA to avoid excessive target heating. Due to the nature of forward momentum transfer by ion implantation, an interstitial-rich region takes shape in the region close to the end of projectile range deep in the substrate, leaving behind a vacancy-rich region close to the free surface. Upon heating prior to the GaN growth, the vacancies would coalesce to give a corrugated surface that forms the basis of nano-capillaries. While this method may produce the nanorods more easily, growth conditions can be set, especially the flow rate ratio of Ga:N to achieve the nanorod structure.

In any case, the nanorods grow faster than the surrounding matrix area via VLS growth mechanism at the early stage, while intrinsic polarity might have also contributed to the protrusion of the nanorods [109]. The sample preparation followed typical cleaning by HF etching as a precaution for further ultrahigh vacuum (UHV) processing, in which the Si substrates were transferred into an MBE growth chamber with a base pressure of ~10^{-10} Torr. The samples went through an 810°C preheating for 50 min before deposition in order to get rid of any HF residue. A buffer layer of GaN (0001) was first grown under an equivalent N/Ga pressure ratio ~100 at 550°C. The ensuing growth of nanorods and the accompanying matrix thin film on the GaN buffer was carried out at a N/Ga equivalent pressure ratio ~30. The N_2

plasma power was ~500 W and the substrate temperature was ~720°C during the film deposition [110].

The FE properties of the obtained nanostructured samples were studied as a function of various exposures of O_2, CO_2, H_2O, and N_2 gases from 65 to 6,500 L. The experimental setup and procedure were as described in the previous section in this chapter on the effects of gases on ZnO nanorods grown using the arc discharge method, except that the anode–sample distance was approximately 3.7 μm. As shown in Fig. 7.11a, after 6,500 L of O_2 exposure, the turn-on voltage increased significantly by approximately 100% from 170 to 340 V. The slope of the FN curve also increased indicating an increase in work function of 60%, assuming β remained the same. As shown in Figs. 7.11b and 7.12a, after 6,500 L of CO_2 and H_2O exposure, the turn-on voltage increased by approximately 35 and 38%, respectively; from 200 to 270 V and 240 to 330 V, respectively. The work function increased significantly by 50% after CO_2 exposure, but remained approximately the same after H_2O

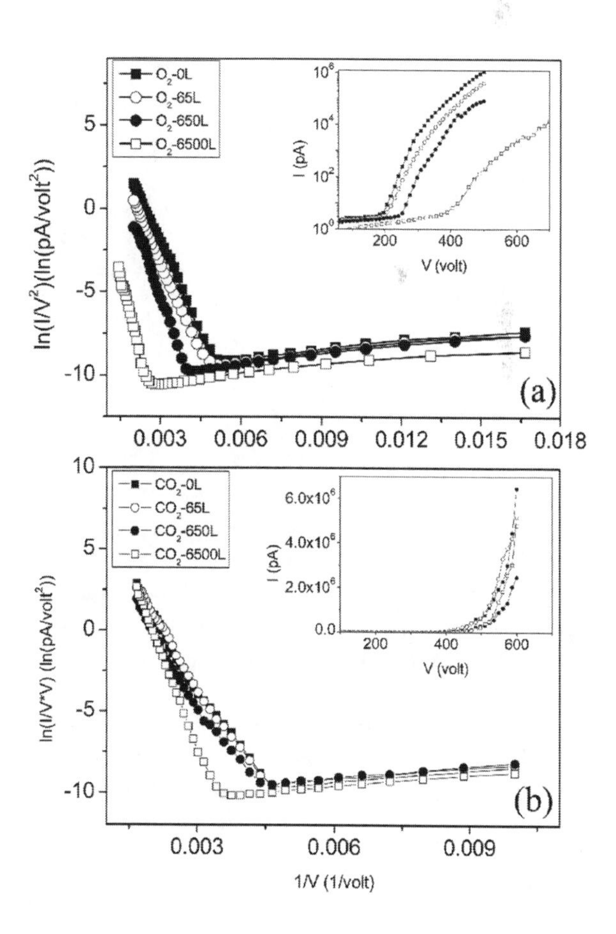

Fig. 7.11 Fowler–Nordheim curves for the GaN nanorods after exposures of 0, 65, 650, and 6,500 L at 3×10^{-7} Torr of (**a**) O_2 and (**b**) CO_2

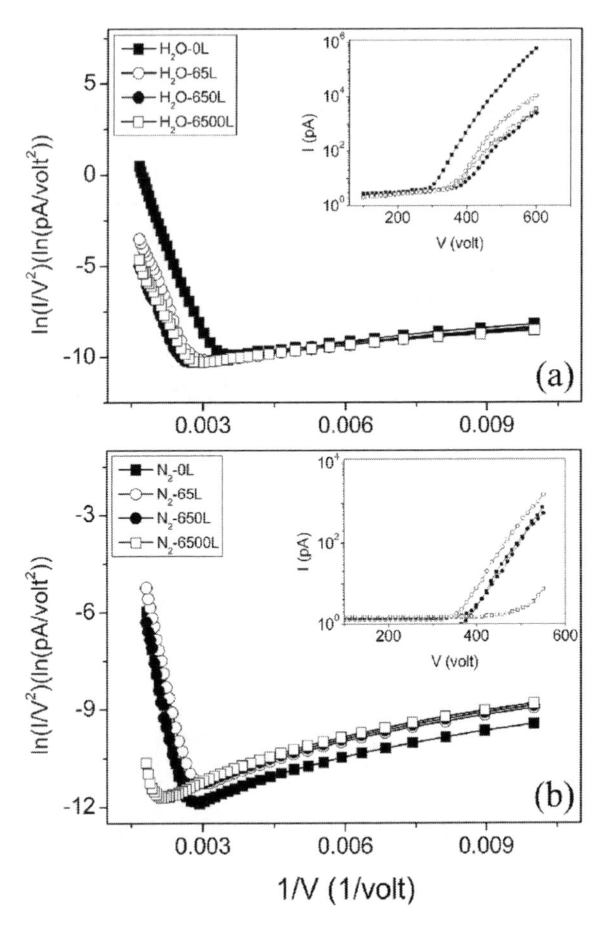

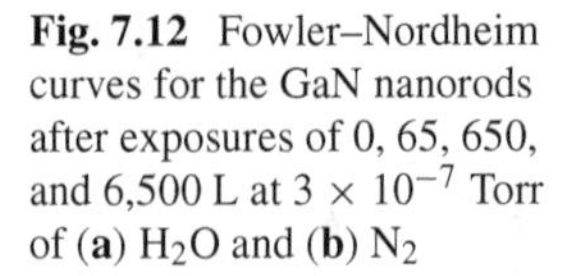
Fig. 7.12 Fowler–Nordheim curves for the GaN nanorods after exposures of 0, 65, 650, and 6,500 L at 3×10^{-7} Torr of (**a**) H_2O and (**b**) N_2

exposure. After 6,500 L of N_2 exposure, the turn-on voltage increased by approximately 35%, from 340 to 460 V, and the work function remained approximately the same, as shown in Fig. 7.12b.

We also investigated the effects of Cs deposition on the FE properties of the GaN nanorods. The sample was placed in a UHV chamber at a base pressure $<10^{-10}$ Torr. A Cs-metal dispenser from SAES Getters [111] was used to deposit approximately one monolayer of Cs on the sample. As shown in Fig. 7.13, Cs deposition resulted in a decrease in turn-on voltage of approximately 30%, from 370 to 260 V. The work function decreased by approximately 60%, assuming β remained the same. We propose that the improvement in FE properties is due to a Cs-induced space-charge layer at the surface that reduces the barrier for FE such as by producing an NEA surface [112–115]. It has been reported that Cs deposition on GaN films produces an NEA surface [116–118]. It would be interesting to investigate if Cs deposition improves the FE properties of ZnO, ZnS, and other wide bandgap nanostructures.

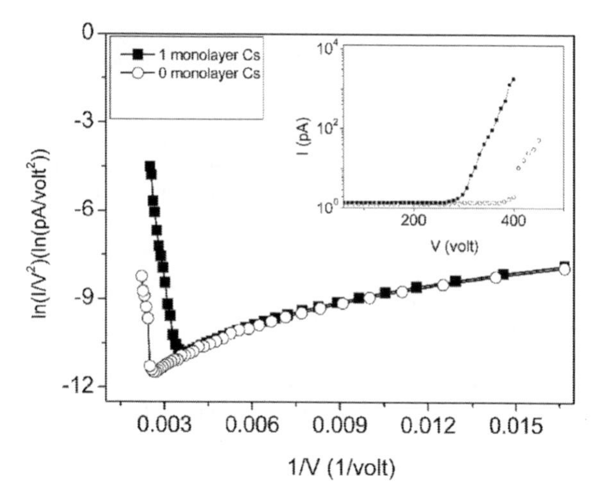

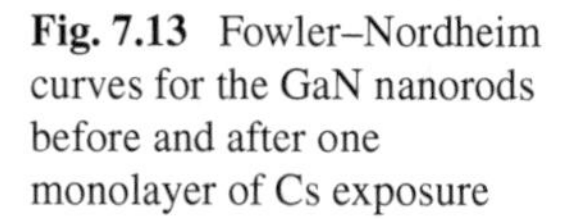
Fig. 7.13 Fowler–Nordheim curves for the GaN nanorods before and after one monolayer of Cs exposure

We note that in the same family of the III-nitrides, InN has recently been reported to show record-low turn-on voltages; however, its long-term performance stability as compared to GaN, which may last as long as 8 h at 1 mA FE current, is still unclear [119]. It remains for future studies to understand what dictates the turn-on characteristics and to what extent one can work to reduce it for practical device applications when the fundamental surface physics or chemistry is better known.

Acknowledgments J.J. Schwartz and M.H. Lynch thank the National Science Foundation Research Experience for Undergraduates program for support. This work was supported, in part, by a Faculty Research Grant from the University of North Texas. Support by the National Science Council of Taiwan is acknowledged. This work was supported by the Center for Advanced Research and Technology at the University of North Texas.

References

1. Lee, C.J., Lee, T.J., Lyu, S.C., Zhang, Y., Ruh, H., Lee, H.J.: Field emission from well-aligned zinc-oxide nanowires grown at low temperature. Appl. Phys. Lett. **81**, 3648–3650 (2002)
2. Dong, L.F., Jiao, J, Tuggle, D.W., Petty, J.M., Ellis, S.A., Coulter, M.: ZnO nanowires formed on tungsten substrates and their electron field emission properties. Appl. Phys. Lett. **82**, 1096 (2003)
3. Zhu, Y.W., Zhang, H.Z., Sun, X.C., Feng, S.Q., Xu, J., Zhao, Q., Xiang, B., Wang, R.M., Yu, D.P.: Efficient field emission from ZnO nanoneedle arrays. Appl. Phys. Lett. **83**, 144 (2003)
4. Li, Y.B., Bando, Y., Goldberg, D.: ZnO nanoneedles with tip surface perturbations: excellent field emitters. Appl. Phys. Lett. **84**, 3603 (2004)
5. Tseng, Y.K., Huang, C.J., Cheng, H.M., Lin, I.N., Liu, K.S., Chen, I.C.: Characterization and field-emission properties of needle-like zinc oxide nanowires grown vertically on conductive zinc oxide films. Adv. Funct. Mater. **13**, 811–814 (2003)
6. Xu, C.X., Sun, X.W.: Field emission from zinc oxide nanopins. Appl. Phys. Lett. **83**, 3806 (2003)

7. Li, S.Y., Lin, P., Lee, C.Y., Tseng, T.Y.: Field emission and photofluorescent characteristics of zinc oxide nanowires synthesized by a metal catalyzed vapor-liquid-solid process. J. Appl. Phys. **95**, 3711 (2004)
8. Jo, S.H., Lao, J.Y., Ren, Z.F., Farrer, R.A., Baldacchini, T., Fourkas, J.T.: Field emission studies on thin films of zinc oxide nanowires. Appl. Phys. Lett. **83**, 4821–4823 (2003)
9. Wan, Q., Yu, K., Wang, T.H., Lin, C.L.: Low field electron emission from tetrapod-like ZnO nanostructures synthesized by rapid evaporation. Appl. Phys. Lett. **83**, 2253–2255 (2003)
10. Zhao, Q., Zhang, H.Z., Zhu, Y.W., Feng, S.Q., Sun, X.C., Xu, J., Yu, D.P.: Morphological effects on the field emission of ZnO nanorod arrays. Appl. Phys. Lett. **86**, 203115 (2005)
11. Ye, Z.Z., Yang, F., Lu, Y.F., Zhi, M.J., Tang, H.P., Zhu, L.P.: ZnO nanorods with different morphologies and their field emission properties. Solid State Commun. **142**, 425 (2007)
12. Xu, C.X., Sun, X.W., Chen, B.J.: Field emission from gallium-doped zinc oxide nanofiber array. Appl. Phys. Lett. **84**, 1540 (2004)
13. Kim, D.H., Jang, H.S., Lee, S.Y., Lee, H.R.: Effects of gas exposure on the field-emission properties of ZnO nanorods. Nanotechnology **15**, 1433–1436 (2004)
14. Jang, H.S., Kang, S.O., Nahm, S.H., Kim, D.H., Lee, H.R., Kim, Y.I.: Enhanced field emission from the ZnO nanowires by hydrogen gas exposure. Mater. Lett. **61**, 1679–1682 (2007)
15. Yeong, K.S., Meung, K.H., Thong, J.T.L.: The effects of gas exposure and UV illumination on field emission from individual ZnO nanowires. Nanotechnology **18**, 185608–185611 (2007)
16. Temple, D.:Recent progress in field emitter array development for high performance applications. Mater. Sci. Eng. R **24**, 185 (1999)
17. Schwoebel, P.R., Brodie, I.: Surface-science aspects of vacuum microelectronics. J. Vac. Sci. Technol. B **13**, 13911410 (1995)
18. Chalamala, B.R., Wallace, R.M., Gnade, B.E.: Surface conditioning of active field emission cathode arrays with H_2 and helium. J. Vac. Sci. Technol. B **16**, 2855–2858 (1998)
19. de Heer, W.A., Chatelain, A., Ugarte, D.: A carbon nanotube field-emission electron source. Science **270**, 1179 (1995)
20. Gulyaev, Y.V., Chernozatonskii, L.A., Kosakovskaja, J.Z., Sinitsyn, N.I., Torgashov, G.V., Zakharchenko, Y.F.: Field emitter arrays on nanotube carbon structure films. J. Vac. Sci. Technol. B **13**, 435 (1995)
21. Wang, Q.H., Setlur, A.A., Lauerhaas, J.M., Dai, J.Y., Seelig, E.W., Chang, R.P.H.: A nanotube-based field-emission flat panel display. Appl. Phys. Lett. **72**, 2912 (1998)
22. Choi, W.B., Chung, D.S., Kang, J.H., Kim, H.Y., Jin, Y.W., Han, I.T., Lee, Y.H., Jung, J.E., Lee, N.S., Park, G.S., Kim, J.M.: Fully sealed, high-brightness carbon-nanotube field-emission display. Appl. Phys. Lett. **75**, 3129–3131 (1999)
23. Dean, K.A., Chalamala, B.R.: The environmental stability of field emission from single walled carbon nanotubes. Appl. Phys. Lett. **75**, 3017–3019 (1999)
24. Lim, S.C., Choi, Y.C., Jeong, H.J., Shin, Y.M., An K.H., Bae, B.J., Lee, Y.H., Lee, N.S., Kim, J.M.: Effect of gas exposure on field emission properties of carbon nanotube arrays. Adv. Mater. **20**, 1563–1567 (2001)
25. Wadhawan, A., Stallcup, R.E., Stephens, K.F., Perez, J.M., Akwani, I.A.: Effects of O_2, Ar, and H_2 gases on the field-emission properties of single-walled and multiwalled carbon nanotubes. Appl. Phys. Lett. **79**, 1867–1869 (2001)
26. Sheng, L.M., Liu, P., Liu, Y.M., Qian, L., Huang, Y.S., Liu, L., Fan, S.S.: Effects of carbon-containing gases on the field-emission current of multiwalled carbon-nanotube arrays. J. Vac. Sci. Technol. A **21**, 1202 (2003)
27. Yi, G.C., Wang, C., Park, W.I.: ZnO nanorods: synthesis, characterization and applications. Semicond. Sci. Technol. **20**, 22–34 (2005)
28. Wei, A., Sun, X.W., Xu, C.X., Dong, Z.L., Yu, M.B., Huang, W.: Stable field emission from hydrothermally grown ZnO nanotubes. Appl. Phys. Lett. **88**, 213102 (2006)

29. Wagner, R.S., Ellis, W.C.: Vapor-liquid-solid mechanism of single crystal growth. Appl. Phys. Lett. **4**, 89 (1964)
30. Huang, M.H., Wu, Y.Y., Feick, H., Tran N., Weber E., Yang, P.D.: Catalytic growth of zinc oxide nanowires by vapor transport. Adv. Mater. **13**, 113 (2001)
31. Li, S.Y., Lee, C.Y., Tseng, T.Y.: Copper-catalyzed ZnO nanowires on silicon (100) grown by vapor-liquid-solid process. J. Cryst. Growth **247**, 357 (2003)
32. Ding Y., Gao, P.X., Wang, Z.L.: Catalyst-nanostructure interfacial lattice mismatch in determining the shape of VLS grown nanowires and nanobelts: a case of Sn/ZnO. J. Am. Chem. Soc. **126**, 2066 (2004)
33. Konenkamp, R., Boedecker, K., Lux-Steiner, M.C., Poschenrieder, M., Zenia, F., Clement, C.L., Wagner, S.: Thin film semiconductor deposition on free-standing ZnO columns. Appl. Phys. Lett. **77**, 2575 (2000)
34. Liu, X., Wu, X., Cao, H., Chang, R.P.H.: Growth mechanism and properties of ZnO nanorods synthesized by plasma-enhanced chemical vapor deposition. J. Appl. Phys. **95**, 3141 (2004)
35. Park, W.I., Kim, D.H., Jung, S.W., Yi, G.C.: Metalorganic vapor-phase epitaxial growth of vertically well-aligned ZnO nanorods. Appl. Phys. Lett. **80**, 4232 (2002)
36. Park, W.I., Yi, G.C., Kim, M.Y, Pennycook, S.J.: ZnO nanoneedles grown vertically on Si substrates by non-catalytic vapor-phase epitaxy. Adv. Mater. **14**, 1841 (2002)
37. Kim, K.S., Kim, H.W.: Synthesis of ZnO nanorod on bare Si substrate using metal organic chemical vapor deposition. Physica B **328**, 368 (2003)
38. Maejima, K., Ueda, M., Fujita, S., Fujita, S.: Growth of ZnO nanorods on a-plane (120) sapphire by metal-organic vapor phase epitaxy. Japan. J. Appl. Phys. **42**, 2600 (2003)
39. Park, J.Y., Oh, H., Kim, J.J., Kim, S.S.: Growth of ZnO nanorods via metalorganic chemical vapor deposition and their electrical properties. J. Cryst. Growth **287**, 145–148 (2006)
40. Park, W.I., Kim, D.H., Jung, S.W., Yi, G.C.: Metalorganic vapor-phase epitaxial growth of vertically well-aligned ZnO nanorods. Appl. Phys. Lett. **80**, 4232 (2002)
41. An, S.J., Park, W.I., Yi, G.C., Kim, Y.J., Kang, H.B., Kim, M.: Heteroepitaxial fabrication and structural characterizations of ultrafine GaN/ZnO coaxial nanorod heterostructures. Appl. Phys. Lett. **84**, 3612 (2004)
42. Heo, Y.W., Varadarajan, V., Kaufman, M., Kim, K., Norton, D.P., Ren, F., Fleming, P.H.: Site-specific growth of ZnO nanorods using catalysis-driven molecular-beam epitaxy. Appl. Phys. Lett. **81**, 3046–3048 (2002)
43. Pearton, S.J., Tien, L.C., Norton, D.P., Hung-Ta, W., Ren, F.: Nucleation control for ZnO nanorods grown by catalyst-driven molecular beam epitaxy. Appl. Surf. Sci. **253**, 4620–4625 (2007)
44. Li, Y., Meng, G.W., Zhang, L.D., Phillip, F.: Ordered semiconductor ZnO nanowire arrays and their photoluminescence properties. Appl. Phys. Lett. **76**, 2011 (2000)
45. Park, W.I., Yi, G.C., Kim, M., Pennycook, S.J.: Quantum confinement observed in Zn)/ZnMgO quantum structures. Adv. Mater. **15**, 526 (2003)
46. Geng, B.Y., Wang, G.Z., Jiang, Z., Xie, T., Sun, S.H., Meng, G.W., Zhang, L.D.: Synthesis and optical properties of S-doped ZnO nanowires. Appl. Phys. Lett. **82**, 4791
47. Bae, S.Y., Seo, H.W., Park, J.: Vertically-aligned sulfur-doped ZnO nanowires synthesized via chemical vapor deposition. J. Phys. Chem. B **108**, 5206 (2004)
48. Wan, Q., Li, Q.H., Chen, Y.J., Wang, T.H., He, X.L., Gao, X.G., Li, J.P.: Positive temperature coefficient resistance and humidity sensing properties of Cd-doped ZnO nanowires. Appl. Phys. Lett. **84**, 3085 (2004)
49. Vayssieres, L.: Growth of arrayed nanorods and nanowires of ZnO from aqueous solutions. Adv. Mater. **15**, 464–466 (2003)
50. Wang, Z., Qian, X., Yin, J., Zhu, Z.: Aqueous solution fabrication of large-scale arrayed obelisk-like zinc oxide nanorods with high efficiency. J. Solid State Chem. **177**, 2144–2149 (2004)

51. Lee, J.H., Leu, I.C., Hon, M.H.: Substrate effect on the growth of well-aligned ZnO nanorod arrays from aqueous solution. J. Cryst. Growth **275**, e2069–e2075 (2005)
52. Lin, C.C., Chen H.P., Chen, S.Y.: Synthesis and optoelectronic properties of arrayed p-type ZnO nanorods grown on ZnO films/Si wafer in aqueous solutions. Chem. Phys. Lett. **404**, 30–34 (2005)
53. Tak, Y, Yong, K.: Controlled growth of well-aligned ZnO nanorod array using a novel solution method. J. Phys. Chem. B **109**, 19263–19269 (2005)
54. Wahab, R., Ansari, S.G., Kim, Y.S., Seo, H.K., Kim, G.S., Khag, G., Shin, H.S.: Low temperature solution synthesis and characterization of ZnO nano-flowers. Mater. Res. Bull. **42**, 1640–1648 (2006)
55. Yu, K., Jin, Z., Liu, X., Liu, Z., Fu, Y.: Synthesis of size-tunable ZnO nanorod arrays from $NH_3H_2O/ZnNO_3$ solutions. Mater. Lett. **61**, 2775–2778 (2007)
56. Liu, B., Zeng, H.C.: Hydrothermal synthesis of ZnO nanorods in the diameter regime of 50 nm. J. Am. Chem. Soc. **125**, 4430 (2003)
57. Li, Z.Q., Xiong, Y.J., Xie, Y.: Selected-control synthesis of ZnO nanowires and nanorods via a PEG-assisted route. Inorg. Chem. **42**, 8105 (2003)
58. Choy, J.H., Jang, E.S., Won, J.H., Chung, J.H., Jang, D.J., Kim, Y.W.: Hydrothermal route to ZnO nanocoral reefs and nanofibers. Appl. Phys. Lett. **84**, 287 (2004)
59. Wang, J.M., Gao, L.: Wet chemical synthesis of ultralong and straight single-crystalline ZnO nanowires and their excellent UV emission properties. J. Mater. Chem. **13**, 2551 (2003)
60. Marathe, S.K., Koinkar, P.M., Ashtaputre, S.S., More, M.A., Gosavi, S.W., Joag, D.S., Kulkarni, S.K.: Efficient field emission from chemically grown inexpensive ZnO nanoparticles of different morphologies. Nanotechnology **17**, 1932–1936 (2006)
61. Ahsanulhaq, Q., Kim, J.H., Hahn, Y.B.: Controlled selective growth of ZnO nanorod arrays and their field emission properties. Nanotechnology **18**, 485307–485313 (2007)
62. Chen, J.C., Chai, W., Zhang, Z., Li, C., Zhang, X.: High field emission enhancement of ZnO-nanorods via hydrothermal synthesis. Solid State Electron. **52**, 294–298 (2008)
63. Gudiksen, M.S., Lauhon, L.J., Wang, J., Smoth, D.C., Lieber, C.M.: Growth of nanowire superlattice structures for nanoscale photonics and electronics. Nature **415**, 617–620 (2002)
64. Dai, Y., Zhang, Y., Li, Q.K., Nan C.W.: Synthesis and optical properties of tetrapod-like zinc oxide nanorods. Chem. Phys. Lett. **358**, 83 (2002)
65. Lyu, S.C., Zhang, Y., Lee, C.J., Ruh, H., Lee, H.J.: Low-temperature growth of ZnO nanowire array by a simple physical vapor-deposition method. Chem. Mater. **15**, 3294 (2003)
66. Zhang, Y., Wang, N., Gao, S., He, R., Miao, S., Liu, J., Zhu, J., Zhang, Z.: A simple method to synthesize nanowires. Chem. Mater. **14**, 3564 (2002)
67. Yao, B.D., Chan, Y.F., Wang, N.: Formation of ZnO nanostructures by a simple way of thermal evaporation. Appl. Phys. Lett. **81**, 757 (2002)
68. Gao, P.X., Wang, Z.L.: Self-assembled nanowire-nano-ribbon junctions arrays of ZO. J. Phys. Chem. B **106**, 12653 (2002)
69. Lao, J.Y., Wen, J.G., Ren, Z.F.: Hierarchical ZnO nanostructures. Nano Lett. **2**, 1287 (2002)
70. Gundiah, G., Deepak, F.L., Govindaraj, A., Rao, C.N.R.: Carbothermal synthesis of the nanostructures of AlO_3 and ZnO. Top. Catal. **24**, 137 (2003)
71. Choi, J.H., Tabata, H., Kawai, T.: J. Cryst. Growth **226**, 493 (2001)
72. Obuliraj, S.K., Yamauchi, K., Hanada, Y., Miyamoto, M., Ohba, T., Morito, S., Fujita, Y.: Nitrogen doped ZnO nanomaterials for UV-LED applications. Proceedings of the 2nd IEEE International Conference. NanoMicro Engineered and Molecular Systems, Bangkok, Thailand, 159–162 (2007)
73. Fowler, R.H., Nordheim, L.: Electron emission in intense electric fields. Proc. R. Soc. London, Ser. A **119**, 173–181 (1928)
74. Boscolo, I., Cialdi, S., Fiori, A., Orlanducci, S., Sessa, V., Terranova, M.L., Ciorba, A., Rossi, M.: Capacitive and analytical approaches for the analysis of field emission from

carbon nanotubes in a sphere-to-plane diode. J. Vac. Sci. Technol. B **25**, 1253–1260 (2007).

75. Mo, Y., Neogi, A., Perez, J.M., Fujita, Y.: Effects of residual gases on the field emission properties of ZnO nanorods. Poster presented at the Japan Society for the Promotion of Sciences – University of North Texas Winterschool on Nanophotonics, University of North Texas, Denton, TX, February 14–15, 2008
76. EXFO Burleigh, Quebec, Canada. http://www.exfo-burleigh.com/
77. Ito, F., Tomihari, Y., Okada, Y., Konuma, K., Okamoto, A.: Carbon-nanotube-based triode-field-emission displays using gated emitter structure. IEEE Electron Dev. Lett. **22**, 426 (2001)
78. Liu, J., Zhirnov, V.V., Choi, W.B., Wojak, G.J., Myers, A.F., Cuomo, J.J., Hren, J.J.: Electron emission from a hydrogenated diamond surface. Appl. Phys. Lett. **69**, 4038–4040 (1996)
79. Lim, S.C., Stallcup II, R.E., Akwani, I.A., Perez, J.M.: Effects of O_2, H_2, and N_2 gases on the field emission properties of diamond-coated microtips. Appl. Phys. Lett. **75**, 1179–1181 (1999)
80. Liu, M., Kim, H.K.: Ultraviolet detection with ultrathin ZnO epitaxial films treated with oxygen plasma. Appl. Phys. Lett. **84**, 173–175 (2004)
81. Losurdo, M, Giangregorio, M.M., Capezzuto, P., Bruno, G., Malandrino, G., Blandino, M., Fragala, I.L.: Reactivity of ZnO: impact of polarity and nanostructure. Superlattices Microstruct. **38**, 291–299 (2005)
82. Maki, H., Ichinose, N., Sakaguchi, I., Ohashi, N., Haneda, H., Tanaka, J.: The effect of the nitrogen plasma irradiation on ZnO single crystals. Key Eng. Mater. **216**, 61 (2002)
83. Yamaga, S., Yoshikawa, A., Kasa, H.: Electrical and optical properties of donor doped ZnS films grown by low-pressure MOCVD. J. Cryst. Growth **86**, 252–256 (1988)
84. Ye, C., Fang, X., Li, G., Zhang, L.: Origin of the green photoluminescence from zinc sulfide nanobelts. Appl. Phys. Lett. **85**, 3035–3037 (2004)
85. Vacassy, R., Scholz, S.M., Dutta, J., Plummer, C.J.G., Houriet, R., Hofmann, H.: Synthesis of controlled spherical zinc sulfide particles by precipitation from homogeneous solutions. J. Am. Ceramic Soc. **81**, 2699–2705 (1998)
86. Liu, X., Caia, X., Maob, J., Jinc, C.: ZnS/Ag/ZnS nano-multilayer films for transparent electrodes in flat display application. Appl. Surf. Sci. **183**, 103–110 (2001)
87. Shao, L.X., Chang, K.H., Hwang, H.L.: Zinc sulfide thin films deposited by RF reactive sputtering for photovoltaic applications. Appl. Surf. Sci. **212–213**, 305–310 (2003)
88. Johnstona, D.A., Carlettoa, M.H., Reddyb, K.T.R., Forbesa, I., Miles, R.W.: Chemical bath deposition of zinc sulfide based buffer layers using low toxicity materials. Thin Solid Films **403–404**, 102–106 (2002)
89. Meng, X.M., Liu, J., Jiang, Y., Chen, W.W., Lee, C.S., Bello, I., Lee, S.T.: Structure- and size-controlled ultrafine ZnS nanowires. Chem. Phys. Lett. **382**, 434–438 (2003)
90. Zhu, Y.C., Bando, Y., Xue, D.F.: Spontaneous growth and luminescence of zinc sulfide nanobelts. Appl. Phys. Lett. **82**, 1769–1771 (2003)
91. Lu, F., Cai, W., Zhang, Y., Li, Y., Sun, F., Heo, S.H., Cho, S.O.: Appl. Phys. Lett. **89**, 231928 (2006)
92. Khosravi, A.A., Kundu, M., Jatwa, L., Deshpande, S.K., Bhagwat, U.A., Sastry, M., Kulkarni, S.K.: Green luminescence from copper doped zinc sulphide quantum particles. Appl. Phys. Lett. **67**, 2702–2704 (1995)
93. Zhang, D., Qi, L., Cheng, H., Ma, J.: Preparation of ZnS nanorods by a liquid crystal template. J. Colloid Interface Sci. **246**, 413–416 (2002)
94. Lu, F., Cai, W., Zhang, Y., Li, Y., Sun, F., Heo, S.H., Cho, S.O.: Fabrication and field-emission performance of zinc sulfide nanobelt arrays. J. Phys. Chem. C **111**, 13385–13392 (2007)
95. Ha, B., Seo, S.H., Cho, J.H., Yoon, C.S., Yoo, G-C.Y., Park, C.Y., Lee, C.J.: Optical and field emission properties of thin single-crystalline GaN nanowires. J. Phys. Chem. B **109**, 11095–11099 (2005)

96. Yilmazoglu, O., Pavlidis, D., Litvin, Yu, M., Hubbard, S., Tiginyanu, I.M., Mutamba, K., Hartnagel, H.L., Litovchenko, V.G., Evtukh, A.: Field emission from quantum size GaN structures. Appl. Surf. Sci. **220**, 46–50 (2003)
97. Komirenko, S.M., Kim, K.W., Kochelop, V.A., Stroscio, M.A.: High-field electron transport controlled by optical photon emission in nitrides. Int. J. High Speed Electron. Syst. **12**, 1057–1081 (2002)
98. Goodman, S.A., Auret, F.D., Koschnick, F.K., Spaeth, J.-M., Beaumont, B., Gilbart, P.: Field-enhanced emission rate and electronic properties of a defect introduced in n-GaN by 5.4 mev He-ion irradiation. Appl. Phys. Lett. **74**, 809–811 (1999)
99. Tong, X.L., Jiang, D.S., Li, Y., Liu, Z.M., Luo, M.Z.: Folding field emission from GaN onto polymer microtip array by femtosecond pulsed laser deposition. Appl. Phys. Lett. **89**, 061108-1–061108-3 (2006)
100. Ng, D.K.T., Hong, M.H., Tan, L.S., Zhu, Y.W., Sow, C.H.: Field emission enhancement from patterned gallium nitride nanowires. Nanotechnology **18**, 375707–375711 (2007)
101. Ye, F., Xie, E.Q., Pan, X.J., Li, H., Duan, H.G., Jia, C.W.: Field emission from amorphous GaN deposited on Si by dc sputtering. J. Vac. Sci. Technol. B **24**, 1358– 1361 (2006)
102. Berishev, I., Bensaoula, A., Rusakova, I., Karabutov, A., Ugarov, M., Ageev, V.P.: Field emission properties of GaN films on Si(111). Appl. Phys. Lett. **73**, 1808–1810 (1998)
103. Liu, B.D., Bando, Y., Tang, C.C., Xu, F.F., Hu, J.Q., Golberg, D.: Needlelike bricrystalline GaN nanowires with excellent field emission properties. J. Phys. Chem. B **109**, 17082–17085 (2005)
104. Levitt, Albert P.: Whisker Technology. Wiley, New York, NY (1970)
105. Seo, H.W., Chen, Q.Y., Tu, L.W., Hsiao, C.L., Iliev, M. N., Chu, W.K.: Catalytic nanocapillary condensation and epitaxial GaN nanorod growth. Phys. Rev. B **71**, 235314 (2005)
106. Seo, H.W., Chen, Q.Y., Iliev, M.N., Tu, L.W., Hsiao, C.L., Meen, J.K., Chu, W.K.: Epitaxial GaN nanorods free from strain and luminescent defects. Appl. Phys. Lett. **88**, 153124 (2006)
107. Seo, H.W., Chen, Q.Y., Tu, L.W., Chen, M., Wang, X.M., Tu, Y.J., Shao, L., Lozano O., Chu, W.K.: GaN nanorod assemblies on self-implanted (111) Si substrates. Microelectron. Eng. **83**, 1714 (2006)
108. Chu, W.K., Seo, H.W., Chen, Q.Y., Wang, X.M., Tu, L.W., Hsaio, C.L., Chen, M., Tu, Y.J.: US Patent Pending, 60/696,020
109. Tsai, M.-H., Jhang, Z.-F., Jiang, J.-Y., Tang, Y.-H., Tu, L.W.: Electrostatic and structural properties of GaN nanorods/nanowires from first principles. Appl. Phys. Lett. **89**, 203101 (2006)
110. Tu, L.W., Hsiao, C.L., Chi, T.W.: Self-assembled vertical GaN nanorods grown by molecular-beam epitaxy. Appl. Phys. Lett. **82**, 1601 (2003)
111. SAES Getters, Inc., 1122 E Cheyenne Mountain Blvd, Colorado Springs, CO 80906-4598, http://www.saesgetters.com/
112. Sommer, A.H.: Photoemissive Materials. Wiley, New York, NY (1968)
113. Bell, R.L.: Negative Electron Affinity Devices. Clarendon Press, Oxford (1973)
114. Zhu, Wei (ed.): Vacuum Microelectronics. Wiley, New York, NY (2001)
115. Modinos, A.: Field, Thermionics, and Secondary Electron Emission Spectroscopy. Plenum Press, New York, NY (1984)
116. Martinelli, R.U., Pankove, J.I.: Secondary electron emission from the GaN:Cs-O surface. Appl. Phys. Lett. **25**, 549–551 (1974)
117. Monch, W., Kampen, T.U., Dimitrov, R., Ambacher, O., Stutzmann, M.: Negative electron affinity of cesiated p-GaN(0001) surfaces. J. Vac. Sci. Technol. B **16**, 2224–2228 (1998)
118. Machuca, F., Zhi, L., Sun, Y., Pianetta, P., Spicer, W.E., Pease, R.F.W.: Oxygen species in Cs/O activated gallium nitride (GaN) negative electron affinity photocathodes. J. Vac. Sci. Technol. B **21**, 1863–1869 (2003)
119. Wang, K.R., Lin, S.J., Tu, L.W., Chen, M., Chen, Q.Y., Chen, T.H., Chen, M.L., Seo, H.W., Tai, N.H., Chang, S.C., Lo, I.K., Wang, D.P., Chu, W.K.: InN nanotips as excellent field emitters. Appl. Phys. Lett. **92**, 123105 (2008)

Chapter 8
Growth, Optical, and Transport Properties of Self-Assembled InAs/InP Nanostructures

Oliver Bierwagen, Yuriy I. Mazur, Georgiy G. Tarasov, W. Ted Masselink, and Gregory J. Salamo

Abstract A comprehensive study, including growth, optical characterization and anisotropic transport in quantum well (QW), quantum wire (QWr), and quantum dot (QD) systems, is carried out for the InAs/InP nanostructures grown by gas-source molecular beam epitaxy on InP (001) substrates. Role of substrate misorientation in the self-organized formation, shape, and alignment of InAs nanostructures is investigated. The emission and absorption properties related to interband transitions of the InAs/InP nanostructures are studied by means of polarization-dependent photoluminescence (PL) and transmission spectroscopy. It is demonstrated that the emission wavelength of grown nanostructures extends up to 2 μm, including the technologically important 1.3 μm and 1.55 μm, at room temperature. Polarization-dependent PL and transmission measurements for all QWrs and QDs reveal much similarity in temperature behavior in spite of a qualitatively different character of the one (1D)- and zero (0D)-dimensional density-of-states functions. The in-plane transport of electrons and holes in QWrs, QDs, and QWs is investigated and interpreted in terms of anisotropic two-dimensional carrier systems, and in terms of coupled 1D or 0D systems. Peculiar band structure and carrier relaxation in the InAs/InP nanostructures suggest currently the large application potential for the optical devices – mainly for the telecommunication wavelength of 1.55 μm.

Keywords III-V semiconductors · InAs/InP · Molecular beam epitaxy · Surface morphology · Self-assembly – semiconductor quantum wells · Semiconductor quantum wires · Semiconductor quantum dots · Light transmission · Photoluminescence · Polarization · Magnetoresistance · Weak localization · Electrical conductivity · Electron density · Electron mobility · Electronic switching systems · Field effect devices · Exciton · Valence bands

---

Y.I. Mazur (✉)
Department of Physics, University of Arkansas, 226 Physics Building, Fayetteville, AR 72701, USA
e-mail: ymazur@uark.edu

Z.M. Wang, A. Neogi (eds.), *Nanoscale Photonics and Optoelectronics*, Lecture Notes in Nanoscale Science and Technology 9,
DOI 10.1007/978-1-4419-7587-4_8,

1 Introduction

There are numerous experimental and theoretical studies of the self-assembling process [1–7] in strained heterostructures, which result in the appearance of quantum dots (QDs) or quantum wires (QWrs). The electrons (holes) in these structures are confined in the nanoscale volumes, which leads to the quantization of electronic energy spectra and to the peaked density of states (DOS). Due their characteristic energy spectrum and DOS, such heterostructures develop unique physical properties widely used in low-threshold lasers and optical amplifiers [8], single-photon emitters for quantum information processing [9], and quantum cryptography [10].

For fiber optics the development of nanoscale laser structures operating at the 1.55 μm (0.8 eV) wavelength for minimum losses is crucially needed [11–14]. Using InAs or (In, Ga)As QDs grown on GaAs substrates, such operation is difficult to achieve because of the large lattice mismatch (∼7%) which generally results in a large strain accumulation in the vicinity of QDs and an increase of the InAs band gap energy [15, 16] beyond the required spectral region (∼1.55 μm). In order to reduce the strain in the (In, Ga)As/GaAs system, the metamorphic buffers like InGaAs instead of GaAs [17] or InP substrate can be used as a completely different host material [18], leading in both cases to a reduced lattice mismatch.

Indeed, the growth of self-assembled InAs nanostructures on InP(001) substrates leads to the formation of InAs QDs or QWrs [19] with emission in the spectral range of ∼1.55 μm since the lattice mismatch in the InAs/InP system is only of ∼3.2%, much smaller than the one in the InAs/GaAs system. Several groups [19–23] have demonstrated the growth of the InAs/InP QDs. In this case, the smaller lattice mismatch as compared to the InAs/GaAs system leads to the formation of large QDs with a small QD density and complicates the reproducible formation of the nanometer-sized islands. Recent study of the self-organized InAs nanostructures grown by gas-source molecular-beam epitaxy in an InP matrix on both nominally oriented and vicinal InP(001) surfaces has shown that the off-cut direction of the vicinal substrates determines the morphology of nanostructures, that is, quantum dot, quantum wire, or two-dimensional growth, whereas, on nominally oriented substrates, the morphology is primarily only dependent on the growth conditions [19]. Various island shapes and morphologies are observed, ranging from elongated QDs (quantum dashes) to almost isotropic QDs. Moreover, the islands with and without a capping layer have different dimensions because the As/P exchange reaction takes place [24, 25]. Postgrowth modifications in the InAs QDs are caused by simultaneous action of Group V overpressure and stress field, produced by the InAs nanostructures, resulting in a strong material transport. The direction of this material net current depends on the type of Group V element used for the overpressure flux [24]. Thus the QDs capped with InP might have a smaller height due to As/P exchange-induced decomposition [25].

It was also demonstrated that under controlled epitaxy InAs/InP QWrs can be grown on the substrate surface [26–28]. During the last few years, the growth mechanism and structural and photoluminescence (PL) properties depending on the growth conditions have been explored for this type of wires [29–31]. It has been shown that the control of the growth conditions allows one to produce nanostructures emitting in the range of 1.2–1.9 μm [30]. In order to increase the wire density and improve the size homogeneity, a vertical stacking can be used [31]. In this case the size of the stacked wires is greatly influenced by the thickness of the InP spacing layer between adjacent InAs layers stacks.

While both wavelength and linewidth of the luminescence emission in the InAs/InP nanostructures are determined by their size, composition, and homogeneity, the growth conditions have to be precisely tuned to obtain dense and homogeneous QDs or QWrs. Besides, more accurate geometrical description of the islands requires high-resolution transmission electron microscopy (TEM) measurements rather than atomic force microscopy (AFM). Optical study of the electronic and hole subsystems is also quite necessary to accomplish the characterization of the InAs/InP nanostructures. This provides information about confinement, strain, and piezoelectric fields that alter the symmetry properties of the confinement potential and induce mixing of the valence band wavefunctions [6, 32]. Ultimately, there is an optical anisotropy, which depends on the InAs/InP nanostructure geometry and substrate orientation. This anisotropy can be effectively changed if the inter-nano-block coupling is taken into account. The coupling will change the energy spectra, the DOS structure, and relaxation in the coupled nano-blocks as compared with the isolated one. For example, a further complication of QWr DOS structure is expected in the system of coupled QWRs, both laterally and vertically [33–39]. The strength of coupling can be varied by changing the interwire spacing or the thickness of the spacer layer separating adjacent QWr layers. The change of the coupling strength is immediately reflected in the experimental photoluminescence (PL) decay time shortening due to the exciton wavefunction becoming delocalized along the growth direction. This common property of coupled nanostructures has also been demonstrated for the vertically coupled InAs/GaAs QDs [40]. Lateral coupling of QWrs is also of importance because the system may be seen as lying between 1D and 2D as has been shown by means of magneto-transport for the InAs/InP QWr system [35]. The change of reduced dimensionality appears not only in the details of the physics of the usual scattering mechanisms but also in the quantum mechanical corrections to the conductivity such as weak localization and electron–electron scattering as was revealed at liquid helium temperatures. In this review, we present the data demonstrating the change of dimensionality in dense arrays of InAs/InP nanostructures and compare to the InAs/InP quantum well (QW) system. The change is detected by optical and magneto-transport methods. While at low temperature the carriers can be strongly localized by potential fluctuations in the QWrs and QDs, which can influence the interblock coupling, the temperature is varied from 5 K up to room temperature. In this way the trapped carriers become partly released and facilitate the observation of the expected change of the dimensionality in the coupled nanostructure system. To date, the transport due to lateral coupling between self-assembled nanostructures is not well investigated in general, and in particular no such data exists for InAs nanostructures in InP. The main focus of this review is the detailed investigation of growth conditions, optical properties, and carrier transport, which turns out to be anisotropic, in laterally coupled, self-assembled InAs nanostructures in an InP matrix.

This chapter is organized as follows: the preparation of InAs/InP nanostructures and their structural properties are discussed in Section 2, in particular with respect to the effect of substrate misorientation on the nanostructure shape; optical properties of the coupled InAs/InP nanostructures, particularly the interband emission and absorption, are discussed in Section 3 in terms of thermally induced change of dimensionality; investigation of the anisotropic carrier transport in coupled InAs/InP nanostructures, interpreting the transport of electrons and holes from the point of view of coupled 1D (one-dimensional) or 0D (zero-dimensional) nanostructures is presented in Section 4; a peculiar electronic switching, that is based on the anisotropic transport properties of coupled InAs QWrs, is discussed in Section 5; finally, the results are summarized in Section 6.

2 Growth of InAs/InP Nanostructures

2.1 P–As Exchange Process

Here we present details of molecular beam epitaxy (MBE) of InAs onto InP substrate. In contrast to most of the other III–V heterointerfaces, a characteristic feature of the InAs/InP interface is the P–As exchange (PAsX) during the growth. It has been shown that annealing of InP surfaces under an As flux in a solid-source MBE system produces a thin pseudomorphic InAs layer of maximum 2 monolayer (ML) thickness due to the replacement of desorbing P by impinging As [41]. At an As flux of 1.5×10^{-5} torr the thickness of the InAs layer formed due to the PAsX process increases with increasing substrate temperature from 1.25 ML at 450°C to 2.6 ML at 575°C [42]. It is also shown [43] that the exchange starts at 350°C, and that an InAs film is of 7.3 Å thickness. A cross-sectional STM study of InAs QDs grown on InP(311)B determines 2 ML of additional InAs due to PAsX [25]. In Ref. [44], the amount of excess InAs formed by PAsX is determined to be 4–5 ML. The same study shows that significantly less excess InAs forms on $In_{0.53}Ga_{0.47}As$ and $In_{0.52}Al_{0.48}As$ buffers as these contain no P to be replaced by As. The formation of InAs QWr exclusively by PAsX has been demonstrated with solid-source MBE in Ref. [45]. A subsequent reduction of the As flux resulted in the decomposition of the QWrs into QDs. The amount of excess InAs due to PAsX was determined to be 2.1 ML based on the volume of QWrs measured with STM [46].

The limitation to the thickness of the InAs formed by PAsX is understood in terms of protecting the InP from further P desorption by the formed InAs layer. However, if strain-driven transport of In ad-atoms on the surface plays a role, the amount of excess InAs due to PAsX is not limited. Desorbing P atoms are replaced by the impinging As. The InAs formed in this way is laterally strained to match the underlying InP layer. If the strain in the surface is nonhomogeneous, it is energetically favorable for the InAs to form in regions with a lateral lattice constant close to that of relaxed InAs. Since In atoms from the surface are mobile, they can migrate to the energetically favorable regions and form InAs with the impinging As. The energetically unfavorable regions continue supplying In from the InP layer because no protective InAs layer forms there to limit the PAsX. As a result of this described mechanism, the excess InAs due to PAsX during the growth of InAs QDs by MOCVD at high growth temperatures and AsH_3 flux has been shown to be 17 ML [47]. The formation of QDs with heights in the range of 100 nm, in this case, has been explained by the migration of In ad-atoms to the dot apex. Lower growth temperatures and AsH_3 flux have been shown to result in less excess InAs. The insertion of an $In_{0.53}Ga_{0.47}As$ layer underneath the InAs effectively suppresses the formation of excess InAs. Similarly, the insertion of a 1-ML-thick GaAs layer prior to InAs deposition has been used to suppress PAsX and optimize the optical properties of InAs QDs [48]. On the other hand, the formation of QDs and QWrs only by PAsX on InP buffers with a built-in strain field induced by buried InAs nanostructures has been observed and explained by strain-driven In migration in MBE growth [31].

During capping, the reverse process of As–P exchange (AsPX) reduces the total amount of InAs by transforming it into InP. It has been shown [49] that at an AsPX time of 10 s the amount of excess InAs is 2 ML for 15 s PAsX and 3 ML for 120 s PAsX. The results for the comparably long time of 120 s suggest a saturation of the PAsX at 3 ML due to the protective effect of the InAs layer, whereas the results from the sample with thickness of 2 ML at relatively short 15 s suggest a quick PAsX of the first 2 ML. This value is in agreement with most of the aforementioned results from other groups. The PAsX times of 15 s and AsPX times of 10 s as standard parameters have been used for sample growth. Therefore, an excess of 2 ML InAs is assumed for all InAs/InP samples in our studies. The morphology of InP surfaces with nominally oriented (N) (001) substrate, vicinal (A) substrate, off-cut toward [110], and vicinal (B) substrate, off-cut toward [−110] after a long anneal in As has been measured with AFM. During the annealing, reflection high-energy electron diffraction (RHEED) observations show a 2D–3D transformation only on the B-type substrate after 7 min, whereas the RHEED pattern on the other substrates indicate planar layers. The morphologies shown in Fig. 8.1 reveal very different results for different substrate orientations. The layer formed on N substrate (Fig. 8.1a) shows a smooth surface with large monolayer terraces. The fact that no nanostructures have formed shows that the thickness of the InAs is below the critical thickness for Stranski–Krastanow (SK) growth suggesting a limited PAsX. On the A-type

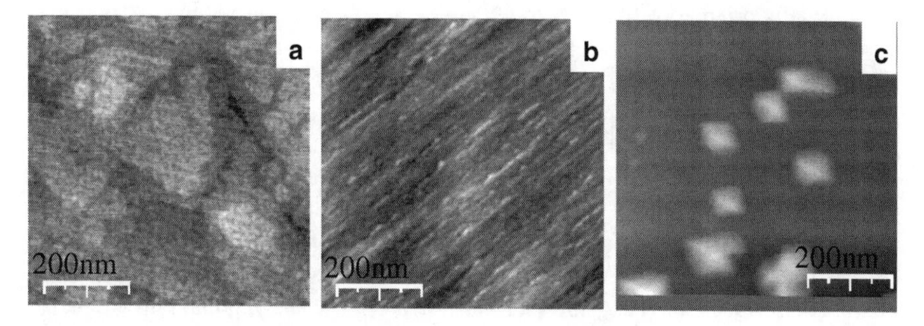

Fig. 8.1 AFM images of nominally oriented (001)InP (**a**), A-type vicinal InP (**b**), and B-type vicinal InP (**c**) after an annealing under As flux for 15 min [49]. The lateral size is 1 μm × 1 μm for all images. The height scales are 1.2 nm (**a,b**), and 80 nm (**c**) [49]

substrate, a surface modulation with a wire-like morphology (along the direction of the surface steps) is visible. In comparison to InAs nanostructures grown by InAs deposition on this type of substrate, the surface modulation formed by PAsX is low, suggesting a limited PAsX. In marked contrast, the InAs on the B-type substrate forms giant InAs dots of typically 100 nm base length and 50 nm height. These dots look similar to those of Ref. [47] that formed at the highest temperature and AsH_3 flux used in that study. The volume of our giant InAs dots clearly demonstrates unlimited PAsX due to surface diffusion.

Thus, the PAsX process typically forms a 2 ML thick InAs layer on the expense of the underlying InP through the replacement of desorbing P by impinging As. This process can be suppressed by growth temperatures below the P desorption temperature (360°C). Nonhomogeneous strain distribution can drive a lateral material transport resulting in unlimited P–As exchange in regions with unfavorable strain for InAs and accumulation of this formed InAs in regions with favorable strain. On B surfaces, the PAsX process is unlimited due to spontaneous formation of 3D islands and the strain-driven transport.

2.2 Role of Vicinal Substrates

The most commonly used growth technique for InAs/InP nanostructures is a solid-source MBE, for which the shape of self-organized nanostructures can depend on buffer [26] or even surface reconstruction [45]. The miscut of vicinal substrates plays an important role influencing the type and arrangement of self-organized InAs nanostructures. Off-cuts from InP(001) can be made toward [110] (A surface), toward [−110] (B surface), and toward [100] (C surface), usually with a magnitude of 2°. This off-cut angle results in monolayer steps every 8.4 nm. QWr formation was reported on A surfaces grown by gas-source MBE (GSMBE) [50]. Using MBE, QWrs were found on N surfaces and on A surfaces, while QDs were observed on C surfaces [51]. Our systematic studies [19, 49] demonstrate that self-organized nanostructures grown by GSMBE on vicinal InP(001) substrates occur as either QWrs or QDs depending on the off-cut direction and generally independently of other growth conditions. This result is in contrast to the growth of nanostructures on nominally oriented substrates for which the details of the growth conditions play a determining role on the resulting morphology. It is shown by varying InAs deposition rate and deposition temperature that for nominally oriented substrates, the growth conditions determine the type of nanostructure, whereas for vicinal substrates, it is only the off-cut which is important. Thus, the use of vicinal substrates represents a robust way to determine the nanostructure type in the InAs/InP system. Figure 8.2 shows AFM images of the samples with 2 ML, 4 ML, and 6 ML of deposited InAs, all using a substrate temperature of $T = 450°C$. One can see that the deposition of 2 ML InAs onto the N surface results in a 2D layer with large monolayer terraces; on the A surface, short, flat, closely packed quantum wires parallel to the surface steps formed; the 2 ML onto the B surface results in quantum dots with low areal density, roughly aligned parallel to the surface steps. The height of the QWrs is about 0.6–1.2 nm (measured at the edge between QWrs and flat

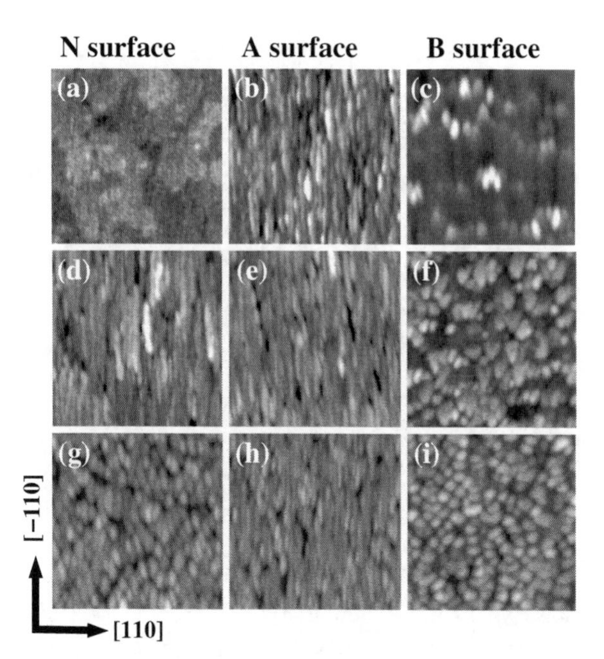

Fig. 8.2 AFM images of 0.5 μm × 0.5 μm InAs/InP with different deposited InAs thicknesses. The height scales are given in parentheses. 2 ML deposited: (**a**) "N" (12 Å), (**b**) "A" (12 Å), (**c**) "B" (70 Å). 4 ML deposited: (**d**) "N" (10 Å), (**e**) "A" (20 Å), (**f**) "B" (70 Å). 6 ML deposited: (**g**) "N" (70 Å), (**h**) "A" (40 Å), (**i**) "B" (160 Å) [19]

areas between the QWrs) and the QDs are 3–7 nm high. These results demonstrate that misoriented substrates lead to an earlier onset of islanding (3D growth) than nominal surfaces. When the deposited InAs is increased to 4 ML, 3D growth also begins on N surfaces. The onset of 3D growth with a critical thickness (between 2 and 4 ML of deposited InAs in our case) indicates a nanostructure formation in the SK growth mode. The AFM images show long, flat, and closely packed quantum wires on both the N surface (0.6–0.9 nm high) and the A surface (0.9–1.5 nm high), but dots (typically 3–5 nm high) of higher areal density and with random alignment on the B surface. The QD areal density of 5×10^{10} cm^{-2} on the B surface is as high as that obtained in Ref. [52] on (311)B InP substrates which have been intentionally used to obtain a high density of QDs. Similar morphologies for the N and A surfaces were obtained using solid-source MBE [51]. Cross-sectional transmission electron microscopy (TEM) images (see Fig. 8.3) of a stack of 15 InAs (4 ML)/InP(20 nm) layers grown under the same growth conditions confirm that the capped InAs nanostructures are wire-like on A surfaces and dot-like on B surfaces. (The InP spacer between the InAs layers is thick enough to suppress any influence due to the strain fields of the previous InAs layer.) The height of the wires (2.4 nm) is bigger than the height determined with AFM, due to tip-sample convolution. The precise cross-sectional shape of the QWrs cannot be resolved with these images. The height of the dots (3 nm), on the other hand, is smaller than that determined by AFM (4 nm), suggesting a shrinking of the InAs nanostructures due to AsPX when exposed to a P flux before capping. Owing to AsPX and/or tip-sample convolution of AFM, the width of the capped QDs at their base is approximately 15 nm (see TEM image in Fig. 8.3c) as opposed to the apparent width of uncapped QDs of 50 nm (see AFM image in Fig. 8.2f). When the InAs deposition is further increased to 6

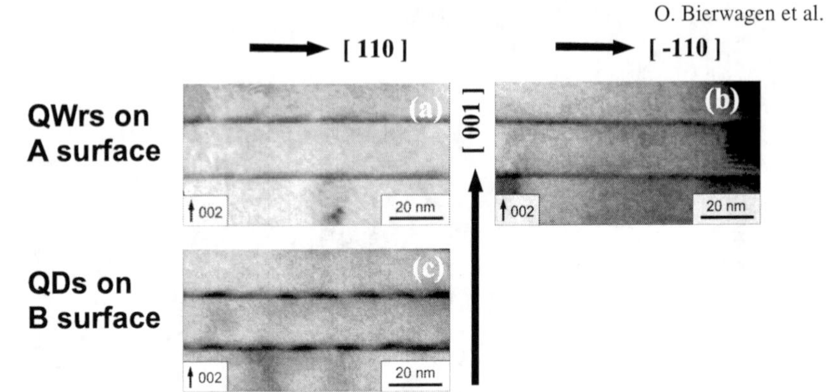

Fig. 8.3 Cross-sectional TEM 002 dark-field images of (**a, b**) quantum wires on A surface and (**c**) quantum dots on B surface. The *black areas* indicate InAs and the *gray areas* indicate InP [49]

ML, the QDs on the N surfaces elongate in the [−110] direction (quantum sticks, quantum dashes), typically 3–5 nm high; the A surface results in longer quantum sticks and QWrs (typically 2.5 nm high); and the B surface results in the growth of dots (typically 5–10 nm high).

The influence of surface steps on the shape of the nanostructures is an interesting issue not yet fully explained. It is of importance that the InAs A steps (forming the long edge of these islands) are energetically more favorable than steps along [110] (B steps), resulting in straight A steps and rough B steps [45, 53]. The energetically favorable A steps determine the direction of the QWrs. Furthermore, it is known that ad-atoms tend to attach themselves to existing steps during growth. The formation of QWrs along the [−110] direction on N and A surfaces can then be understood based on these points: On A surfaces, InAs preferably nucleates at the A steps resulting in a higher local InAs coverage along the steps than on the terraces between steps. This higher local coverage leads to an earlier 2D–3D transition in comparison to N surfaces with a uniform nucleation of InAs. The same argument holds for steps on B and C surface. The steps on B (and C) surfaces, on the other hand, impede the formation of wires in the [−110] direction. Therefore it is observed experimentally that instead of forming QWrs in the [110] direction, the InAs forms high QDs, even for low InAs deposition. Kinetically, the roughness (kinks) along the B steps provides preferred nucleation sites for the ad-atoms leading to a nonuniform nucleation along the B steps. This nonuniformity results in the formation of local InAs clusters that later undergo a 2D–3D transition. Thermodynamically, dots incorporate lower strain energy than do wires. This argument may explain that it is now energetically favorable for InAs to nucleate on top of existing InAs islands forming dots (less strain) than to form uniformly along B steps. In addition, the more dot-like shape on N surface at a deposition of 6 ML compared to 4 ML suggests an energetic trade-off by lowering the strain energy (dot shape) and increasing the surface and step energy (incorporating more B steps). To theoretically confirm this explanation, more sophisticated kinetic Monte Carlo simulations that include the local strain distribution would be necessary.

Thus, vicinal substrates result in an onset of 3D growth at lower deposited InAs thickness than nominally oriented (001) substrates. Vicinal surface misoriented by 2° toward [110] (A surfaces) support the formation of the type of QWrs that also grow on exactly oriented (001) surfaces (N surfaces). Vicinal (001) surface misoriented by 2° toward [−110] (B surfaces) lead to the formation QDs. Similar QDs form on C surfaces (misoriented by 2° toward [100]). In addition, the QDs on C surfaces are roughly aligned along the surface steps ([010] direction) demonstrating the potential of in-plane alignment of QDs through vicinal substrates.

2.3 Influence of Growth Parameters on InAs Nanostructures

In order to get more information about growth kinetics and dynamics, we have varied parameters that influence the ad-atom surface diffusion: growth temperature, growth rate, and As flux. The effect of growth temperature on the morphology of self-organized nanostructures was investigated by depositing 4 ML of InAs at different temperatures (including 450°C discussed above) whereas all other growth conditions were maintained (PAsX for 15 s, InAs growth rate 0.4 ML/s, AsH_3 flux 0.6 sccm, postdeposition annealing for 10 s). At a low substrate temperature of 380°C (to reduce ad-atom diffusivity), high growth rate, and without annealing (to reduce ad-atom diffusion time), the 2D–3D transition is mainly suppressed. On the nominally oriented substrate, the InAs surface is found to be quite smooth without InAs nanostructures. Monolayer terraces similar to those on the 2 ML sample are observed. This sample serves as reference QW sample with the same nominal InAs coverage (4 ML) as the nanostructure-containing samples compared to it. On the B surface, a rougher layer formed without discernable nanostructures, indicating preferred nucleation at B step kinks but insufficient surface transport for a 2D–3D transition. Increasing the substrate temperature to 400°C, we find a relaxation of the kinetic restrictions. On the A surface, shallow wires are discernable, whereas the nominal surface contains no wires. The B surface shows QDashes oriented along [−110]. These morphologies suggest a surface diffusion sufficient to reach the steps of the vicinal substrate. At 450°C, QWrs form on N and A surfaces, and QDs with an average height of 4.5 nm form on the B surface. The effect of the higher temperature like 485°C on the N surface is that low areal density quantum dots form instead of the wires seen at $T \sim 450°C$; these QDs are roughly 10 nm in height.

The As flux was varied for fixed growth conditions (PAsX for 15 s, 4 ML of InAs deposited, growth temperature 450°C, InAs growth rate 0.4 ML/s, postdeposition annealing for 10 s). The actual As content of the surface, that determines the kinetics, is a function of the As flux/In flux ratio during the InAs deposition. For each AsH_3 flux, the morphologies on N and A surfaces are very similar, but differ markedly from the morphology on B surfaces (QDs at all AsH_3). At AsH_3 fluxes of 0.3 and 0.6 sccm, narrow, closely spaced QWrs form on N and A surfaces. These QWrs are shorter at the lower AsH_3 flux. Similarly, at a higher surface As content (AsH_3 flux of 1.5 sccm or growth rate 0.13 ML/s at 0.6 sccm), thicker, higher, less ordered QWrs form similar to those formed at higher substrate temperature on the A surface. Hence, with increasing diffusivity due to increasing surface As content, the morphology on N and A surfaces develops from short narrow wires, over long narrow wires, to thick, high, and less ordered wires. This development suggests thick QWrs to be energetically preferred to thin QWrs on N and A surfaces. The narrow or thick QWrs may be seen as large or small 2D islands (depending on As flux) that underwent the 2D–3D transition. On the B surface, QD height drastically increases with increasing AsH_3 flux, whereas the areal density of QDs decreases only slightly.

Thus, growth temperature and AsH_3 flux influence the nanostructure shape through their impact on growth kinetics. Low growth temperatures can be used to suppress the 2D–3D growth transition and grow a QW sample with the same nominal InAs thickness as the QWr and QD samples. At high growth temperatures, QDs form also on N surfaces, and the height of QDs on B surfaces is increased. A high AsH_3 flux creates thicker, less ordered QWrs on N and A surfaces and higher QDs on B surfaces.

2.4 Postdeposition Modifications of InAs Nanostructures

The final shape of InAs nanostructures on InP can be significantly modified by surface transport of In after the deposition of the InAs layer: The QWrs discussed in Ref. [54] have been formed during the 2D–3D transition by in situ annealing an MBE-deposited 2D InAs layer at elevated temperature. The influence of group V overpressure on the final shape of CBE-grown InAs nanostructures has been investigated in Ref. [24]. The authors point out that a cooldown of samples under either As or P overpressure modifies the nanostructure shape due to the combination of PAsX or AsPX and strain-induced surface transport. A cooldown in vacuum resulting in an In-rich surface is shown to preserve the nanostructure shape. Optimized sample cooling under As overpressure has been shown to trigger the ripening of MBE-grown QDashes into QDs of low areal density [55]. Our data [49] show that the morphologies on N surfaces with and without 10 s annealing at growth temperature after the InAs deposition are different: the 2D–3D transition happens during the annealing. On the unannealed surface, shallow islands with the shape and arrangement of the QWrs are visible. The capping of InAs nanostructures by InP at lower temperatures may be necessary to reduce their decomposition by AsPX. The (fast) cooldown in vacuum instead of As overpressure results in better-defined nanostructures. The shape, areal density, and arrangement, however, do not change significantly. A slow cooldown under As overpressure (temperature drop of 30°C during 260 s) transforms the small nanostructures into larger, higher ones of smaller areal density and regular shape. This process can be seen as a step of the ripening described in Ref. [55]. This ripening takes place only at lower temperatures. It suggests the As content of the surface and/or the reconstruction to be the decisive factor, and not the temperature. Indeed, Ref. [55] describes the disappearance of the (2×4) reconstruction during the cooling. This reconstruction has been suggested to favor more anisotropic nanostructures [56], which is in agreement with the argument of a lower energy for A steps (steps along the $[-110]$ direction) of Ref. [53]. It thus follows that the ripening is not simply kinetically enhanced by high temperature and consequently surface diffusivity but it is rather due to the thermodynamics of what nanostructure shape is energetically favorable at different surface reconstructions. Finally, this argument would also explain the elongated dots on the B surface after a longer annealing at growth temperature with a (2×4) reconstruction.

Thus, the annealing at growth temperature (450°C) for 10 s is necessary for the 2D–3D transition that forms the InAs nanostructures through surface In diffusion. A slow cooling under As overpressure leads to a coalescence of the nanostructures (ripening) whereas a longer annealing at growth temperature under As overpressure does not change the nanostructures significantly. The ripening is tentatively attributed to changed surface reconstruction at lower temperatures under As overpressure which makes the narrow long wires energetically less favorable. Rapid cooling in vacuum preserves the shape of surface InAs nanostructures.

2.5 Growth of InAs/InP Quantum Dots

Let us consider the height and areal density (D_a) of QDs grown on B surfaces for a fixed amount of deposited InAs (4 ML) under different growth conditions. It is of interest to establish whether ripening or PAsX process determines the final QD D_a and height. It has been demonstrated [49] that a low AsH_3 flux results in a larger D_a value for lower QDs in comparison to the QDs grown at conventional conditions. Higher AsH_3 flux or growth temperature T_g or a slow cooling with a AsH_3 flux, on the other hand, leads to the formation of higher QDs with a smaller D_a. The increased height, however, is not compensated by a lower areal density which results in a larger total InAs volume in the QDs. Assuming the same wetting layer (WL) thickness for all samples, this larger volume strongly suggests the incorporation of excess InAs formed by P–As exchange. Also, excess InAs is incorporated into the tall QDs with a large D_a value formed at a lower growth rate and a longer annealing time. The increased dot height observed at varied growth conditions is primarily an effect of excess InAs formed by PAsX accompanied by the strain-driven surface transport of In to the QDs. Thus, on B surfaces, a deviation from the conventional growth conditions to higher growth temperature, higher AsH_3 flux, a slow cooling under As overpressure, or low growth rate increases the QD height. Excess InAs formed due to P–As exchange supplies the material for the height increase of the QDs.

2.6 As–P Exchange Process During Capping

Capping of the grown InAs nanostructures is a necessary step for most of the applications that involve their optical and transport properties. However, during the capping of InAs nanostructures with InP, the exchange of As by impinging P (AsPX) transforms parts of the InAs into InP. A net surface indium transport actively disassembles the nanostructures and forms InP between the nanostructures at areas with a lattice constant close to that of relaxed InP as described in Ref. [24]. This strain-driven net In transport makes the AsPX faster on nanostructures than on smooth InAs quantum wells. In our experiments [49], we observed the As–P exchange in situ with RHEED through disappearance of wedges in the RHEED image upon annealing of 3D InAs nanostructures under a P flux at growth temperature. Furthermore, InAs nanostructures were capped with InP at different growth temperatures in order to control the AsPX. In the reference sample, the growth temperature was not changed for the capping. The temperature reduction, performed in the other samples, by approximately 60°C after the formation of the nanostructures (at typically 450°C) is, however, prone to change the nanostructure morphology due to the In surface migration. In order to suppress this reorganization of the surface, a rapid temperature reduction was performed before capping. A 5 nm InP cap was grown immediately after the temperature reduction at 390°C to protect the InAs nanostructures, followed by a substrate heating to the original growth temperature and growth of the remaining cap at 450°C. Comparing the results of AFM, TEM, and PL studies for various capping regime, we conclude that much faster AsPX process takes place with QDs on B surfaces than with QWrs on N or A surfaces. This leads to a height of QDs capped at the InAs growth temperature close to the height of the QWrs. Overgrown QWrs and QDs at nominal growth conditions are 2.4 and 3.0 nm high. The faster As–P exchange on B surfaces coincides with the faster P–As exchange.

Typically, the QDs have an areal density of 5×10^{10} cm^{-2} and a height of 5 nm. After capping, their height is reduced to 3 nm, and their base length is approximately 15 nm. The QDs are likely to have an anisotropic base with the longer axis in the [−110] direction. This height reduction is a result of the As–P exchange process during capping of the InAs with InP.

2.7 Double-Cap Technique of Growth of InAs/InP Quantum Dots

For the InAs/InP QDs grown via the SK growth mode, it is difficult to control the size, density, and position of the QDs. In order to reduce the QD height dispersion and, hence, to decrease drastically the PL linewidth, the double-cap procedure can be used [18, 57–61]. The double-cap procedure includes two steps that are separated by growth interruption under a phosphorus flux. During first step, InAs QDs are obtained in the SK growth mode on the substrate with further capping them with a thin InP layer. While the elastic relaxation of the tallest InAs QDs makes them energetically unfavorable for the planar growth of the InP layer, this layer accumulates predominantly between the islands. The thickness of this first InP layer is chosen to be smaller than the height of the tallest QDs. Thus, the tops of tall dots are sticking out above the first cap layer. During the growth interruption, As–P exchange takes place. This exchange plays a vital role in the double-cap technique. Due to the AsPX process the tops of QDs become truncated prior to the growth of the remaining cap. The second step of the procedure is the growth of the final barrier layer. As a result the height of the tallest capped islands is determined to a great extent by the thickness of the first capping layer and consequently the QD height distribution should be substantially reduced. This allows a control of the InAs/InP QDs emission wavelength and its tuning to 1.55 μm. TEM and PL experiments show that the InAs/InP QD structure is still present after growth interruption under phosphorus flux and that the PL linewidth at 1.55 μm is reduced from 120 to 50 meV due to the double-cap procedure [18].

The QD density appears approximately similar under double-cap procedure and conventional capping without growth interruption [18]. The TEM measurements indicate that the double-cap procedure primarily results in a QD height reduction whereas the lateral QD dimensions are obviously only slightly changed. In contrast, the island height changes drastically from 40 Å to around 25 Å. To improve the density and positioning of the QDs, a selective area growth combined with the double-cap technique can be used [59]. In the metal–organic vapor phase epitaxy selective area growth, the substrate is covered by SiO_2 mask and the growth of the semiconductor material is performed selectively in the areas free of the SiO_2. In these areas the growth precursors appear which cause lateral vapor diffusion and surface migration. Due to these diffusion mechanisms, the layer thickness and composition will be fitted to the geometry of the SiO_2 masks. The supplied material diffuse to the areas uncovered with mask from the wide mask. The QD size in each area changes following the amount of material supplied. Thus, using both the double-cap and the selective area growth techniques, it is possible to obtain uniform height QDs in patterned waveguides emitting in wide spectral range. Recently, it is realized 120 nm emission wavelength range in 16 array, five layer InAs QD waveguides [59].

Concluding this section, we established that the surface steps on vicinal InP(001) surfaces play a decisive role in the self-organized formation, shape, and alignment of InAs nanostructures during MBE growth. The tendency of the QDs to align along the steps might be exploited for realization of QD chains by optimizing the growth conditions and varying the off-cut angle. Thus, in contrast to the nominally oriented substrates, vicinal substrates provide a rather growth-condition-independent template for the robust realization of defined QDs or QWrs. In addition to the deposited InAs, a 2 ML thick InAs layer forms due to the P–As exchange process. On substrates off-cut to [−110] this process seems not to be limited to 2 ML due to strain-induced surface transport of In. As a result, the volume and height of QDs grown with the same amount of deposited InAs depend strongly on the impact of growth conditions on the P–As exchange. Thus, areal densities between 1×10^{10} and 8×10^{10} cm^{-2} and average QD heights from 3 to 11 nm are found. The QWrs are closely spaced with a lateral period of 20 nm, and a height after overgrowth of typically 2.4 nm. Their length is varying with typical lengths of 200–600 nm. (Some QWrs are longer than 1 μm.) These QWrs are grown on nominally oriented substrates and on substrates off-cut toward [110]. The formation is related to the (2 $\times$ 4) surface reconstruction. The QDs are generally higher than QWrs that are grown on a different substrate orientation under the same growth conditions. Typically, the QDs have an areal density of 5×10^{10}cm^{-2} and a height of 5 nm. After capping, their height is reduced to 3 nm, and their base length is approximately 15 nm. The QDs are likely to have an anisotropic base with the longer axis in the [−110] direction. The height reduction is a result of the As–P exchange process during capping of the InAs with InP.

3 Optical Properties of InAs/InP Nanostructures

3.1 Photoluminescence and Absorbance of InAs/InP Quantum Dot Samples

Currently, the largest application potential of InAs nanostructures grown on InP substrates is considered to be optical devices – mainly for the telecommunication window of 1.55 μm which can easily be reached in this material system. In view of this, optical properties of InAs/InP nanostructures are of great interest and have to be thoroughly studied. Emitting systems normally exploit the high-density arrays of InAs/InP QDs. Therefore, we focus on such arrays in our further analysis [62]. The samples for this study were grown in a Riber 32P GSMBE system on semi-insulating InP(001) substrates. We used the vicinal InP(001) surface because on this vicinal surface the formation of QDs always takes place and it provides a rather growth-condition-independent template for the robust realization of defined QDs. The total absorption of the nano-size samples was significantly increased by growing the 30-stack InAs/InP QDs of identical InAs layers separated by 20-nm-thick InP spacers. The morphology of the grown samples is demonstrated in Fig. 8.4, representing an AFM image of the corresponding uncapped nanostructures (Fig. 8.4a), and the capped QDs seen in Fig. 8.4b by means of cross-sectional TEM. Typically, the QDs have an areal density of 5×10^{10} cm^{-2} and a height of 5 nm which after capping reduces to ~3 nm. As it was mentioned in Section 2 this height reduction results from the As–P exchange during capping of the InAs with InP. The QDs have an anisotropic base with the longer axis in the [−110] direction. From the AFM images (Fig. 8.4a) we derived the 43 nm [−110] × 35 nm [110] QD lateral dimensions that are typical for our samples. Similar results showing elongated InAs dots (though MOVPE grown) have been presented in Ref. [63]. These dots grown on the N-surfaces of InP possessed typically 45 nm × 35 nm dimensions. Slightly elongated shape of InAs/InP QDs was observed also for the QDs grown on the N

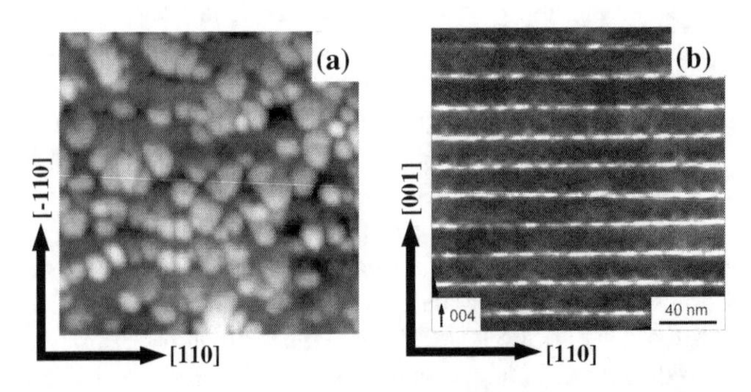

Fig. 8.4 (**a**) AFM image showing 500 × 500 nm^2 of the uncapped QDs and (**b**) TEM image of the 30-period superlattice samples of InAs/InP QDs used in the transmission and PL measurements [62]

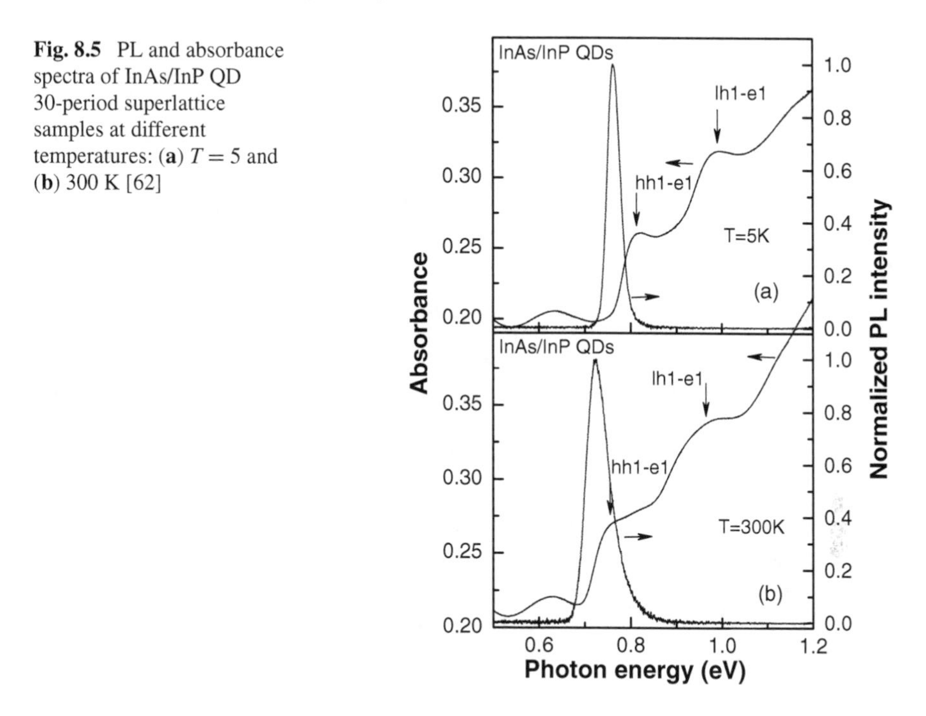

Fig. 8.5 PL and absorbance spectra of InAs/InP QD 30-period superlattice samples at different temperatures: (**a**) $T = 5$ and (**b**) 300 K [62]

substrates [64] and deduced from the PL polarization measurements for the QDs grown on the C substrate [51].

We used Fourier transform infrared spectroscopy (FTIS) and PL measurements to detect the transition energies in the InAs/InP QD systems. At low temperature we observe the excitonic states both in absorption and emission [62]. PL and absorbance spectra of the 30 period superlattice samples of the InAs/InP QDs on the B-type substrate measured at (a) $T = 5$ K and (b) $T = 300$ K are shown in Fig. 8.5. The low-temperature PL spectrum (Fig. 8.5a) can be fit with a symmetric Gaussian distribution by energies centered at 762 meV with a full width at half maximum (FWHM) $\sim$ 33 meV. This width suggests a very uniform height distribution of the nanostructures because a change in height by 1 ML would result in a peak shift of $\sim$40 meV according to our theoretical estimation. While the low-temperature PL band is clearly seen even under the lowest excitation densities used in our experiments, it is ascribed to the conduction band (e1)–heavy-hole (hh1) transition in InAs/InP QDs. Temperature increase leads to a shift of the PL maximum toward smaller energies and at $T = 300$ K (Fig. 8.5b) the maximum of emission takes place at 0.724 eV with an FWHM of $\sim$59 meV. The short-wavelength asymmetry of PL band evidences the hot carrier contribution at high temperatures.

Three distinct features at 0.636, 0.822, and 0.991 eV are clearly seen in the low-temperature absorbance shown in Fig. 8.5a. The absorbance feature at 0.822 eV definitely correlates with the PL feature at 0.762 eV and can also be ascribed to the lowest hh1–e1 absorption. The nature of two other transitions must be established. Temperature increase from $T = 5$ K up to $T = 300$ K results in a predictable shift (~50 meV) of the 0.822 eV feature toward lower energies (Fig. 8.5b) due to the temperature-induced shrinkage of the InAs band gap. The change of spectral position of the 0.636 feature within this wide temperature interval is insignificant (<5 meV) signaling that this feature cannot be related to the InAs (or InP). Similar absorbance feature in the spectral range from 0.6 to 0.8 eV was also observed in a 12-stack InAs/InP QD sample with thick 120 nm InP spacers grown on a semi-insulating InP(311) B substrate by GSMBE using the SK method [34]. This feature has been ascribed to the Fabry–Perot cavity effects (FPCE), originating from the refractive index difference between the epitaxial layers [34]. Those might be visible in the optical transparent region from 0.60 to 0.80 eV. In order to extract the FPCE contribution from our FTIS data, the propagation matrix technique for multilayer structures is used [65]. For application of this technique the thicknesses of all layers and their refractive indexes must be introduced. Following Ref. [34], the energy-dependent refractive index for InP layer [66] is taken, whereas for the refractive index of the InAs active layers, consisting of QD and wetting layers (WL), a constant refractive index of 3.5 is chosen. Variation of the InAs refractive index with energy does not lead to a noticeable change of the FPCE oscillations within the region from 0.60 to 0.70 eV. While the oscillation period of the FPCE substantially depends on the spacer thickness d_{sp},we use the d_{sp} value as a fitting parameter for the FPCE oscillations in the 0.6–0.7 eV range, and the thickness of InAs active layer is taken to be 3 nm. The results of fit are shown in Fig. 8.6 by dashed line for $d_{sp} = 22$ nm what is in good agreement with the value of 20 nm derived from the cross-sectional TEM analysis. Using the calculated FPCE structure, the corrected low-temperature absorbance data are obtained by subtracting it from direct absorbance measurements as shown in Fig. 8.6. It is seen that the band at 0.636 eV practically vanishes whereas the remaining absorbance features at 0.822 and 0.991 eV become even more pronounced. While the FPCE picture remains almost unchanged with temperature increase, the same correction of the absorbance data is performed for all temperatures.

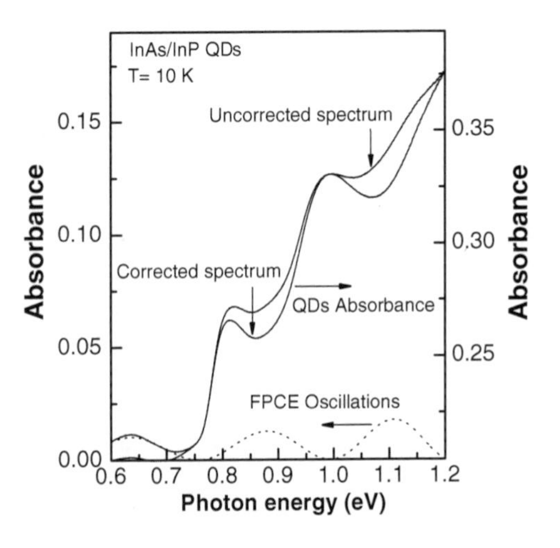

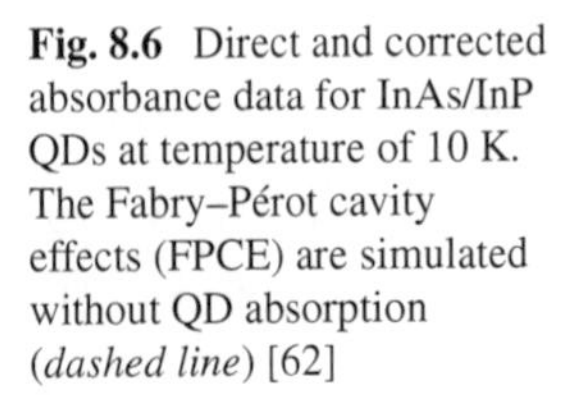
Fig. 8.6 Direct and corrected absorbance data for InAs/InP QDs at temperature of 10 K. The Fabry–Pérot cavity effects (FPCE) are simulated without QD absorption (*dashed line*) [62]

The corrected absorbance spectra $\alpha(E)d$ shown in Fig. 8.6 allow calculation of the QD absorption coefficient $\alpha(E)$ if an effective value for the total thickness d of the absorbing InAs material is known. The samples consist of 30 absorbing layers with approximate nominal InAs thickness ~3 nm. Assuming absorption only in the InAs nanostructures (and not in the InP barriers), the total thickness would be $d = 9 \times 10^{-6}$ cm. At an energy of ~0.820 eV (hh1–e1 transitions), the absorption coefficient of the QD sample is found to be approximately $\alpha = 7 \times 10^3\ \text{cm}^{-1}$. This value is higher than that ($\alpha = 4.4 \times 10^3\ \text{cm}^{-1}$) obtained for the InAs QDs on InP(311)B [34] with similar areal density ($5 \times 10^{10}\ \text{cm}^{-2}$) of QDs. A possible reason of difference is a flatter shape of our QDs resulting in a higher areal coverage.

3.2 Simulation of the InAs/InP Quantum Dots Spectra

In order to interpret the spectral features seen both in PL and absorbance spectra, one has to take into account the rather complicated structure of the InAs/InP QD system. As a result there are several sources for appearance of spectral features. Coming back to the absorbance feature at 0.991 eV seen in the low-temperature transmission measurements, various origins of its appearance might be assumed: (i) WL transitions [6, 34, 61, 67]; (ii) QD excited states transitions of InAs/InP QDs [61]; and (iii) QD light-hole states–conduction band transitions [68]. Therefore, in order to assign the energies of different transitions it is also necessary to treat the WL underneath InAs nanostructures in InP. This WL can be considered to some extent as a biaxially strained InAs quantum well (QW). Different components of strain differently influence to energy band structure. The band-edge energies of unstrained InAs are altered by hydrostatic strain, mainly affecting the conduction band, and by biaxial strain, thus lifting the heavy hole (hh) and light hole degeneracy at the Γ point in the valence band. The subband edges of WL were calculated based on the states in a finite QW (envelope function and effective mass approximation) with the band offsets taken to be 439/281 meV for heavy/light holes, and 514 meV for electrons [69]. The transition energies for InAs/InP QW are calculated for different well widths at $T = 0$ K. It is found that the energies in the region of <1 eV correspond to the thicker than 4 ML QW. Referring this finding to our absorbance data we have to conclude that the width of WL giving the absorption in the range of 0.991 eV must be greater than 4 ML that contradicts the available data: the WL thickness in the InAs/InP QD system is quantified only to $\leq$1 ML by X-ray diffraction [70], to 2 ML by PL [63], and to ~3.0 ML by PL and absorption [34, 61]. Therefore, we rule out the contribution of the WL layer states in the formation of the 0.991 eV absorption resonance in our experiments.

Before starting further calculations one has to pay attention that the TEM images (Fig. 8.4b) give the height of the overgrown nanostructures in our samples to be in the range of ~3 nm. The lateral dimensions are somewhat one order bigger, i.e., the aspect ratio (the QD height/width) is ~0.1. Besides, the capped QDs are assumed to be truncated with a flattened upper surface [71]. Therefore, we model the deformation of capped QDs on the (001) substrate surfaces assuming that they are equivalent

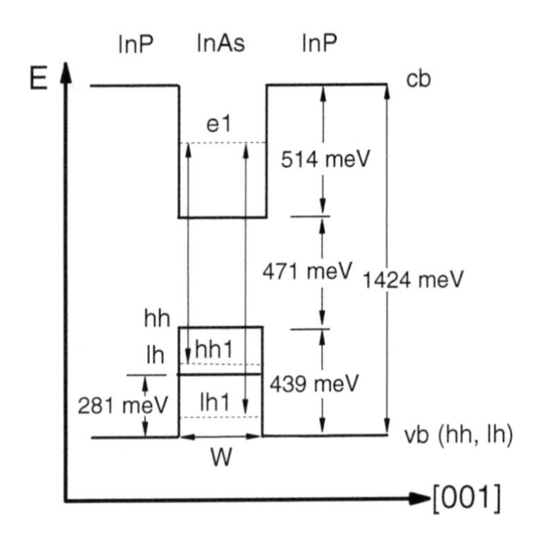

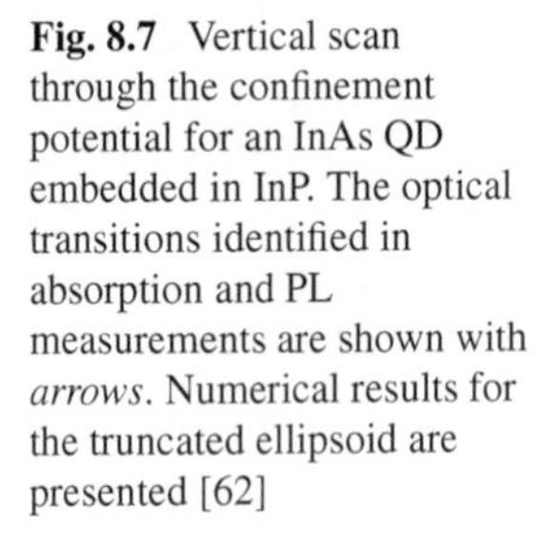
Fig. 8.7 Vertical scan through the confinement potential for an InAs QD embedded in InP. The optical transitions identified in absorption and PL measurements are shown with *arrows*. Numerical results for the truncated ellipsoid are presented [62]

to epitaxial layers with a coherent interface [71, 72]. Indeed, the confinement energy is dominated by the 3 nm dot height, whereas the confinement energy associated with the long dimensions of the dot is an order of magnitude smaller than that related to the dot height [69]. A biaxial compression in the InAs/InP system results in the tetragonal deformation of the InAs lattice. This deformation is assumed to be uniform inside the QD that is reasonable for flat and wide QDs. A one-band effective mass approximation can be used both for the conduction band and for the valence band. For flat and wide QDs, the hh and lh potentials are well separated. Therefore, one can consider each of the valence bands separately with renormalization by strain potential and effective masses. The same assumption is made with respect to the conduction band. The shape of the QD is taken to be a truncated ellipsoid [71] with a height of 3 nm and a diameter of 35 nm. To compute the electronic structure of InAs islands embedded in InP, the calculation technique described in Ref. [71] is exploited. The material parameters for InP and InAs are taken from Ref. [73]. The one required material parameter, which is not directly tabulated, is the band alignment between the InP and InAs. Since strain affects the band alignment as well, we are referring here to the contribution to the band offset which is independent of strain. We have treated the strain-independent offset as a fitting parameter and take it to be 0.4 for unstrained valence band [69]. After inclusion of strain the valence band offset increases. Figure 8.7 shows the potential profile along the growth axis in the InAs/InP QD system used in the calculations. The results of calculations in the framework of the one-band kp approximation are shown in Fig. 8.8. The energies of the ground and excited state of the exciton transitions for the hh band together with the ground-state exciton transition for the lh band are presented as a function of the QD height varying in the range from 2.5 to 4.5 nm. On the one hand this interval of the QD height variation covers the actual QD height expected in our capped system, and on the other hand in this range of QD heights the results of the one-band

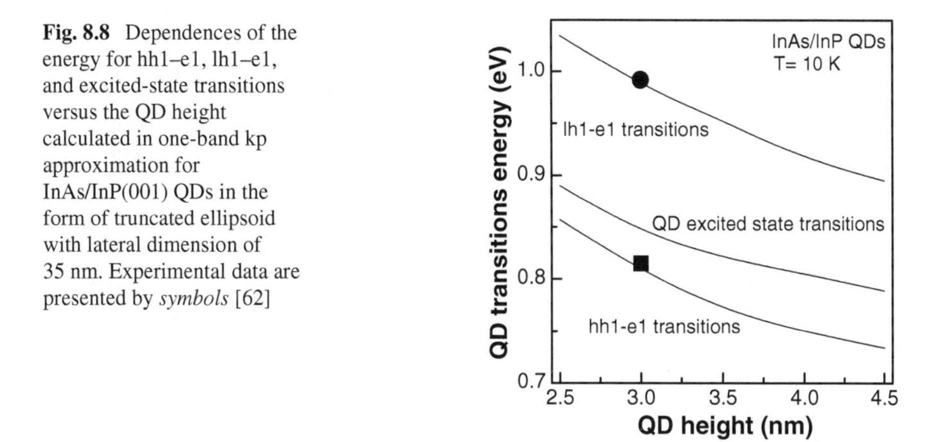

Fig. 8.8 Dependences of the energy for hh1–e1, lh1–e1, and excited-state transitions versus the QD height calculated in one-band kp approximation for InAs/InP(001) QDs in the form of truncated ellipsoid with lateral dimension of 35 nm. Experimental data are presented by *symbols* [62]

and eight-band kp calculations, as it has been shown in Ref. [6], do not differ significantly. The ground-state transition energy grows by ~120 meV under reduction of the QD height from 4.5 nm down to 2.5 nm for the QD shape like a truncated ellipsoid. It is also seen from Fig. 8.8 that the energy of the excited-state transition is closer to the energy of the ground-state transition for flatter QD (smaller QD height) compared to the higher QD. This is explained by the different aspect ratio of considered QDs [6]. In the case of flat QDs, the weak lateral confinement due to large lateral dimensions influences predominantly the energy of the p-like (excited) QD states, whereas the strong vertical confinement (in growth direction) mainly impacts the energy of the s-like ground QD state. Therefore, the energies of ground- and excited-state transitions go closer for flatter QDs. What is important is that the energy distance between ground and excited states does not exceed 70 meV for the whole range of the QD height values (from 2.5 to 4.5 nm). For the same range of QD heights, the energy of the lh1–e1 transition varies from 893 meV at the height of 4.5 nm to 1034 meV at the height of 2.5 nm. The energies of two absorption resonances found from the transmission measurements are plotted in Fig. 8.8 as well. Comparing these experimental energies with the calculated dependence, we find a remarkable fit for the QD height of 3 nm in case of hh1–e1 and lh1–e1 transitions. Concerning other transitions involving excited states in the InAs/InP QD system, they have been found fairly weak [74] due to the similar symmetry of the hole and electron wavefunctions for the relevant transitions. Only at an energy of about 80 meV above the ground-state transition, a significant absorption probability caused by the different symmetry of the electron and hole wavefunctions involved in the transitions has been calculated. However, this spectral range is still far below the vicinity of the 0.991 eV absorption, which is remote by ~170 meV from the energy of the ground-state transition in our experiment. Similar calculations for the InAs quantum dashes grown on the InP(001) substrate also give numerous transitions, but many of them vanish because of a slight overlap of the wavefunctions [75]. The energies of the four remaining transitions possessing the highest optical dipole moments in case of the highly nonsymmetrical dashes cover only the energy region of about 100 meV above the energy of ground-state transition. Basing on these estimations and the results of our calculations we rule out the contribution of the QD excited states in the formation of the 0.991 eV resonance in absorbance. Thus, the energy scheme and the optical transitions shown in Fig. 8.7 are the most plausible for the InAs/InP QD system under investigation. The calculated ground-state energies for the electrons and heavy holes are found to be 1.214 and 0.392 eV, respectively, where the zero is taken to be the top of the InP valence band, as in Fig. 8.7. These values correspond to binding energies of 0.210 eV for electrons and 0.392 eV for heavy holes, which will be compared with the experimental values derived from the temperature dependences of QD absorbance and PL.

3.3 Temperature Effects in Photoluminescence and Transmission of the InAs/InP Quantum Dot Samples

In order to uncover the nature of the transitions seen either in PL or in the absorption, thorough temperature measurements have to be carried out in the range from 5 K up to room temperature. Figure 8.5 compares the PL and absorbance measured at low temperature $T = 5$ K (Fig. 8.5a) and at high temperature $T = 300$ K (Fig. 8.5b). Carefully analyzing each spectral dependence of absorbance at a given temperature, one can notice that two spectral features at hh and lh absorption edges possess well-pronounced maximums. These maximums are accompanied by a flat part of the spectral dependence on the short wavelength side of the features. The length of the flat part increases with increasing temperature, transforming finally into the well-pronounced spectral step typical for the density of states of 2D electron system. Taking the minimum on the right side of the absorbance feature (or the flat part of the spectrum if this part is already clearly developed) as an origin we can consider the left-side part of the absorbance towering over the local origin as a manifestation of the excitonic enhancement in the QD absorption spectrum. Subtracting the contribution of absorption step from the left side of the absorbance we get pure excitonic features. Figure 8.9 demonstrates the behavior of these excitonic features seen in the absorbance (Fig. 8.5) at the hh (Fig. 8.9a) and lh (Fig. 8.9b) band edges versus temperature. The excitonic resonance seen at 0.822 eV (hh exciton) survives with temperature increasing up to 170 K, whereas the resonance at 0.994 eV (lh exciton) decays already at $\sim$120 K. This is reasonable while the hh exciton is more strongly bound than the lh exciton due to its larger mass. What is important is immediately after the exciton decay the absorbance demonstrates an ideal step-like shape typical for a 2D electron gas density of states. This finding allows the model of studied QDs to be the 2D quantum well with excitonic coupling. Temperature behavior of the PL line shape also corroborates this model. The variation of the integrated PL (Fig. 8.10a) with the inverse temperature demonstrates two linear parts. One of them allows determination of activation energy E_a^1, $\sim$12 meV, whereas another one gives the E_a^2 value of $\sim$200 meV. Following the behavior of the hh excitonic feature shown in Fig. 8.9a, we conclude that E_a^1 corresponds to the hh exciton binding

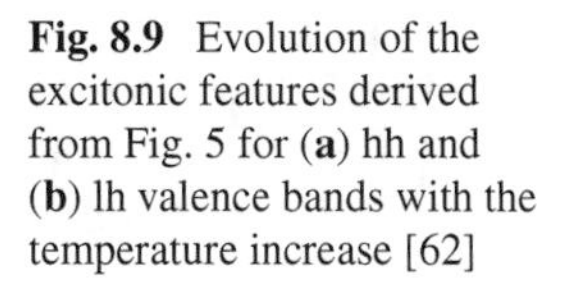

Fig. 8.9 Evolution of the excitonic features derived from Fig. 5 for (**a**) hh and (**b**) lh valence bands with the temperature increase [62]

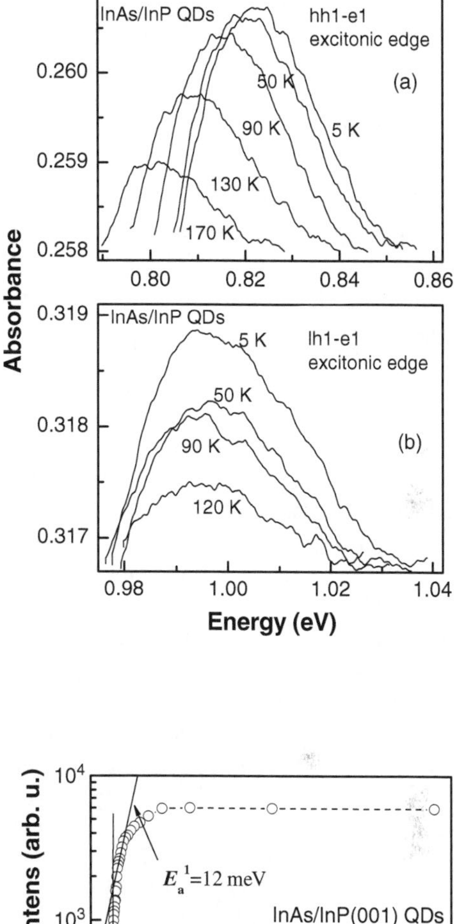

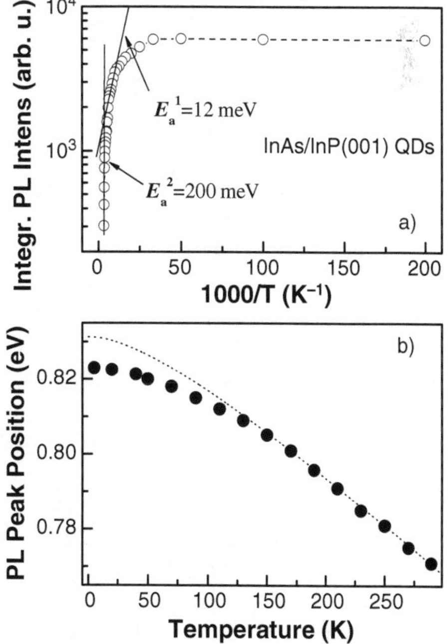

Fig. 8.10 (**a**) Integrated PL intensity in the InAs/InP QDs for different temperatures. Two activation energies of $E_a^1 = 12$ meV and $E_a^2 = 200$ meV are determined. (**b**) Variation of the InAs energy gap with temperature. Varshni law dependence is plotted by *dashed line*, whereas experimental data derived from the absorbance measurements are shown by *dots* [62]

energy. At temperature ~150 K, the excitonic coupling for the electron–hole pair in a QD decays and the electron and the hole relax independently. In this case we observed significant high-energy asymmetry of the PL band (Fig. 8.5b) which is attributed to the hot carrier thermalization. Assuming a Maxwell–Boltzmann distribution for the hot carriers the high-energy tail of the free carrier band-to-band transitions can be fit quite well. Figure 8.10b shows the variation of the energy gap in the InAs/InP QD system (dots) derived from the temperature dependence of the absorbance. The temperature dependence of the transition energies is mainly determined by the decrease of the band gap E_g (dashed line) with increasing temperature T, which is often described by the empirical Varshni formula [76]:

$$E_g(T) = E_g(T = 0) - \alpha T^2/(T + \beta), \tag{8.1}$$

with the Varshni parameters α and β. The Varshni parameters $\alpha = 0.276$ meV/K and $\beta = 93$ K for InAs [76] correspond to a band-gap reduction of 63 meV for a temperature increase from 0 to 300 K. Least square fit of Eq. (8.1) to the experimental data gives the same set of Varshni parameters for InAs as cited in Ref. [76]. Comparing the calculated dependence $E_g(T)$ with the experimental data we observe excellent fit in the high temperature range, and a pronounced deviation in the low temperature region. The latter can be related to the excitonic effects in QDs and through the maximal deviation of the experimental and calculated data at $T = 5$ K we found the exciton binding energy to be of ~10 meV, similar to the binding energy derived from the temperature dependence of the integrated PL intensity.

As to the lh exciton, the corresponding spectral feature in the absorption spectrum (Fig. 8.6b) is significantly broader (~38 meV at $T = 5$ K) than that for the hh resonance (~26 meV at $T = 5$ K) reflecting the fact of smaller localization of the lh wavefunction in the QD potential well and its spillover into the InP barrier. The temperature increase up to 120 K leads rather to smearing the absorption peak than to its red shift contrary to the hh exciton. As a result only small resulting shift (~3 meV) with a significant enhancement of the low-energy side of the lh absorption band is detected whereas for the hh exciton this shift is about 12 meV.

We find a Stokes shift between the hh excitonic resonance in the absorbance at 822 meV and the QD PL at 762 meV about 60 meV. The origin of this shift can be in the statistical fluctuations of the fundamental band edge due to the alloying, interdiffusion, and different well widths experienced by the excitons [77–80]. This results in the excitonic absorption resonance blue shift as compared to the maximum of the luminescence line. For the case of ultrathin QWs, a well width fluctuation of 1 ML is expected to have a strong influence on the quantized energy levels of the carriers. Within the envelope function model, the total band edge fluctuation corresponds to a maximum Stokes shift of about 55 meV. However, a pseudo-smooth can also result in a smaller Stokes shift as it has been found in (100) InAs single ML QWs grown in bulk-like GaAs [80]. Thus, the Stokes shift can be explained in terms of an ML fluctuation of the well with a lateral extent of the islands less than the exciton radius. One sees that the Stokes shift decreases as temperature increases as a result of ionization of bound excitons and the PL peak approaches the corresponding transmission edge at high temperatures.

3.4 Photoluminescence and Absorbance in InAs/InP(001) Quantum Well and Quantum Wire Systems

InAs/InP QWs and QWrs for this study were grown in a Riber 32P GSMBE system on semi-insulating nominally (001) oriented InP substrates. Figure 8.11 shows AFM images of two samples: (a) InAs/InP QW with 4 ML of deposited InAs and (b) InAs/InP QWrs obtained in the SK growth mode with the same InAs coverage. In order to get the QW sample (Fig. 8.11a), the low substrate temperature of 380°C (to reduce ad-atom diffusivity D_s), high growth rate, and no annealing (to reduce ad-atom diffusion time τ_s) were used, which suppress the SK growth mode. The AsH_3 flux was chosen as low as possible to still have an As-rich (2×4) reconstructed surface. In this case (Fig. 8.11a), the InAs surface is quite smooth without InAs nanostructures. Monolayer terraces can be seen. QWr formation (Fig. 8.11b) occurred under the optimal InAs deposition conditions given by a substrate temperature of 450°C, a deposition rate of 0.4 ML/s, and an AsH_3 flux of 0.6 sccm. In this case the AFM picture shows long, flat, and closely packed QWrs (0.6 to 1.2 nm high), oriented along the $[-110]$ direction. In order to increase the total optical absorption coefficient, samples with 30 periods of identical InAs layers separated

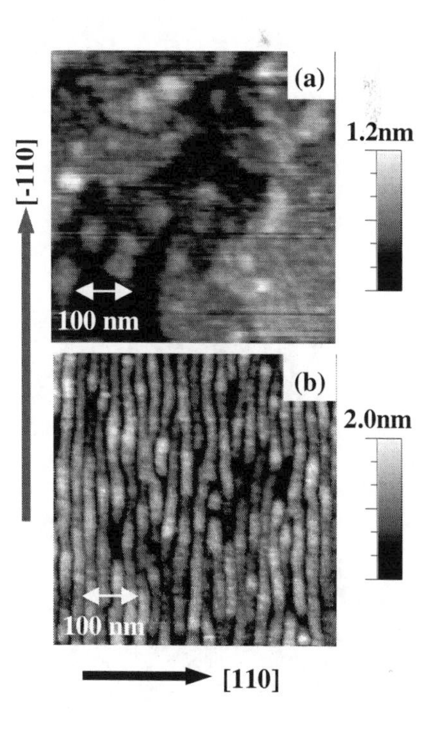

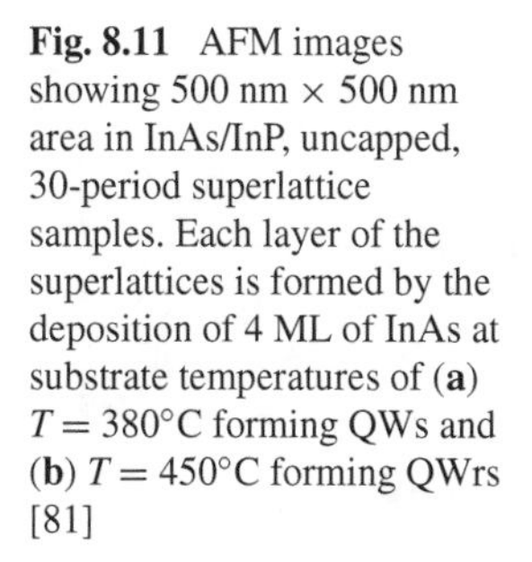

Fig. 8.11 AFM images showing 500 nm × 500 nm area in InAs/InP, uncapped, 30-period superlattice samples. Each layer of the superlattices is formed by the deposition of 4 ML of InAs at substrate temperatures of (**a**) $T = 380°C$ forming QWs and (**b**) $T = 450°C$ forming QWrs [81]

by 20 nm thick InP spacers were grown. The InP spacers between the InAs layers are thick enough to suppress any influence caused by the strain fields of the previous InAs layer. There is no vertical coupling between the QWrs of adjacent layers; thus, only lateral, interwire coupling due to the small interwire distances is expected in the dense system under investigation. For the optical studies the structures were capped with 43 nm of InP.

FTIS measurements were performed with a Bruker IFS 113 IR spectrometer supplied with a liquid-nitrogen-cooled InSb photodetector. The incident intensity was approximately 16 mW cm^{-2} giving a maximum photogenerated electron–hole pair density of 1.25×10^8 cm^{-2}, which ensures the linear absorption regime for the QWrs. The PL measurements were performed in a variable temperature 8–300 K closed-cycle helium cryostat. The 532 nm line from a doubled Nd:YAG laser was used for continuous-wave PL excitation. The laser spot diameter was $\sim$30 μm and the typical optical excitation power was $\sim$2 mW. The PL signal from the sample was dispersed by a monochromator and detected by a liquid-nitrogen-cooled InGaAs photodiode detector array.

To reveal the optically active states and DOS in InAs/InP(001) QWr system and parent QW system shown in Fig. 8.11, PL and FTIS measurements were carried out [81]. We observe a strongly analogous behavior of spectral features developed both in the PL and absorption spectra of QWr and QW nanostructures. Figure 8.12 presents low-temperature PL spectra and absorbance spectra derived

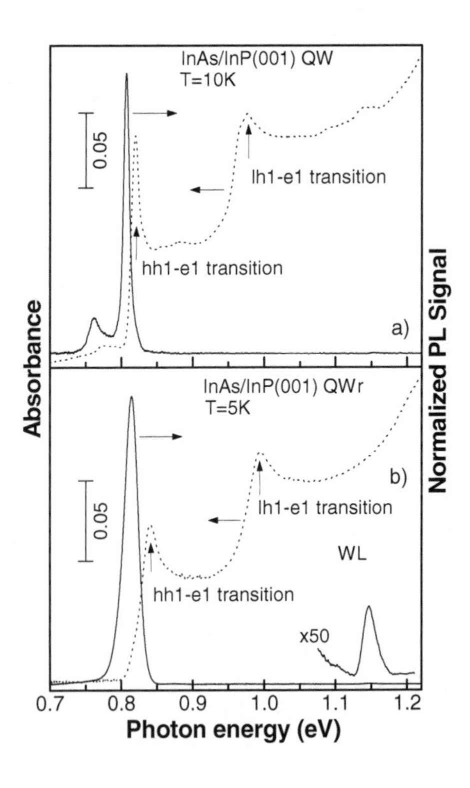

Fig. 8.12 Low-temperature PL and corrected absorbance spectra for InAs/InP(001) (**a**) QWs and (**b**) QWrs. Magnified part of the PL spectrum in (**b**) is ascribed to the WL emission [81]

from the IR transmittance and corrected by subtraction of the FPCE oscillations in the InAs/InP(001) QW structure (Fig. 8.12a) and the similar spectra for the InAs/InP(001) QWr system (Fig. 8.12b). In the PL spectrum of the QW structure, strong emission is observed at ∼0.807 eV accompanied by the low-energy PL band at ∼0.762 eV. The FWHM of ∼9.76 meV for a strong PL feature suggests its excitonic character. The low-energy band is significantly less intensive with an FWHM of ∼18.5 meV. We ascribe the excitonic emission to the e1–hh1 QW transition, whereas the low-energy emission can be ascribed to the InAs/InP QD transition.

Indeed, it is shown in Section 2 that InAs deposition on the nominal surface of InP(001) can result in the appearance of QWs, QWrs, and QDs depending on the growth conditions. The InAs/InP QDs studied above emit exactly at this spectral position ∼0.762 eV. Therefore, we consider a coexistence of vast patches of 2D character and residual 3D islands in QW structure as a possible origin for the appearance of the low-energy PL band in the vicinity of ∼0.762 eV.

The absorbance feature at 0.819 eV (Fig. 8.12a) correlates with the PL feature at 0.807 eV and can be ascribed to the lowest hh1–e1 absorption. The difference of 12 meV between the PL band maximum position and the corresponding absorbance peak gives the value of the Stokes shift for the excitonic transition. The shoulder seen in the absorbance at the 0.778 eV is ascribed to the QD excitation. The value of Stokes shift for QD transition is ∼16 meV.

The absorbance feature at 0.975 eV is broad and we consider it as a compound band. The energy difference between the two main absorbance peaks shown in Fig. 8.12a is ∼156 meV. We ascribe this significantly large energy difference to the valence band states that form due to the strong vertical confinement and strain in the QW. The possible nature of the higher energy state visible in the absorbance spectrum may be attributed to the excited hh2 state or the ground state of the light hole (lh1). While the hh2–e1 transition is parity forbidden for the QW, we might expect to see the lh1–e1 transition. The latter transition is not forbidden by parity but is a factor of 3 weaker than the hh1–e1 transition. At the same time, it should be noted that the selection rule prohibiting the hh2–e1 transition might be strongly relaxed in a real system. Therefore, in order to identify the detected hole state we have calculated the energy splitting for a strained InAs QW embedded in InP in the envelope function approximation developed by Bastard [82]. For this calculation we include in the initial Hamiltonian the perturbations due to strain [83] along with the perturbation due to band coupling. For the calculation of the subband edges in a finite QW we take the band offsets to be 439/281 meV for heavy/light holes, and 514 meV for electrons [69]. A solution of the Schrödinger equation gives the confined levels in the conduction, light-hole, and heavy-hole bands. For a QW width of 1.2 nm, as in our experiment, we calculate a hh–lh splitting equal to 157 meV which is in good agreement with the experimentally observed value of 156 meV. Also, for such a narrow QW only one confined level in the hh potential well exists. Therefore, the excited hh2 state has to be withdrawn from further consideration. Based on this analysis, the high-energy feature in the QW absorbance is attributed to the lh1–e1 transition.

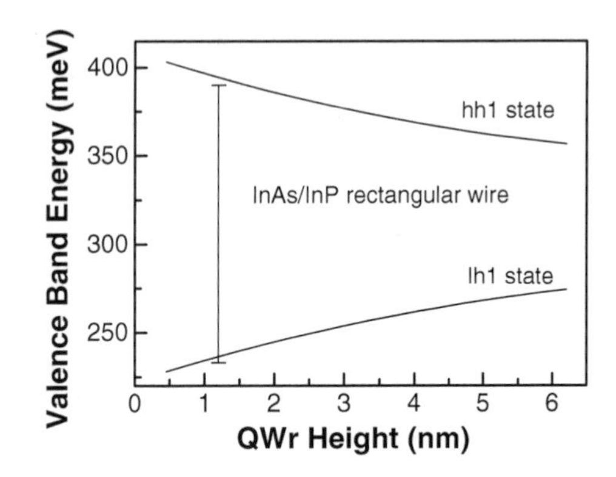

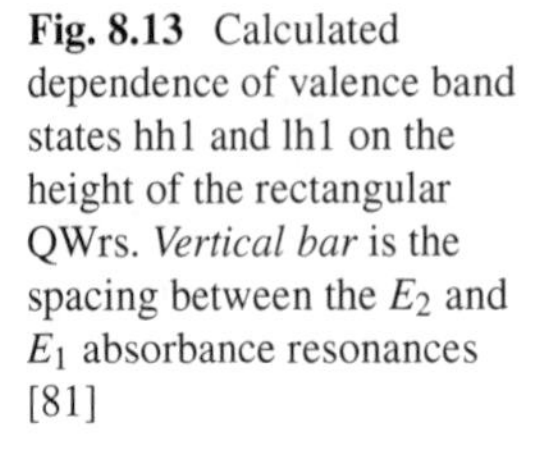
Fig. 8.13 Calculated dependence of valence band states hh1 and lh1 on the height of the rectangular QWrs. *Vertical bar* is the spacing between the E_2 and E_1 absorbance resonances [81]

Figure 8.12b represents the low temperature ($T = 5$ K) PL and absorbance spectra measured for the InAs/InP(001) QWrs system grown with the same InAs coverage (4 ML) as the QW heterostructure but with different growth conditions (see Fig. 8.11b). The PL band is centered at 0.814 eV with an FWHM of ~23 meV. The low energy broadening can be related to a distribution of QWr sizes. Low-intensity emission is also observed at ~1.146 eV. The absorbance spectra also develop two main features at $E_1 \sim 0.840$ and $E_2 \sim 0.994$ eV. The E_1 feature is ascribed to the hh1–e1 transition in InAs/InP QWr and the difference between the energy position of E_1 and the position of the PL maximum (~26 meV) represents a Stokes shift for the exciton in the QWr. The E_2 feature is shifted to the blue from E_1 by 154 meV. Following the analysis given above for the QW system, one may assume that the E_2 resonance can be related to the lh1–e1 transition in QWrs or alternatively to the WL absorption. In order to find the energy position of both the hh and lh states we calculate the energy spectrum of 2D rectangular quantum boxes with height h and width w [84]. We assume that the conduction band minimum and the valence band maximum occur at the Γ point of the wire, and that the conduction band and the valence band are decoupled. However, hh–lh mixing is taken into account. The hh and lh confinement potentials are obtained from the Pikus–Bir strain Hamiltonian by its value in the center of the wire [83]. The results of this calculation for the valence band states are presented in Fig. 8.13. Here, we find that the experimentally observed spacing between two absorbance features in the case of 1.2 nm high QWrs (vertical bar) is best fit by the calculated spacing for the same height of the rectangular QWrs, 154 meV experimental versus 157 meV theoretical. Therefore, we ascribed the E_2 resonance to the lh1–e1 transition in InAs/InP QWrs. Thus the structure of the absorbance spectra both of QW and QWr samples is very similar. This similarity becomes even more pronounced in the temperature-dependent behavior of the distinguished spectral features.

The change of the PL line shape under temperature variation in the QW and QWr samples is shown in Fig. 8.14. In general, the PL spectra shift to the red with increasing temperature. Partly, this can be assigned to the reduction of the InAs band gap and thermally induced carrier redistribution between nanostructures of different

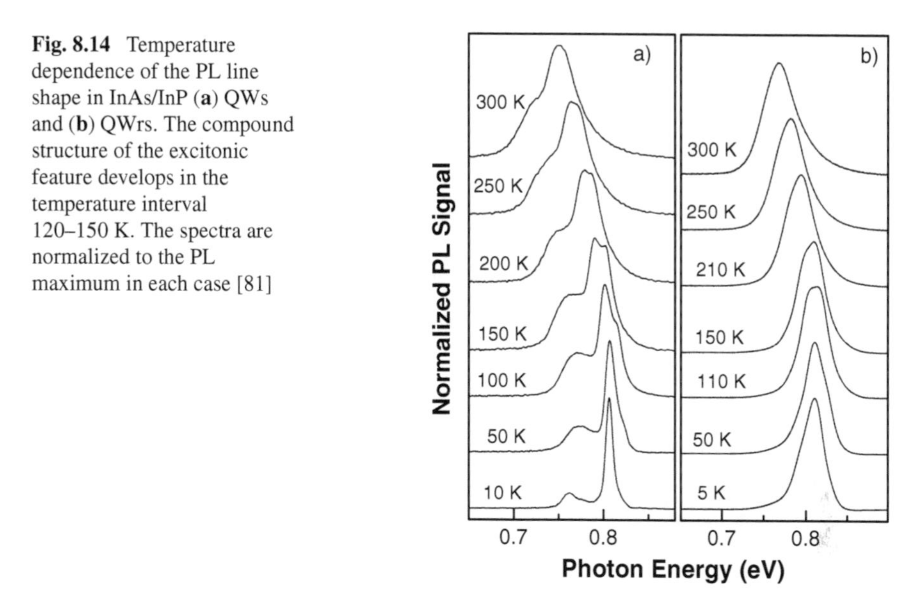

Fig. 8.14 Temperature dependence of the PL line shape in InAs/InP (**a**) QWs and (**b**) QWrs. The compound structure of the excitonic feature develops in the temperature interval 120–150 K. The spectra are normalized to the PL maximum in each case [81]

sizes. The shift amounts to ~57 meV for QW sample and ~42 meV for QWr sample if the temperature varies from 5 to 300 K. The excitonic bands both in the QW and QWr PL spectra consist of a set of peaks and at least two contributions separated by ~10 meV can be easily distinguished beginning in the range of temperatures from 120 to 150 K.

In order to quantify the thermally induced modification of the PL line shape in the QW sample (Fig. 8.14a) one has to take into account a coexistence of the QD and QW contributions in this sample: the low-temperature PL band at ~0.762 eV ascribed to the residual QDs overlaps with a significantly stronger PL band at ~0.807 eV ascribed to the excitonic transitions in the QW. These two contributions were distinguished using a two-Gaussian fit in the QW PL spectrum and Fig. 8.15a represents the temperature-assisted change of the energy and the FWHM of the QW excitonic PL. It is seen that the energy of PL maximum $E_{\max}(T)$ jumps onto another branch of temperature dependence at the temperature ~150 K. This abrupt blue shift is caused by the thermal activation of the excitonic ground states of higher energies. Because the states of higher energies decay with elevated temperature quicker, further elevation of the temperature leads to a return of the PL to the lower branch of the $E_{\max}(T)$ dependence. This low-energy branch follows the Varshni law for bulk InAs at higher temperatures [76]. As can be seen from Fig. 8.15 the temperature dependence of the FWHM mirrors the peculiar variation of the $E_{\max}(T)$. A plateau in the FWHM(T) correlates with the range of $E_{\max}(T)$ switching from the low-energy to the high-energy branches. Similar temperature-dependent processes take place for the QD PL band in the QW sample.

The $E_{\max}(T)$ and FWHM(T) dependences for the QWr sample are shown in Fig. 8.15b. The switching of the $E_{\max}(T)$ dependence from the low-energy branch

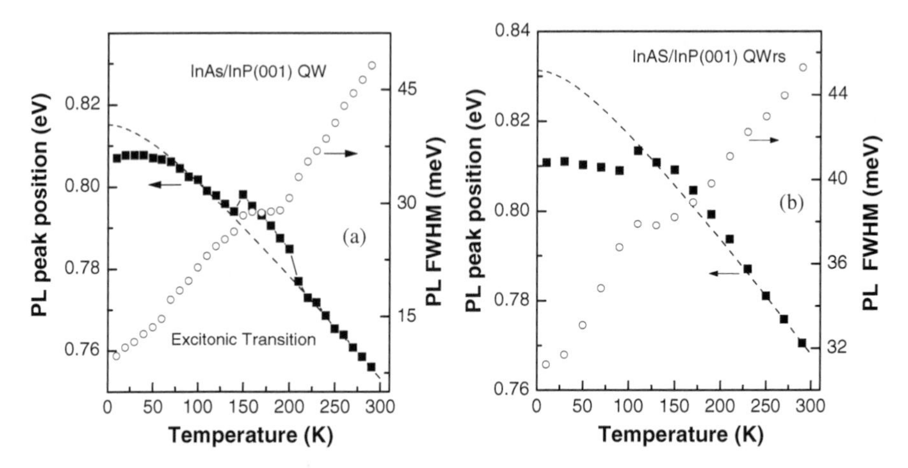

Fig. 8.15 Temperature dependence of the PL peak position (*dark squares*) and the FWHM value (*empty circles*) in the InAs/InP(001) QW sample (**a**) and QWr sample (**b**) for the excitonic transition. The variation of the InAs energy gap by Varshni law is shown with *dashed line* [81]

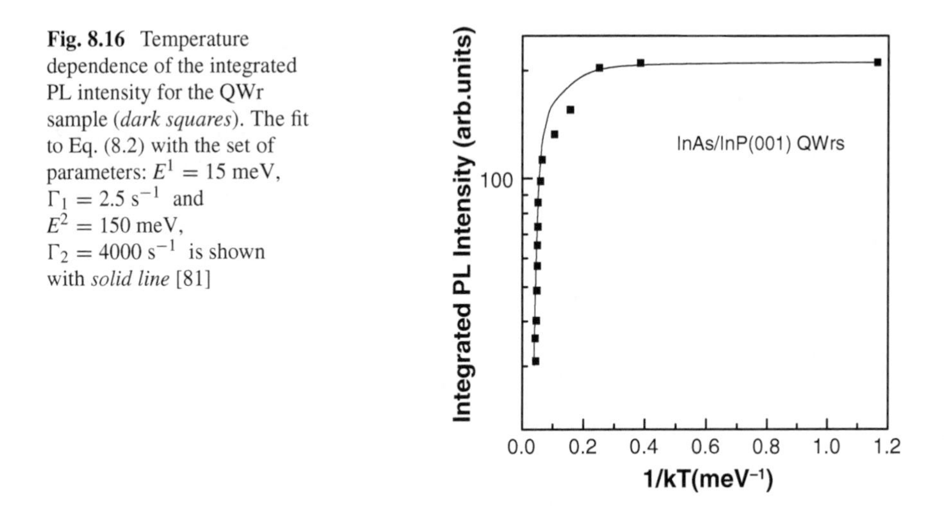

Fig. 8.16 Temperature dependence of the integrated PL intensity for the QWr sample (*dark squares*). The fit to Eq. (8.2) with the set of parameters: $E^1 = 15$ meV, $\Gamma_1 = 2.5\ \mathrm{s}^{-1}$ and $E^2 = 150$ meV, $\Gamma_2 = 4000\ \mathrm{s}^{-1}$ is shown with *solid line* [81]

onto a higher energy one with the accompanied change in the FWHM(T) is also evident here. Thus, the temperature dependences of the PL spectral characteristics reveal similar relaxation processes acting both in the QW and QWr samples. The temperature dependence of the integrated PL intensity for the QWr sample is shown in Fig. 8.16. This dependence also reflects the $E_{\max}(T)$ switching and can be fit outside this temperature range by a Boltzmann model for excitonic recombination with two quenching mechanisms [39]: (i) Carrier recombination through impurities in the QWr or InAs/InP interfaces. This process is characterized with low activation energy E^1 and determines the low temperature PL quenching. (ii) Intrinsic nonradiative recombination mechanism in self-assembled QWr related to a thermal escape of carriers into the barrier material. This process is characterized by higher activation energy E^2 and determines the high temperature PL quenching.

In total, both these mechanisms quench the integrated PL intensity $I_{PL}(T)$ by [85]

$$I_{PL}(T) = I_0/\{1 + \tau_0\,[\Gamma_1 \exp(-E^1/kT) + \Gamma_2 \exp(-E^2/kT]\}, \quad (8.2)$$

where I_0 is the integrated PL intensity at $T = 0$ K, τ_0 is the radiative recombination time for excitons in the QWr which is taken to be 1 ns, and Γ_1 and Γ_2 are two scattering rates. The best fit of the whole PL band quenching, shown in Fig. 8.16, is given by $E^1 = 15$ meV, $\Gamma_1 = 2.5\ \mathrm{s}^{-1}$ and $E^2 = 150$ meV, $\Gamma_2 = 4000\ \mathrm{s}^{-1}$.

Concerning the PL feature in the region of $E \sim 1.146$ eV seen in the PL spectrum of InAs/InP QW sample (Fig. 8.12b) we attribute it to the transitions in the InAs WL. Two factors serve in favor of the WL recombination: (i) absence of such recombination in the QW sample (see Fig. 8.12a) and (ii) the temperature dependence of this recombination. The maximum of this emission shifts to the blue by ~7 meV as the temperature increases from 10 up to 180 K. After 180 K, this emission reduces to the PL background. Temperature-induced change of the line shape substantiates the presence of two components in the formation of this PL response. We assume that these components represent the excitonic states localized due to potential relief of the WL. With increased temperature the excitons can be released from deeper potential wells and trapped by shallower potential wells, thus shifting the PL maximum position to the blue.

Thus, we observed the blue PL shift for various nanostructures and in different spectral regions. The increase of the energy of excitonic PL can also result from a combined action of the phase-space filling effects and the screening of the internal piezoelectric field by free carriers in the QWr system [84]. However, it is unreasonable to expect a strong thermally enhanced density of free electrons in our nanostructure system. Therefore, based on the development of the PL band structure at elevated temperature, we conclude that the thermally induced filling of the higher energy states realized in the size-distributed nanostructures is a dominating mechanism for the observed blue shift and switching of PL maximum energy in InAs/InP(001) samples. This conclusion is also supported by the results of the temperature studies of the absorbance spectra both for the QW and QWr samples. Indeed, Fig. 8.17 demonstrates the absorbance spectra for the InAs/InP(001) QWs (Fig. 8.17a) and QWrs (Fig. 8.17b) measured in the temperature range from 5 to 300 K. The resonance at 0.810 eV in absorbance spectra of QW is of excitonic nature. By increasing the temperature to 300 K, it shifts toward the red to 0.757 eV. It is worth noting that the development of excitons at room temperature is very specific. In this case the absorption of a photon creates an exciton which immediately scatters by a longitudinal optical phonon thereby producing a free electron–hole pair. This sequence of events causes unusual dynamics in the ultrafast nonlinear

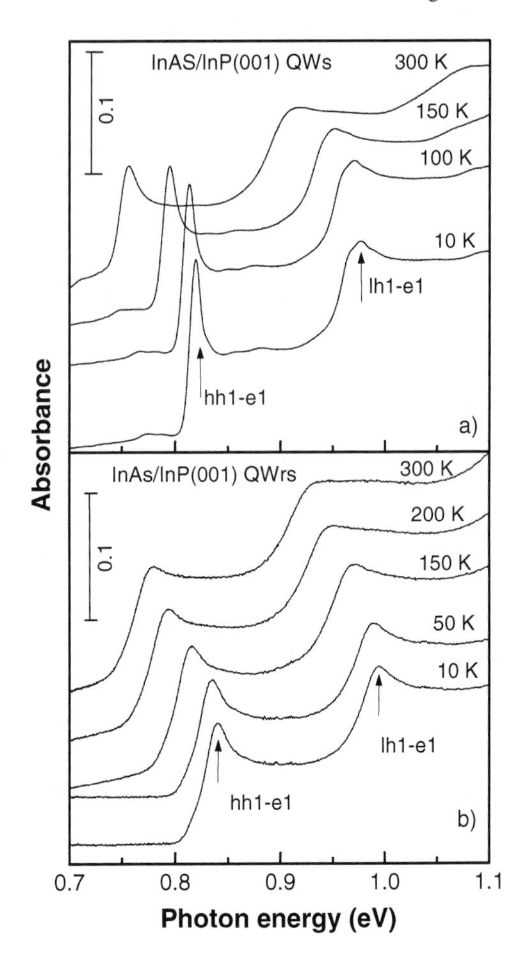

Fig. 8.17 Temperature-induced change of the absorbance for the InAs/InP(001) (**a**) QW sample and (**b**) QWr sample. *Curves* are labeled with the temperature for each [81]

response of QW structures [86]. The shift of the excitonic energy with temperature increase is monotonic in contrast with the jump observed in the $E_{\max}(T)$ dependence in QW PL spectrum. The resonance at 0.957 is a compound band and we ascribe it to the lh1–e1 exciton transition. While the exciton binding energy for the lh exciton is smaller than that for the hh exciton, the temperature increase dissolves the excitonic state and a step-like absorption, typical for 2D system, is clearly observed.

3.5 Thermally Induced Change of Dimensionality in Coupled InAs/InP Quantum Wire Nanostructures

The QWr behaves like a quasi-1D system in optical and transport phenomena characterized by a 1D band structure, 1D density of states, and reduced phase space for scattering. The DOS function develops a strong peak near the band edge for QWrs, allowing, for example, a very high gain in QWr lasers at very low injection. The narrowing of the DOS leads to a lower excitation threshold for phase-space filling in QWRs [84, 87], thus enhancing nonlinear optical effects important for applications in optical communications.

The ideal 1D DOS is given by $\rho(E) = \left[(E - E_i)/2\pi^2\hbar^2\right]^{-1/2}$ with the energy E_i being the quantization energy due to the 2D confinement in the wire. If the wire is perfect, the wavefunction along the axis of the wire (x axis) is represented by Bloch functions $\Psi_{\mathrm{ik}}(x) = U_{\mathrm{ik}} \exp(ikx)$, where U_{ik} is the band edge wavefunction for the bulk taken at the cell center. The singularity in the joint density of states at the energy of the QWr band edge is expected to appear above the band edge, in the interband absorption spectra. Therefore, it is of great importance to study the linear absorption spectra both below and above the band edge in order to reveal the characteristic features of the 1D system. In what follows, we examine the absorption spectra of the InAs/InP(001) QWrs and compare them with the 2D InAs/InP(001) quantum well system.

The DOS and the wavefunction given above are strictly valid only for a perfect wire system. However, in a 1D system, any disorder can cause localization and drastically change the density of states. Indeed, it was demonstrated experimentally that localization of the 1D exciton is the main channel of radiative processes taking place in QWRs at low temperatures [88, 89]. In order to describe the fine details of the experimental data, the surface roughness and composition fluctuations have to be taken into account [90]. The disorder-induced tail of localized states of the exciton center of mass (lying below the ground state of the ideal 1D exciton) gives rise to the observed broadening of the exciton lines. Analysis of the disorder arising from surface roughness and alloy disorder [91] allows to conclude that the changes of DOS of the 1DEG in a wire reduce to a tail of localized states far below the subband edge.

Further complication of QWr DOS structure is expected in the system of coupled QWRs, both laterally and vertically [33–39]. The strength of coupling can be varied by changing the interwire spacing or the thickness of the spacer layer separating adjacent QWr layers. According to Ref. [38] for the system of vertically coupled InAs/InP QWrs, depending on the InP spacer layer thickness d(InP), three regimes of coupling can be realized: weak electron coupling and negligible hole coupling for $d(\mathrm{InP}) > 10$ nm, intermediate electron coupling and weak hole coupling for 5 nm $\leq d(\mathrm{InP}) \leq 10$ nm, and strong electron coupling and moderate hole coupling for $d(\mathrm{InP}) < 5$ nm. The change of the coupling strength is immediately reflected in the experimental PL decay time shortening due to the exciton wavefunction becoming delocalized along the growth direction. Lateral coupling of QWrs is of importance also because the system may be seen as lying between 1D and 2D. This will also be shown in Section 4 by means of magneto-transport for the InAs/InP QWr system. The change of reduced dimensionality in optical processes we investigate in the dense InAs/InP QWrs system together with InAs/InP QW system by means of PL and FTIS. At low temperature, the carriers are partly localized by potential fluctuations in the QWrs and their wavefunctions undergo additional local confinement, which can influence the interwire coupling. Therefore, the temperature is increased in our experiment from 5 K up to room temperature in order to release the trapped carriers and facilitate the observation of the expected change of the dimensionality in the QWr system caused by the enlarged overlapping of the wavefunctions for carriers in adjacent QWrs.

The temperature-dependent evolution of the QWr absorption is shown in Fig. 8.17b. The peculiarities caused by the 1D DOS for QWrs are observed at the band edges for the hh and lh states. With increasing temperature the excitonic edges smear and the flat spectral regions typical for the 2D system are developed at the room temperature. At low temperature, the inhomogeneous broadening of the exciton transitions due to the disorder caused by surface roughness and composition fluctuations alludes to the 1D nature of the QWrs. This inhomogeneous broadening can be introduced into the calculation of QWr DOS in case of rectangular QWr as a fitting parameter.

Figure 8.18 shows the results of this calculation using a 3 meV Gaussian broadening for low temperature. However, at room temperature such a broadening can be enhanced on the one hand due to the electron–phonon coupling but on the other hand due to the e–h plasma which is present at room temperature. The results taking into account both these contributions together with a Gaussian broadening are shown in Fig. 8.18b. It can be seen that the calculated 1D DOS reproduces the

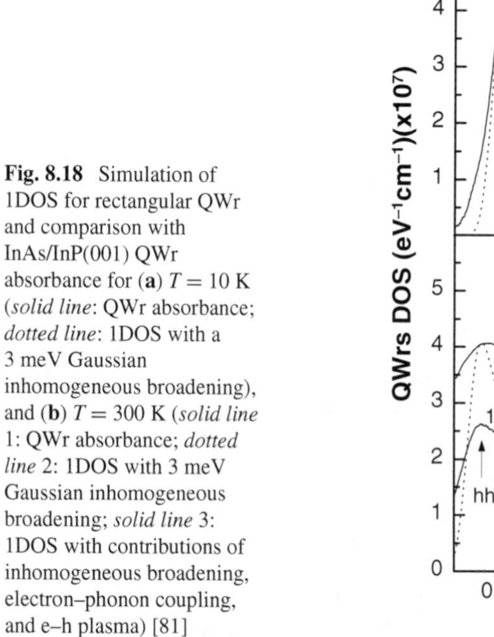

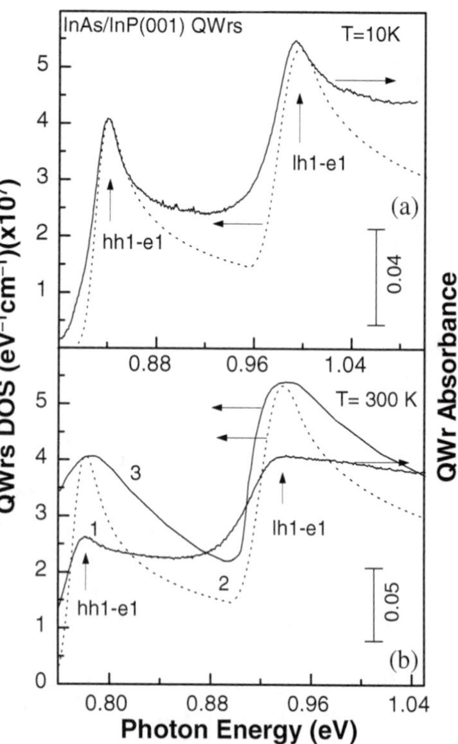

Fig. 8.18 Simulation of 1DOS for rectangular QWr and comparison with InAs/InP(001) QWr absorbance for (**a**) $T = 10$ K (*solid line*: QWr absorbance; *dotted line*: 1DOS with a 3 meV Gaussian inhomogeneous broadening), and (**b**) $T = 300$ K (*solid line* 1: QWr absorbance; *dotted line* 2: 1DOS with 3 meV Gaussian inhomogeneous broadening; *solid line* 3: 1DOS with contributions of inhomogeneous broadening, electron–phonon coupling, and e–h plasma) [81]

absorbance line shape fairly well for low temperatures ($T = 10$ K) (Fig. 8.16a), whereas it fails to fit the absorbance data at room temperature even if the e–h correlation effects are taken into account (curve 3 in Fig. 8.18b). This discrepancy has a certain physical source: at low temperature the carriers are localized in fluctuation minima. Properties of the QWrs which are more dependent upon the DOS such as optical transitions will still see an enhancement due to the peak at the band edge in the density of states. Wavefunctions of carriers are strongly localized in the QWr area and the overlap with the wavefunctions of neighboring QWrs is of negligible value, corresponding to an approximation of isolated QWrs. As the temperature increases the carriers release from the local traps and transfer to extended states along the wire axis and become more delocalized in the plane perpendicular to the QWr axis. Such a delocalization results in the spillover of the carrier wavefunction into neighboring QWrs and in fact is a change in the dimensionality of QWr system approaching an anisotropic 2D system in both optical [81, 92] and transport [35] properties.

3.6 *Polarization-Dependent Transmittance and Photoluminescence of InAs/InP Nanostructures*

The distinctly visible hh–lh splitting in absorbance spectra of the InAs/InP QD and QWr systems can result in the anisotropic absorption. The in-plane polarization of light propagating along the [001] direction arises due to transitions between a mixture of lh and hh valence states, and electrons in the conduction band. The hh–lh admixture can be caused by anisotropic confinement due to shape anisotropy or due to the symmetry change caused by the piezoelectric field [84]. Our InAs nanostructures grown on the B surface appear to be elongated in the [−110] direction, which would suggest a polarization in that direction. Figure 8.19c shows the polarization-dependent absorbance in InAs/InP QD system measured at 10 and 300 K. One can see that the maximal differences in absorbance for the polarizations parallel and

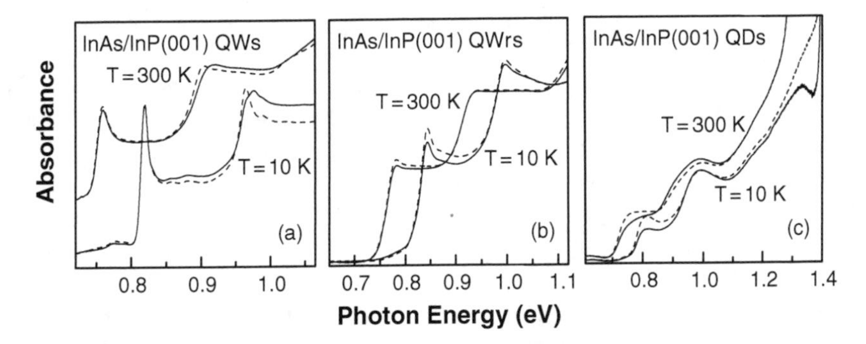

Fig. 8.19 Polarized absorption of InAs/InP QW (**a**), InAs/InP QWrs (**b**), and InAsInP QDs (**c**) 30-period superlattice samples for two different temperatures, $T = 10$ and 300 K. Parallel (*dashed lines*) and perpendicular (*solid lines*) polarizations relative to the [−110] direction

orthogonal to [−110] occur at the excitonic maximums. The qualitative properties of the polarization-dependent absorption are independent of the model used for the lateral confinement [93]. The anisotropy increases with decreasing QD base length due to increased hh–lh mixing. At fixed lateral QD sizes, the transitions that originate from different valence bands exhibit different magnitudes of the anisotropy. Let us introduce absorption coefficients α_x and α_y of the indicated transitions for the polarization parallel to the x axis (along the [−110] direction) and to the y axis (perpendicular to the [−110] direction). Positive $(\alpha_x - \alpha_y)$ anisotropy identifies a transition that involves a valence band of dominant lh character, whereas negative $(\alpha_x - \alpha_y)$ indicates that the contribution of the hh valence band prevails. It is just what we see in Fig. 8.19c. The change of the anisotropy sign when the energy changes from the hh states to the lh states unambiguously corroborates the correctness of the absorption feature assignment.

It has been mentioned above that the hydrostatic strain just causes a constant shift between the electron energy level and all the hole levels. The shear strain for the (001)-oriented zinc-blende crystal adds only to the diagonal terms of the Luttinger Hamiltonian. Therefore, strain does not induce any extra spin mixing but effects only the relative positions of the hh and lh energy bands. The dominant contribution to the anisotropy of the strain distribution arises from the anisotropy of the QD shape. Therefore, the polarization dependence of absorption and/or emission allows to get the information about the anisotropy of the QD and consequently about the anisotropy of confinement in the QD system. Using the results of theoretical investigation of the effect of asymmetry on optical properties and electronic structure of SK QDs modeled with anisotropic parabolic confinement potential, we estimate such anisotropy for our InAs/InP quantum dots [94]. If one introduces the in-plane confinement like anisotropic parabolic potential

$$V_{xy}(x, y) = (\alpha_x x^2 + \alpha_x y^2)/2, \tag{8.3}$$

a parameter of anisotropy $A = (\omega_x - \omega_y)/(\omega_x + \omega_y)$ with

$$\omega_x^{hh(lh)} = \sqrt{\alpha_x / m_{hh(lh)}} \qquad \text{and} \qquad \omega_y^{hh(lh)} = \sqrt{\alpha_y / m_{hh(lh)}}$$

can be defined. It has been shown [94] that the degree of linear polarization is zero for all the allowed states if $A = 0$. This means the absence of preference for a particular orientation of linearly polarized light in case of a symmetric QD. For the transition from the lowest hole state, the degree of polarization monotonically increases with increasing anisotropy parameter A. The value of band offsets also affects the degree of linear polarization, while it influences the mixing between the different contributing states. It is well known that transitions from pure hh or lh states lead to circularly polarized emission and/or absorption, whereas linear polarization arises from the states with an admixture of hh and lh states. Thus, if the coupling between the different hh and lh states weakens, the degree of linear polarization decreases. Defining the polarization ratio to be 0.18 in our QD system, we determine the anisotropy parameter A to be equal to 0.3 that corresponds to a 45 nm elongation in the [−110] direction versus a 35 nm elongation in the perpendicular direction what is in good agreement with the dimensions derived from the AFM analysis (43 nm versus 35 nm).

It is of interest to note that there exist spectral points of complete in-plane optical isotropy. These points arise due to the solution of equation $(\alpha_x(\omega) - \alpha_y(\omega)) = 0$ and are seen in Fig. 8.19c as crossing points of the absorbance curves for the parallel and orthogonal polarizations.

Outside of the vicinity of the isotropy points the QD structure can rotate the polarization of the linearly polarized light [95]. Indeed, if $I_x(d) = I_x^{(0)} \exp\{-\alpha_x d\}$ and $I_y(d) = I_y^{(0)} \exp\{-\alpha_y d\}$ are the intensities of the linearly polarized light transmitted normally to the surface of the QD structure of thickness d with polarization parallel and perpendicular to the QD axis (the [−110] direction) the change of arbitrary linear polarization in (xy) plane can be treated as the change of the angle Θ between the E vector and the x axis. Put $E_x = \sqrt{I_x} = \sqrt{I}\cos(\Theta)$ and $E_y = \sqrt{I_y} = \sqrt{I}\sin(\Theta)$ for normal incidence of light. Then, $\tan(\Theta) = \tan(\Theta_0)\exp\{(\alpha_x d - \alpha_y d)/2\}$ connects the Θ value for transmitted light and the Θ_0 value for the incident light. Taking into account the absorbance difference $(\alpha_x d - \alpha_y d)$ in our experiment does not exceed (see Fig. 8.19c) 0.02, the change of the polarization angle $(\Theta - \Theta_0)$ is expected to be $\sim 0.6°$ that is beyond our recent experimental accuracy. It is also seen from Fig. 8.19c that the lh states are affected by the barrier states at low temperature thus changing the anisotropy value in the range of the lh excitonic transitions.

Figure 8.19a,b shows the linearly polarized absorbance spectra measured with the light polarization parallel and perpendicular to [−110] direction for InAs/InP QW system and InAs/InP QWrs system, respectively. Figure 8.19a shows that the QW system is isotropic in the region of hh1–1h1 transition but demonstrates significant polarization dependence in the region of the lh1–e1 transition. The absorbance peak at 0.975 eV splits into two components: 0.975 and 0.965 eV. One of these components at 0.975 is polarization independent whereas the other one at 0.965 is polarization dependent. The nature of this feature at 0.965 eV is not yet clear and we can assume the existence of regions of uniaxial strain in the QW structure. Upon increasing the temperature to 300 K the lh1–e1 branch of the absorbance demonstrates a perfect 2D step. Polarization-dependent absorbance in the QWr sample is shown in Fig. 8.19b. At low temperature, maximal change of the absorbance is observed in the regions of the hh1–e1 and lh1–e1 resonances. Again, positive $(\alpha_x - \alpha_y)$ anisotropy identifies a transition that involves a valence band of dominant lh character, whereas negative $(\alpha_x - \alpha_y)$ indicates that the contribution of the hh valence band prevails. The change of the anisotropy sign when the energy changes from the hh states to the lh states unambiguously corroborates the accuracy of the absorption feature assignment. With an increase in temperature, the polarization dependence disappears for the region of lh transition, and significantly reduces for the hh transition. This agrees with the idea of temperature-induced dimensionality change in the QWr system outlined above. Strong polarization dependence typical for 1D wires fades and the 1D DOS resembles more the 2D DOS at high temperature.

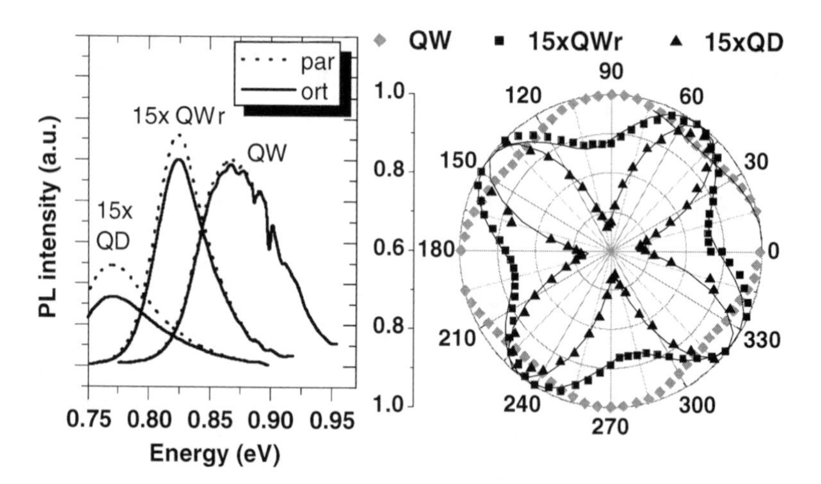

Fig. 8.20 Polarization-dependent PL of QW, QWr, and QD samples measured at room temperature. *Left*: PL spectra taken at room temperature for linear polarization of the PL signal along [−110] "par" and [110] "ort" (dips in the QW spectrum are due to parasitic absorption). *Right*: Dependence of PL maximum on polarization direction. The direction of maximum intensity is [−110] for all samples [49]

Finally, PL and FTIS measurements were carried out for InAs/InP(001) nanostructures that included QW, QD, and QWR samples. All samples have been grown by MBE depositing four monolayers of InAs on different vicinal surfaces of InP substrate. Many similarities are observed in temperature dependences of PL and FTIS spectra of InAs/InP nanostructures caused by peculiarities of the excitonic thermal activation in the ensemble of coupled nanoparticles. It demonstrates a significant change of the excitonic absorption spectrum (flat plateau formation) at high temperatures for both QWr and QD structures. Such a behavior is interpreted as thermally induced change of the dimensionality from 1D (0D) to anisotropic 2D. FTIS measurements with polarized light also indicate the change of dimensionality through a substantial reduction of difference between the absorption spectra in orthogonal polarizations in the regions of the hh1–e1 and lh1–e1 transitions in InAs/InP 0D and 2D samples.

Polarized PL spectra taken for polarization of the PL signal along the [−110] and [110] directions, and the direction-dependent PL intensity for a QW sample, a QWr 15 period superlattice sample, and a QD 15 period superlattice sample are shown in Fig. 8.20.

The lowest polarization degree is obtained for the QW sample as expected. The QWr and QD show larger polarization. The QD sample, however, shows a significantly larger polarization than the QWr sample. This relation holds true for all samples measured. The degree of linear polarization P is defined as $P = (I_{\max} - I_{\min})/(I_{\max} + I_{\min})$ with maximum and minimum PL intensity $I_{\max}$ and $I_{\min}$ in the mutually perpendicular polarization directions determined to be [−110] and [110]. The polarization direction with higher intensity is [−110] for all samples.

This observation also holds true for the QDs elongated into the [100] direction (grown on C surfaces). With decreasing temperature, the degree of polarization increases for QWrs but not for QDs. The fact that P is larger for the QDs than for the QWrs is counterintuitive as the QWrs are much more elongated than the almost symmetric QDs. Furthermore, P in our QWr samples seems lower than the values published by other groups. It was reported that $P = 30\%$ at 18 K [96], $P = 37\%$ at 4.2 K [97] against $P = 6\%$ at $T = 10$ K for our QWrs. The polarization degree of QWrs and QDs grown on N, A, and B substrates are given as 40% (QWr) and 16% (QD) at 10 K in Ref. [51]. Whereas the published results on P for the QD samples are compatible with our results, a marked difference exists for the results on QWr samples. One reason for the lower degree of linear polarization in our investigated QWr samples is their flat shape that makes them structurally closer to a QW as it follows from our absorbance data.

4 Transport in Coupled InAs/InP Nanostructures

In this section, the in-plane transport properties of large ensembles of InAs/InP nanostructures are investigated and discussed. Similar studies have been done in other material systems such as GaAs/AlGaAs QWrs [98, 99] or InAs/GaAs QDs [50, 100]. In the InP-based material system, in-plane transport has been measured in $In_{0.53}Ga_{0.47}As$ QWs with optional insertions of an InAs quantum well [101, 102]. A distinct feature of our arrays of InAs nanostructures is the highly anisotropic transport.

4.1 The Temperature-Dependent Magneto-Transport Experiments

Temperature-dependent transport measurements were used to clarify the nature of lateral coupling between the nanostructures and to determine the main scattering mechanisms. These measurements determine the sheet carrier concentration n (for electrons) or p (for holes) in (cm^{-2}), and the carrier mobility μ (in cm^2/V s). The change of mobility with temperature or carrier concentration can be used to identify the scattering mechanisms.

A widely used experimental technique to determine carrier concentration and mobility is the van der Pauw–Hall measurement [103]. This experimentally simple technique can normally be only applied to samples with isotropic transport. Recently, however, it has been demonstrated how van der Pauw–Hall measurements can also be used to determine anisotropic transport properties [49, 104].

To verify the effect of nanostructures on transport, a set of reference samples was studied first. These samples are an undoped InP layer, an 8 nm doped InP layer inserted into the undoped InP layer, and a modulation-doped InAs QW (sample nQW) with modulation-doping parameters given in Table 8.1. Before deposition of the InP layer, a $In_{0.52}Al_{0.48}As$ layer was grown directly on the substrate to suppress

Table 8.1 Modulation-doping parameters and measured carrier concentration n at $T = 10$ K of n-type QWr, QD, and QW samples

Sample	N_d (cm^{-3})	d (nm)	Substrate: nanostructure	n (10^{11} cm^{-2})
nQW	1.0×10^{18}	5	N : QW	14
nQWr1	1.0×10^{18}	5	A : QWr	9.4
nQWr2	1.0×10^{18}	**10**	N : QWr	3.8 (6.3)
nQWr3	$\mathbf{1.7 \times 10^{18}}$	5	N : QWr	15
nQD1	1.0×10^{18}	5	B : QD	6.2
nQD2	1.0×10^{18}	**10**	B : QD	3.1 (4.9)
nQD3	$\mathbf{1.7 \times 10^{18}}$	5	B : QD	20

Notes: The parameters are donor concentration N_d of the 8 nm InP:Si layer and thickness d of the InP spacer between InAs and doped InP. Deviations from the standard parameters are marked with bold numbers. Carrier densities in parenthesis were measured after illumination (persistent photoconductivity, ppc).

parasitic parallel conductivity at the substrate interface (see [49] and references therein). The measurements of the temperature-dependent Hall mobility of these samples allow to conclude: (i) nonintentionally doped samples are basically isolating (i.e., their conductivity is orders of magnitude lower than the conductivity of intentionally doped samples) which allows to exclude parasitic parallel conductivity effects; and (ii) a typical transport sample represented by a capped, n-doped, 8 nm-thick InP layer has a typical mobility characteristics for highly doped InP. Taking the temperature dependence of the mobility as $\mu \propto T^r$ we find that the exponent r changes if the temperature changes from high to low (from 310 to 9 K). The mobility increases slightly ($r \approx -0.2$) due to phonon scattering but decreases at lower temperatures ($r \approx 0.5$) due to ionized impurity scattering of the electrons at the ionized donors (Si atoms on group III sites). At 9 K, the mean free path for electrons is of $l_e = 9$ nm, which agrees quite well with the average distance of 10 nm between the donors. The mobility of this sample grown on nominal N substrate, and the same structure grown on vicinal A, B, and C substrates occurs to be completely isotropic. This confirms that the transport anisotropy observed in other samples is not due to the doping layer. The insertion of a nominally flat (2D) InAs layer underneath the doped InP separated by an undoped InP spacer layer results in a modulation-doped InAs QW. The spatial separation of donors (in the InP) and electron gas (in the InAs) leads to a reduction of ionized impurity scattering due to the weaker remote impurity scattering. As a result, the exponent r of increasing mobility at high temperatures is high (about 0.9 – 1.0), and no reduction of mobility occurs at low temperatures. Depending on sample and transport direction, the mobility saturates with decreasing temperatures in the range of 10000 – 20000 cm^2/V s. The mobility-limiting scattering is remote impurity scattering and interface roughness scattering. The slight transport anisotropy suggests a slightly anisotropic interface roughness that, however, does not lead to qualitatively different mobility characteristics for both transport directions. The tendency of vicinal substrates (particularly B

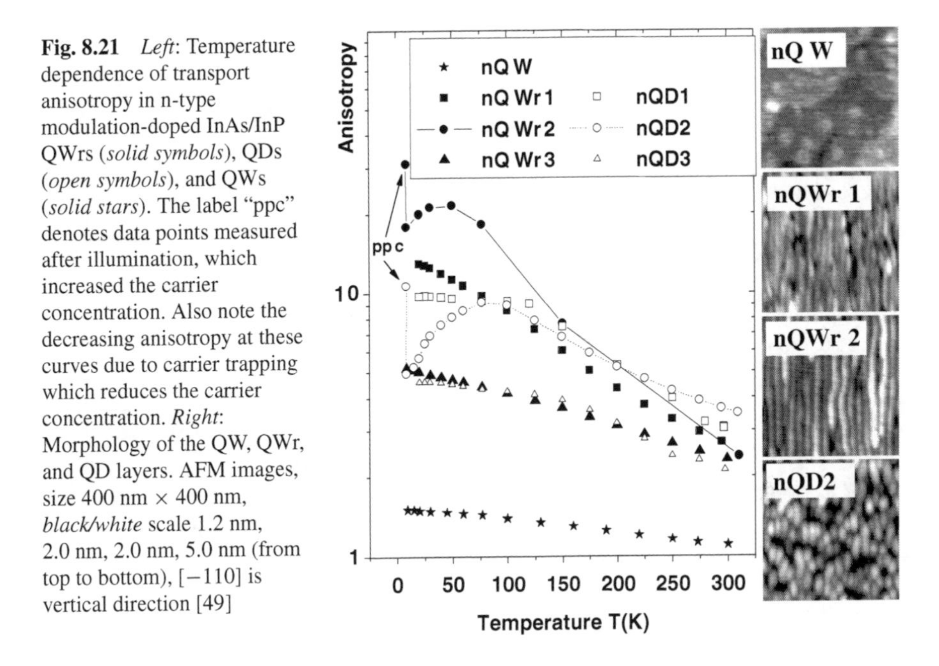

Fig. 8.21 *Left*: Temperature dependence of transport anisotropy in n-type modulation-doped InAs/InP QWrs (*solid symbols*), QDs (*open symbols*), and QWs (*solid stars*). The label "ppc" denotes data points measured after illumination, which increased the carrier concentration. Also note the decreasing anisotropy at these curves due to carrier trapping which reduces the carrier concentration. *Right*: Morphology of the QW, QWr, and QD layers. AFM images, size 400 nm × 400 nm, *black/white* scale 1.2 nm, 2.0 nm, 2.0 nm, 5.0 nm (from top to bottom), [−110] is vertical direction [49]

surfaces) to form rougher layers leads to a higher transport anisotropy. Anisotropic interface roughness scattering is theoretically treated in Ref. [105].

Three QWr transport samples with the same nominal InAs layer morphology and thickness were investigated. The nominal parameters of the modulation doping, i.e., remote donor concentration and spacer thickness were varied as shown in Table 8.1. These parameters provide different strengths of remote impurity scattering ranging from the strongest scattering (higher ionized donor concentration) to the weakest scattering (larger distance from electron gas to ionized donors). In contrast to the QW samples, marked transport anisotropy is present in the QWr samples (Fig. 8.21), which is intuitively expected due to their elongated shape. The high-mobility direction is parallel to the QWrs (along [−110]). The increasing transport anisotropy with decreasing temperature suggests a thermally activated coupling between the nanostructures (thermionic emission). A similar conclusion was drawn in [98] for a corrugated AlGaAs/GaAs heterointerface, and a lateral potential modulation of a few meV for electrons was determined.

A detailed analysis of the temperature-dependent mobilities in the principal transport directions is performed (Fig. 8.22). The following observations are made: (i) The mobility orthogonal to the QWrs (orthogonal mobility) shows the same qualitative behavior for all three QWr samples. At room temperature, the orthogonal mobility is approximately one-third the mobility parallel to the quantum wires (parallel mobility) corresponding to a transport anisotropy of 3. (ii) With decreasing temperature, the orthogonal mobility decreases to only approximately 2/3 its

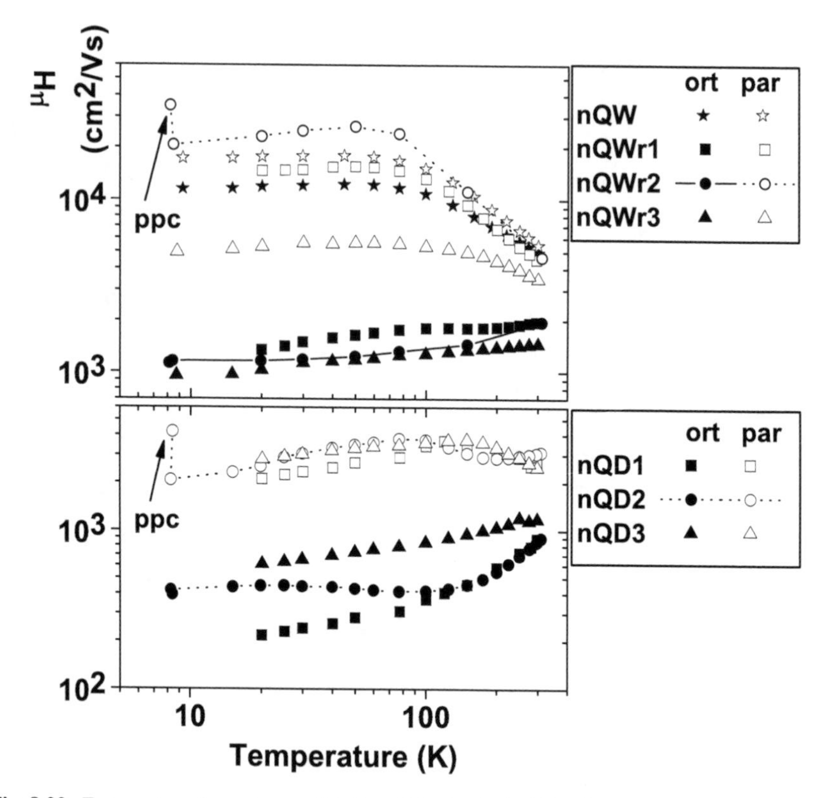

Fig. 8.22 Temperature dependence of Hall mobility in the [−110] direction ("par," *open symbols*) and [110] direction ("ort," *solid symbols*) of the n-type QWr and QD samples. The mobility of the QW (*stars*) is given as Ref. [49]

room temperature value at 20 K. This almost temperature-independent characteristic would correspond to an exponent of $r = 0.1$. It does not depend on the varying parameters of modulation doping either. These observations demonstrate that the transport perpendicular to the QWrs is neither limited by phonon scattering nor by remote impurity scattering. At 10 K, the orthogonal mobility is roughly 1000 cm^2/V s with mean free paths of $l_e = 12 - 21$ nm. (iii) The parallel mobility shows the same qualitative behavior for all three QWr samples and coincides with the behavior of the QW samples. From high to low temperatures it increases drastically, and saturates at about 50 K. The further cooling down leads to a slight degradation of the mobility in some samples. The quantitative details change with the variation of modulation-doping parameters, which shows that the mobility-limiting mechanism is the remote impurity scattering. (iv) The mobility of sample, grown with the same modulation-doping parameters as the QW references, is in quantitative agreement with the QW references. The mobility increases to a maximum of 3.4 times the room temperature value. (v) A thicker spacer layer that increases the spatial separation of ionized donors and electron gas results in a higher maximum mobility (5.4× the room temperature value) and an exponent of $r = -1.1$ at high temperatures. The mobility degradation below 50 K is due to a reduction of the carrier concentration by trapping. Retrieving these carriers at 8 K by a short illumination significantly increases the mobility to a value of approximately 35000 cm^2/V s, which results in a maximum anisotropy of 31. Interpolation to the mobility at 50 K suggests no mobility degradation at decreasing temperature if the carrier concentration would remain constant (without trapping). (vi) A higher donor concentration results in a significantly lower maximum mobility of only $1.6\times$ the room temperature value. The stronger degradation of mobility below 50 K suggests a situation at the transition from phonon and remote impurity scattering to phonon and ionized impurity scattering. (vii) At the lowest temperature (9–20 K), the mean free path of the QW samples in the parallel direction is 210 and 270 nm. In comparison, $l_e = 240$ nm in the QWr sample with the same modulation-doping parameters, which agrees well with that of the QW samples. Modification of these parameters results in a lower l_e of 102 nm (nQWr3, higher donor concentration) or higher l_e of 210/450 nm (nQWr2 thicker spacer, before/after illumination).

Following these observations, we conclude that the transport in our structures is in the diffusive regime as the mean free path is much smaller than the sample dimension (no ballistic transport) and $k_F l_e > 1/(2\pi)$ with $k_F \approx 0.25$ nm^{-1} (no hopping transport according to Ioffe–Regel criterion). The increasing transport anisotropy with decreasing temperature is an effect of the reduction of phonon scattering in the high mobility direction. This behavior is opposite to the observation in GaAs/AlGaAs heterojunctions on periodic multiatomic step arrays of Ref. [106], in which the change of anisotropy is mainly due to decreasing mobility in the low mobility direction. In those heterojunctions, an activation energy of 5 meV for transport across the steps has been observed. In contrast, Arrhenius plots of the orthogonal mobility of our samples (not shown) yield a barrier height of less than 0.6 meV, which is much lower than the results for the corrugated AlGaAs/GaAs heterointerface from Ref. [98]. In the parallel direction, comparison of the l_e value to QW samples and the dependence on the modulation-doping parameters yields that the mobility-limiting mechanism is remote impurity scattering. The l_e of 450 nm in the sample with weakest remote impurity scattering is close to the average wire length of ~600 nm in this sample. Hence, the transport within the wires is quasi-ballistic. In the context of coupled QWrs, this corresponds to a very efficient longitudinal wire–wire coupling. Further optimization of the modulation-doping parameters and the comparison to QW samples of the same nominal thickness would be required to test if there is a mobility limit due to longitudinal coupling. This comparison may also help to test the predicted higher mobility in wires compared to QWs due to reduced backscattering [107]. In the orthogonal direction, the l_e value is less or equals the average lateral wire–wire spacing, which is much smaller than the mean free path in QW samples. The transport is limited due to the strong lateral wire–wire scattering, which corresponds to a weak lateral coupling in the context of coupled QWrs. The coupling, however, is not thermally activated because the empirically determined barrier of less than 0.6 meV is too small to explain the low mobility at room temperature ($kT = 25$ meV), and is much less than the estimated barrier $\Phi_B \sim 200$ meV. In the context of anisotropic 2D systems, the scattering mechanisms for the parallel mobility (phonon scattering, remote impurity scattering) are isotropic and, hence, also effective for the orthogonal mobility. Interface roughness scattering dominates the orthogonal mobility.

Similar to the QWr samples, the QD samples also show pronounced transport anisotropies up to 10 (see Fig. 8.21) with the same high-mobility direction. (In layers containing self-assembled InAs QDs grown on GaAs lower transport anisotropies between 1 and 2 have been observed [108, 109]. These anisotropies have been attributed to the formation of dot-chains forming along steps, and in terms of dislocations.) In our QD samples the temperature dependence of mobility has striking similarities to that of the QWr samples. The absolute mobilities, however, are lower (around 3000 cm^2/V s parallel, 300 cm^2/V s orthogonal) than in the QWr samples. Also, the mobility in the high mobility direction ("parallel") decreases more pronounced with decreasing temperature after its maximum. Comparison to samples with the same modulation-doping parameters but a QWr morphology demonstrates that the QD morphology limits the parallel mobility. The mobilities are still large enough to be described by diffusive transport. Hence, interface roughness scattering is assumed the mobility-limiting scattering for both transport directions in the QD samples. Thermal activation seems to play no role for the same reasons. Given the random spatial distribution of almost isotropic-appearing QDs the transport anisotropy cannot be explained with the apparent InAs morphology. The high mobility direction is the same as for the QWr, suggesting the same underlying physics. In terms of coupled nanostructures, a stronger coupling in the parallel direction than in the orthogonal direction may be responsible for the anisotropic transport. Plan-view TEM images (out of the scope of this work) could give more detailed information of the shape of the covered dots, which contribute to transport.

A variation of the electronic properties by keeping the same basic structural properties was done by growing modulation-doped QWrs in an $In_{0.53}Ga_{0.47}As$ matrix. These wires are structurally similar to the wires in an InP matrix. Electronically, however, the carrier confinement is reduced due to the smaller conduction band offset of 0.25 eV compared to 0.5 eV in InP. In addition, the smaller effective mass in $In_{0.53}Ga_{0.47}As$ results in a weaker localization of the carriers in the InAs, and in higher total mobilities. The estimation of effective barrier $\Phi_B = 15$ meV between adjacent InAs nanostructures suggests much stronger coupling than in an InP matrix. The temperature-dependent principal mobilities and corresponding transport anisotropy of InAs QW in $In_{0.53}Ga_{0.47}As$ reveal the characteristics qualitatively similar to InAs QW in InP. There is also a small transport anisotropy which may be attributed to anisotropic interface roughness scattering. The mobilities, however, are higher than in the InAs/InP structures. In contrast to InAs/InP, the mobilities parallel and orthogonal to InAs QWrs in $In_{0.53}Ga_{0.47}As$ show qualitatively the same behavior and the transport anisotropy is only about of 2. It is concluded that the potential barrier between InAs and $In_{0.53}Ga_{0.47}As$ does not provide sufficient confinement of the electrons to the InAs wires.

The transport of holes in coupled InAs/InP nanostructures, on the other hand, allows to realize the opposite regime of a very strong confinement. Even though

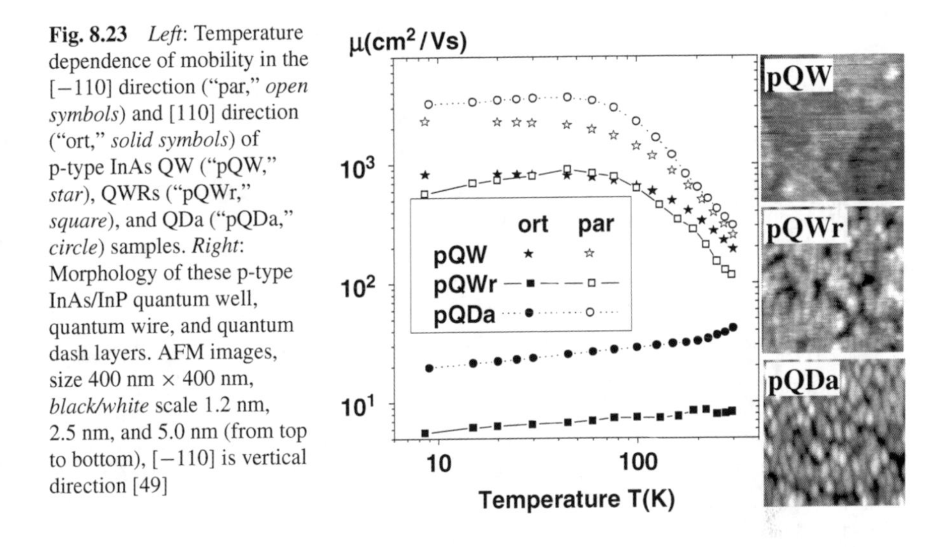

Fig. 8.23 *Left*: Temperature dependence of mobility in the [−110] direction ("par," *open symbols*) and [110] direction ("ort," *solid symbols*) of p-type InAs QW ("pQW," *star*), QWRs ("pQWr," *square*), and QDa ("pQDa," *circle*) samples. *Right*: Morphology of these p-type InAs/InP quantum well, quantum wire, and quantum dash layers. AFM images, size 400 nm × 400 nm, *black/white* scale 1.2 nm, 2.5 nm, and 5.0 nm (from top to bottom), [−110] is vertical direction [49]

the band discontinuity for holes is lower than for electrons, the net confinement for holes is larger than for electrons due to their significantly higher effective mass (by a maximum factor of 5). In particular, the higher effective mass leads to a generally stronger localization of the holes in comparison to electrons as the tunneling probability and the wavefunction dimensions (e.g., Bohr radius) decrease with increasing effective mass. Since the mobility μ is inversely proportional to the effective mass m^* ($\mu = e\tau/m^*$, τ is the momentum relaxation time [elastic scattering time]), lower mobilities than in n-doped samples are expected in the p-doped samples (if the same scattering times are assumed).

The *p*-Type modulation-doped InAs/InP nanostructures, i.e., a quantum well ("pQW"), quantum wire ("pQWr"), and quantum dash ("pQDa") sample, were grown and investigated (Fig. 8.23). The nominal doping was in the range of 1–2 × 10^{12} cm^{-2}. Hall measurements yielded consistent carrier concentration results for the QW sample (p_{H} varying between 1.25 and 1.38 × 10^{12} cm^{-2}). The other samples, however, did not allow measuring the carrier concentration in the entire temperature range. Assuming the same qualitative behavior as the QW sample, i.e., mainly constant carrier concentration, a pseudo-mobility was calculated based on the measured carrier concentration at 275 K. The temperature-dependent principal mobilities are given in Fig. 8.23.

We find that, qualitatively, the holemobilities show the same characteristics as the electron mobilities. An analysis of the quantitative details yields the following results. The p-type QW sample pQW shows transport anisotropies (attributed to anisotropic interface roughness scattering) from 1.3 to 2.8 (300 to 9 K), which is larger than 1.1 to 1.5 in the corresponding n-type QW sample nQW. This larger transport anisotropy indicates stronger confinement of holes to potential fluctuations caused by interface roughness. Along this trend, the transport anisotropy of the QWr samples is significantly larger for holes than for electrons. At low temperatures, values in excess of 100 are obtained. At room temperature, transport anisotropies of 14 and 7 are already found in the QWr and QDash containing samples pQWr and pQDa. The maximum parallel mobility of pQW and pQDa in the range of 2000–4000 cm^2/V s is approximately 1/5 the maximum mobility of comparable n-doped samples and can thus be explained by the higher effective mass but comparable scattering times τ. The relative mobility increase with decreasing temperature (by up to 12 times) is higher than that for electrons. The calculated mean free paths at $T = 9$ K in the parallel direction of $l_e = 45$, 70, and 13 nm for pQW, pQDa, and pQWr, respectively, are consistent with diffusive transport.

On the other hand, the calculated mean free paths in the orthogonal direction of $l_e = 16$, 0.4, and 0.08 nm for pQW, pQDa, and pQWr suggest diffusive transport only in the QW sample. According to the Ioffe–Regel criterion, the hole wavefunctions of pQWr and pQDa samples are localized laterally, and the transport along the orthogonal direction is in the hopping regime. In the context of coupled nanostructures, the longitudinal coupling is very high, an effect in particular visible with the structurally short quantum dashes realized in the pQDa sample. Laterally, however, the quantum wires and quantum dashes can be considered quasi-decoupled.

In the hopping regime, the mobility has no physical meaning. Instead, the conductivity or resistivity should be discussed. Nevertheless, the pseudo-mobility is discussed to have a direct comparison to other samples and to the parallel transport direction. Due to the assumed constant carrier concentration, the mobility is proportional to the conductivity that makes the discussion of mobility equivalent to the discussion of conductivity. The temperature dependence of the orthogonal hole mobility shows the same qualitative behavior as the corresponding electron mobility. From 300 to 9 K, it decreases only to a value between 1/2 and 2/3 its 300 K value. The hole mobility, however, is approximately one to two orders of magnitude lower than the electron mobility which is beyond the effect of the effective mass. At 10 K the orthogonal mobility ranges from 5 to 20 cm^2/V s. Arrhenius plots of the orthogonal conductivity give activation energy below 0.5 meV. Due to the high barrier, hopping via activation into extended states (with $\Phi_B > 180$ meV) is unlikely, either. The small relative change in conductivity with temperature observed in our samples is in marked contrast to published results describing hopping conductivities that change by orders of magnitude [110–112]. Consequently, nearest-neighbor hopping can be considered as the dominant hopping path. One reason for the small temperature dependence may be the formation of bands for the parallel transport. Laterally adjacent bands should have an energy overlap (due to their bandwidth) resulting in a disappearing activation energy for hopping and the conductivity only depends on the lateral overlap of adjacent bands.

4.2 Controlled Transport Anisotropy and Interface Roughness Scattering

In addition to the temperature dependency of mobility, further information on the transport can be obtained by a variation of the Fermi energy. With the Fermi energy, the effective coupling barrier Φ_B can be modified which, in the context of coupled nanostructures, allows one to investigate the longitudinal and lateral coupling as a function of effective barrier height. Furthermore, the Fermi wavelength can be tuned to find Bragg reflections of the electrons at the lateral periodicity of the quantum wires (lateral superlattice). Finally, in the context of anisotropic 2D systems, additional information on the scattering mechanisms is gained due to their characteristic carrier-concentration dependence.

A change of Fermi energy is equivalent to a change of the carrier concentration. Typically, the carrier concentration can be changed by 1×10^{12} cm^{-2}. This change of carrier concentration results in a variation of Fermi energy of less than 80 meV. Taking into account band nonparabolicity the actual value will be even lower. Comparison to the effective barrier of $\Phi_B \approx 200$ meV yields that it will not be possible to occupy extended states in the wetting layer or InP, nor to thermally activate carriers into these states. Hence, the carrier will still remain localized in QWrs. The Bragg reflections of the electrons in a lateral superlattice formed by the QWrs would result in a conductivity minimum if the lateral periodicity a (roughly 20 nm) coincides with half the Fermi wavelength, $a = \lambda_F/2$ [113]. This coincidence is expected at a carrier concentration of 3.9×10^{11} cm^{-2}. The mobility dependence on carrier concentration in case of remote impurity scattering is approximated by $\mu \propto n^{\gamma}$. Numerical calculations [114, 115] obtain exponents in the range of $\gamma = 1$ to $\gamma = 1.7$ depending on the thickness of the undoped spacer that separates the electron gas from the remote donors. The positive exponent is a result of the better screening of the remote impurity potential with higher n. The relevant facts about interface roughness scattering are summarized following Ref. [116]. The roughness of the interface is typically described by a Gaussian fluctuation of the interface $\left\langle \Delta\left(r\right) \Delta\left(r'\right)\right\rangle = \Delta^2 \exp\left[-\left|r - r'\right| / \Lambda^2\right]$ with the lateral size Λ of the Gaussian fluctuation and roughness Δ. The <> brackets denote an ensemble average. The mobility dependence on carrier concentration for interface roughness scattering is governed by the product $k_F \Lambda$. The mobility has a minimum for $\lambda_F = 4\Lambda$ at which the electrons are most effectively scattered and is monotonically increasing mobility to both sides of the minimum. In our samples, the top interface of the InAs gives the major contribution to the roughness since the bottom interface (to the underlying InP layer) is comparably smooth. Consequently, Λ can be determined from AFM images of the QWr surface morphology. An analysis of our AFM data yields $\Lambda =$ 7 nm, which means that the scattering is most effective at a Fermi wavelength of $\lambda_F = 28$ nm or a carrier concentration of 8×10^{11} cm^{-2}. In contrast for the parallel direction $\Lambda^{par} = 34$ nm, the strongest scattering would occur at 3.5×10^{10} cm^{-2}. For the carrier concentrations between 3×10^{11} and 2×10^{12} cm^{-2} in our samples, strong interface roughness scattering is theoretically expected for the orthogonal direction but not for the parallel direction.

A theory for interface roughness scattering at an anisotropic roughness described by two perpendicular lateral fluctuation sizes Λ^{par} and Λ^{ort} is given in Ref. [105]. For a roughness anisotropy $\Lambda^{par}/\Lambda^{ort} = 2$, this theory yields a transport anisotropy A in the range of 2–3 for our λ_F. This transport anisotropy is much lower than our experimentally obtained A for QDs with shape anisotropy $\Lambda^{par}/\Lambda^{ort}$ between 1 and 2. In addition, the significantly larger shape anisotropy of our QWrs is out

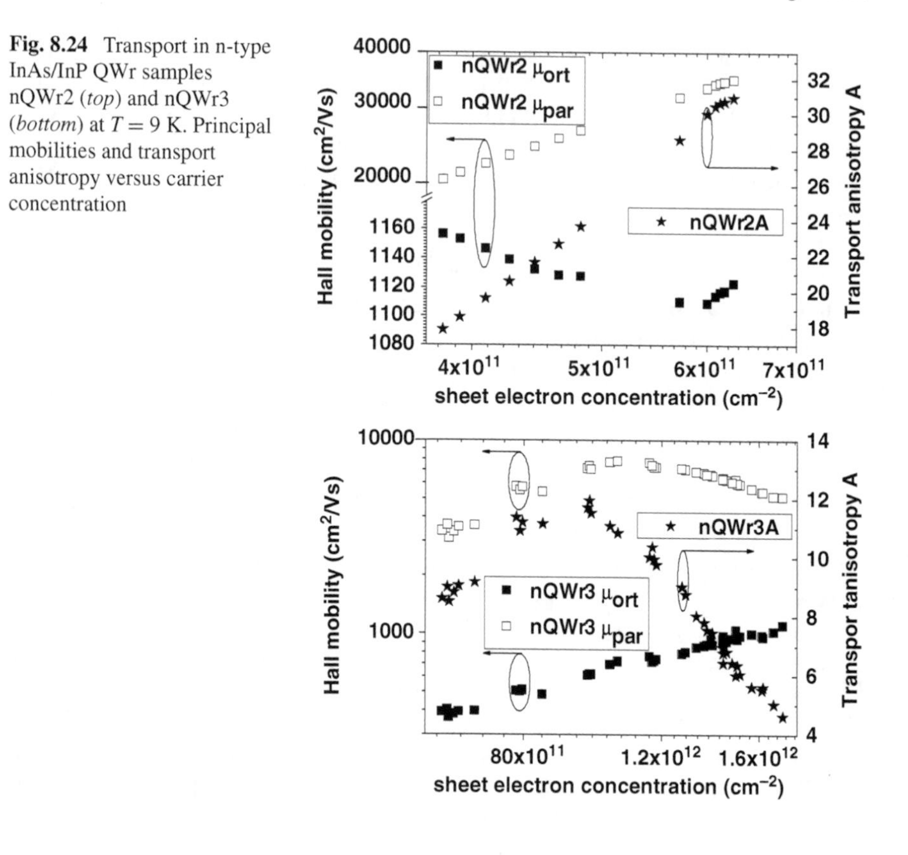

Fig. 8.24 Transport in n-type InAs/InP QWr samples nQWr2 (*top*) and nQWr3 (*bottom*) at $T = 9$ K. Principal mobilities and transport anisotropy versus carrier concentration

of the range treated in Ref. [105]. We note that the theory of anisotropic interface roughness scattering from Ref. [105] cannot explain the mobility anisotropy of our structures.

In two n-type modulation-doped QWr samples, nQWr3 and nQWr2 (see Table 8.1), the carrier concentration is varied at a constant temperature of 9 K. In nQWr3, a gate electrode is used to control the carrier concentration. In nQWr2, the ppc effect was used to increase the electron concentration stepwise by taking advantage of existing traps. The dependence of the principal mobilities on the carrier concentration, and the resulting transport anisotropies are shown in Fig. 8.24 for both samples. In sample nQWr2, the carrier concentration could be increased from $n = 3.8 \times 10^{11}$ up to $n = 6.3 \times 10^{11}$ cm^{-2} which corresponds to a decrease in Fermi wavelength from 40.5 to 31.5 nm. In this range, the parallel mobility increases from 21000 to 35000 cm^2/V s with an exponent $\gamma = 1.0$. This behavior indicates remote impurity scattering. The experimentally obtained exponent is, however, somewhat lower than the numerically calculated values. In contrast, the orthogonal mobility is almost constant over the entire range of n. Even though strong Bragg reflection of the electrons at the lateral periodicity of the QWrs is expected at

$n = 3.9 \times 10^{11}$ cm^{-2}, no increasing mobility is observed as the carrier concentration increases (which drives the Fermi wavelength away from the Bragg condition). Hence, these data give no evidence for a lateral superlattice. Also in terms of effective coupling barrier an increasing mobility would be expected with increasing carrier concentration. Rather, the mobility is slightly decreasing from 1160 to 1110 cm^2/V s up to a carrier concentration of $n = 60 \times 10^{11}$ cm^{-2} and increases slightly beyond. This behavior strongly indicates interface roughness scattering. The carrier concentration at minimum mobility corresponds to a Fermi wavelength of 32 nm suggesting a Gaussian roughness lateral size of 8 nm which is in good agreement with the value of 7 nm obtained from our AFM data. The transport anisotropy increases in the entire range of carrier concentration, mainly due to the increase of parallel mobility. The regime of higher carrier concentrations is probed with nQWr3 that differs from nQWr2 in the parameters of modulation doping but has the same InAs morphology. Using a gate electrode, the carrier concentration was varied in the range of $0.6 - 1.7 \times 10^{12}$ cm^{-2} corresponding to a variation of λ_F between 32 and 19 nm. The orthogonal mobility of nQWr3 increases monotonically with increasing carrier concentration (decreasing Fermi wavelength), which is consistent with the results from nQWr2. The parallel mobility is increasing up to a carrier concentration of 1.1×10^{12}cm^{-2} and decreases beyond. The increasing mobility can be explained with remote impurity scattering. The decrease at higher carrier densities remains unexplained so far. The drastic change of transport anisotropy at carrier concentrations above 1×10^{12} cm^{-2} is a joint effect of the decreasing parallel mobility and the increasing orthogonal mobility. This behavior is beneficial to the device application.

Thus, the mobility dependence on carrier concentration indicates dominant interface roughness scattering for the orthogonal transport. The quantitative details are consistent with the roughness in this direction due to our QWrs as seen in the AFM images. The mobility dependence of the parallel direction, in contrast, agrees with remote impurity scattering as mobility-limiting mechanism. In the context of coupled nanostructures, the lateral coupling is not enhanced due to the effective decrease of Φ_B with higher n. Electron Bragg reflection at the lateral superlattice formed by the wires was not detected. More detailed investigations in the direction of effects due to lateral superlattices and resonant tunneling may require smaller transport structures than those used in our investigations.

4.3 Shubnikov-de-Haas Oscillations and Quantum Hall Regime

This section gives a brief overview of the results from the magneto- transport measurements at low temperatures and high magnetic fields. A complete discussion of the results and the underlying theories is, however, out of the scope of this chapter. The quantization of the electron gas into Landau levels (LLs, with spacing of $\hbar\omega_c$, where ω_c is the cyclotron frequency) offers additional possibilities to characterize the transport properties. Shubnikov-de-Haas (SdH) oscillations of the longitudinal resistivity ρ_{xx} allow to measure the 2D carrier concentration from the oscillation period in inverse magnetic field, and to determine the transport effective mass from the oscillation amplitudes at different temperatures [117, 118]. The advantage in comparison to Hall measurement is that the resulting carrier concentration is not obscured by a Hall scattering factor, finite contact size effects, or parallel conductivity. Plateaus in transversal resistivity ρ_{xy} due to the quantum Hall effect (QHE) [119] occur at the same magnetic fields as the oscillation minima (with ideally vanishing resistivity). At these magnetic fields, the Fermi energy is located between adjacent LLs. Following the extensive review in Ref. [113], the occupied LLs (i.e., with energy below E_F) are localized within the sample while they form extended states only along the sample boundaries (edge channels). Thus, each LL contributes with the fundamental 1D resistivity h/e^2 to the transverse resistivity. Consequently, the plateaus of transverse resistivity are located at

$$\rho_{xy} = h/(e^2 \upsilon) = 25.818\ \mathrm{k}\Omega/\upsilon, \tag{8.4}$$

with the filling factor υ that indicates the number of occupied LLs. In the absence of spin splitting (all data shown here for InAs/InP nanostructures), these levels have a spin degeneracy of 2. The spatial extension of wavefunction of the Landau level with index $n (n = 1, 2, \ldots)$ is the cyclotron radius r_l which decreases with increasing magnetic field B as $r_l(B, n) = l_m \sqrt{2n - 1}$ with $l_m(B) = \sqrt{\hbar/eB}$ being the magnetic length (~25 nm at $B = 1$ T). Similar to the Fermi wavelength in the interface roughness scattering, this wavefunction extension has to be compared with the characteristic length of the potential variations due to InAs nanostructures for the electrons. Figure 8.25, top left, shows the SdH oscillations of the longitudinal resistivity and QHE plateaus in the transversal resistivity of an InAs QW sample (nQW) at different temperatures. At the lower temperature of 0.3 K, the thermal broadening of the LL is small enough to yield a pronounced QHE plateau and vanishing ρ_{xx} at 10 T. The filling factor $\upsilon = 4 (\rho_{xy} = 6.45\ \mathrm{k}\Omega)$ which corresponds to occupied spin degenerate LLs with $n = 1, 2$. The QWr sample nQWr3 (shown in Fig. 8.25, top right), on the other hand, shows no pronounced QHE plateaus, probably due to the strong remote impurity scattering that does not allow the formation of pronounced LLs. SdH oscillations of ρ_{xx}, however, are already present and show the same qualitative behavior in both transport directions. The SdH and QHE in the InAs QWr sample nQWr2 (with weak remote impurity scattering and high anisotropy up to 38) is shown in Fig. 8.25, bottom. At low carrier concentration (see Fig. 8.25, bottom left) a clear transition from diffusive transport to transport in the QHE regime is observed. At $B \approx 6.5$ T, a QHE plateau with filling factor $\upsilon = 2$ forms, and the SdH oscillation minimum is close to zero in comparison to the zero field resistivity. At zero magnetic field, diffusive transport with longitudinal resistivities of $\rho_{xx}^{ort} = 26\ \mathrm{k}\Omega$ and $\rho_{xx}^{par} = 2.8\ \mathrm{k}\Omega$ is observed. The transport in the QHE regime, in contrast, yields resistivities of 710 and 560 Ω. The filling factor of 2 corresponds to the spin degenerate ground state LL ($n = 1$) which has a cyclotron radius of $r_1 = l_m = 10$ nm. While this value is smaller than the lateral period of the potential due to the InAs QWrs (of 20 nm) it has to allow the pronounced formation of LLs. This assumption is supported by the argument that in disordered conductors,

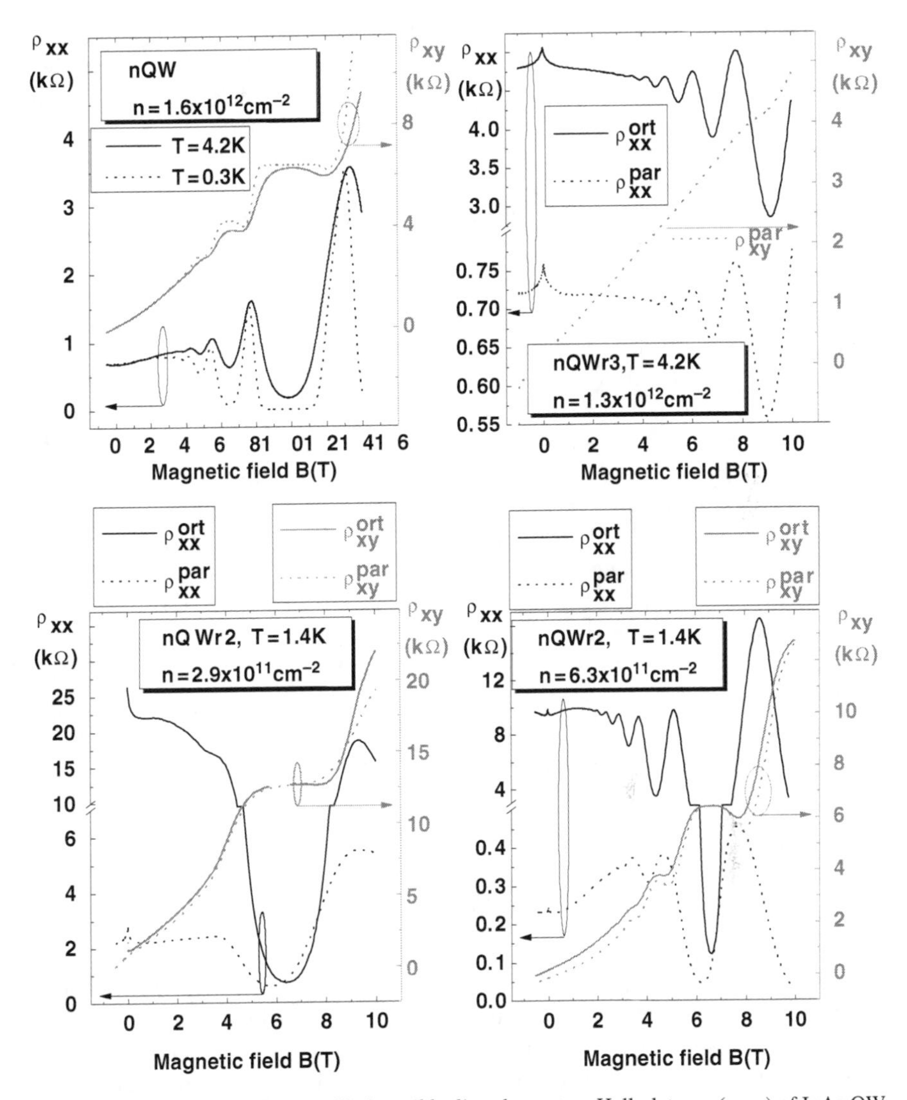

Fig. 8.25 Shubnikov-de-Haas oscillations (*black*) and quantum Hall plateaus (*gray*) of InAs QW and QWr samples. *Top left*: QW sample nQW at different temperatures. *Top right*: QWr sample nQWr3 at $T = 4.2$ K. *Bottom left*: QWr sample nQWr2 at $T = 1.4$ K. *Bottom right*: QWr sample nQWr2 at $T = 1.4$ K and higher carrier concentration [49]

strong potential variations (the InAs nanostructures have strong potential fluctuations leading to the disorder in our samples) should appear on a spatial scale that is large compared to l_m [113]. At a higher carrier concentration, obtained using the ppc effect (see Fig. 8.25, bottom right), the SdH oscillation period is higher, and the QHE plateau at $B \approx 6.5$ T has a filling factor of $\upsilon = 4$, both indicating the higher carrier concentration. While the longitudinal resistance ρ_{xx} is anisotropic, the Hall resistance ρ_{xy} is identical (within the experimental error) for both transport directions. The peak of the longitudinal resistivity at $B = 0$, in Fig. 8.25 bottom, is a macroscopic consequence of quantum mechanical self-interference of the electron – the weak localization.

4.4 Weak Localization in Coupled InAs/InP Quantum Wire Nanostructures

In this section, the magneto-transport properties of QWs, and laterally coupled self-organized InAs QWrs and QDs are investigated with respect to the contributions to conductivity resulting from weak localization. The influence of the type of nanostructure on the weak localization will be compared with focus on QWrs. The weak localization correction shows a directional anisotropy equal to the anisotropy of the total conductivity. This result suggests that even for a conductivity anisotropy ratio as high as 38, the coupling between the QWrs plays a more important role than the quasi-1D nature of the electron transport in describing the quantum mechanical contributions due to the weak localization.

According to the scaling theory of localization [120], at zero temperature in 2D disordered systems the conductivity can be divided into a classical- and a length-dependent term, σ_0 and $\delta\sigma(L)$, with L describing the system's characteristic size and l_e being the elastic scattering length. The total conductivity is given by

$$\sigma_{2D}(L) = \sigma_0 + \delta\sigma \tag{8.5}$$

with

$$\delta\sigma = -\frac{2e^2}{\pi h} \ln\left(\frac{L}{l_e}\right). \tag{8.6}$$

Therefore the conductivity decreases with increasing system size. At finite temperature, inelastic scattering processes cause the electron to lose its phase while traveling a distance of l_Φ (dephasing length). This length is of crucial importance in devices using quantum interference. If the l_Φ value is smaller than L, then $l_\Phi(T)$ replaces L in Eq. (8.6). The logarithmic term $\delta\sigma$ is a correction to the classical conductivity σ_0 resulting from weak localization. Weak localization is a purely quantum mechanical self-interference phenomenon in disordered conductors. At low temperature T, the dephasing length l_Φ becomes large, allowing the coherent self-interference of an electron. Diffusing along a closed path, the electron interferes constructively with its time-reversed path resulting in a higher-than-classical probability of backscattering. This interference is enabled by the non-phase-randomizing nature of the elastic scattering that causes the diffusion. The enhanced backscattering results in a decrease of conductivity whose contribution is of the order of the elementary conductivity e^2/h. Applying a magnetic field perpendicular to the plane of carrier motion destroys the constructive interference as it introduces an

Table 8.2 Transport properties at $T = 1.4$ K of the analyzed samples

Sample	Morphology	A	$n_{\mathrm{H}}(10^{11}$ cm^{-2})	μ^{par} (cm^2/Vs)	μ^{ort}(cm^2/Vs)
nQW	QW	1.5	16	7350	–
nQWr1	QWr	11.2	7.6	6325	565
nQWr2	QWr	13.8(37.9)	2.0(6.0)	4718(38700)	342(1022)
nQWr3	QWr	6.7	13.5	6340	947
nQD2	QD	2.6(13.2)	2.0(6.0)	616(3670)	237(278)
nQD3	QD	8.1	13.0	3102	383

Notes: Hall sheet electron concentration (n_{H}), transport anisotropy (A), Hall mobility parallel (μ^{par}), and (μ^{ort}) to the [−110] direction. Values in parentheses are measured after illumination (ppc).

Aharonov–Bohm phase shift between both time-reversed amplitudes. This is a concurrent process which causes dephasing on the scale of the magnetic length l_{m} as opposed to l_{Φ}. Therefore, the quantum correction $\delta\sigma$ is also dependent on the magnetic field B. In addition to weak localization, electron–electron interaction gives rise to other quantum corrections. These corrections are also magnetic field dependent, but less so than the weak localization. The scaling theory of localization can be further generalized to show that in 2D systems with anisotropic σ, the correction $\delta\sigma$ due to localization or electron–electron scattering is also anisotropic with the same anisotropy as σ [121]. This effect was demonstrated experimentally in the (110)-oriented Si metal-oxide–semiconductor system, in which the 2DEG at the silicon oxide interface shows a directional anisotropy of about 2 [122].

In our experiments [35], it is shown that although the semiclassical transport appears quasi-1D the weak localization is well described as an anisotropic 2D system. Without lateral coupling, the lateral confinement would be defined by the wire width, which is much smaller than the dephasing length l_{Φ}. Therefore, the success of the 2D description signifies that in this system the lateral coupling determines the fundamental behavior of the weak localization. In the transport measurements, samples containing one modulation-doped InAs layer embedded in InP were investigated. Table 8.2 summarizes their transport properties at $T = 1.4$ K along with their morphology. In their magneto-transport characteristics, the resistivity ρ_{xx} exhibits a characteristic peak at $B = 0$. Figure 8.26 shows these peaks of the QWr sample nQWr2 in detail for both transport directions. In the left part, the peak measured at three different temperatures is shown. The right part displays the weak localization peak of the same sample at a higher carrier concentration (after illumination). Additionally, the magneto-resistance in a magnetic field oriented parallel to the InAs layer at higher carrier concentration is shown in Fig. 8.26, right. The corresponding change of conductivity is calculated according to

$$\Delta\sigma(B) = \sigma(B) - \sigma(0) = \rho_{\mathrm{xx}}(B)^{-1} - \rho_{\mathrm{xx}}(0)^{-1} \tag{8.7}$$

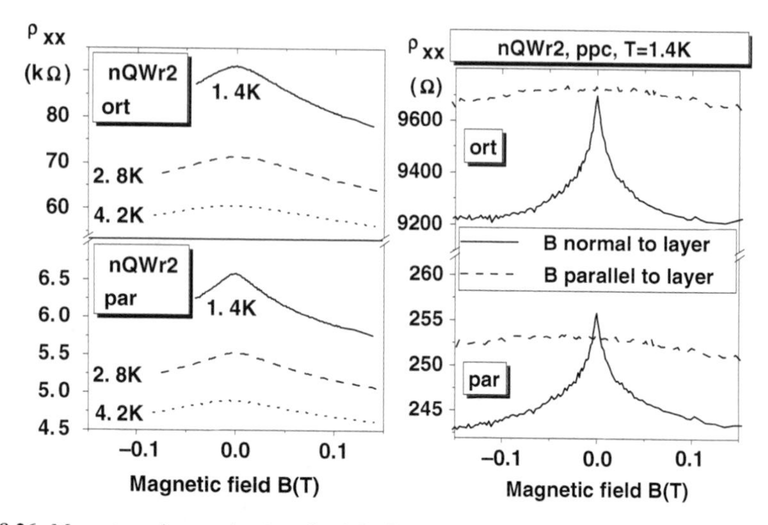

Fig. 8.26 Magneto-resistance (ρ_{xx}) peak of the InAs QWr sample nQWr2 for transport parallel ("par") and orthogonal ("ort") to the wire direction. *Left*: Measured at temperatures $T = 1.4$, 2.8, and 4.2 K with magnetic field normal to the layer. *Right*: Measured at $T = 1.4$ K and higher carrier concentration after illumination ("ppc") with magnetic field normal (*solid line*) and parallel (*broken line*) to the InAs layer [49]

Because for small B, the classical conductivity σ_0 is independent of B (as shown in experiments at $T = 77$ K), the peak in ρ_{xx} is solely a measure of the magnetic field dependence of the quantum correction $\delta\sigma$. Also, the smaller peak height at higher temperatures (see Fig. 8.26, left) indicates a suppression of the quantum correction due to increased inelastic scattering.

The measurement with B parallel to the InAs layer allows one to distinguish between electron–electron interaction and weak localization as a cause of the magneto-resistance peak. Electron–electron interaction yields a peak that consists of a large contribution from spin-splitting, not depending on magnetic field direction, and a contribution due to the perpendicular magnetic field, which is often considerably smaller than the weak localization peak [113]. Consequently, it is the spin-splitting contribution to electron–electron interaction that is likely to distort the weak localization peak. Comparison of the magneto-resistance peaks in perpendicular and parallel B (Fig. 8.26, right) yields that the spin-splitting contribution from electron–electron interaction (measured in parallel magnetic field) is negligible for the measured perpendicular magneto-resistance peak. Thus, the magneto-resistance peaks measured in our samples are mainly related to weak localization.

Using Eqs. (8.5) and (8.7), the experimental quantity $\Delta\sigma$ measures the dependence of the quantum correction to the conductivity on magnetic field. Furthermore, the directional anisotropy of $\Delta\sigma(B)$, defined as $\Delta\sigma^{\mathrm{par}}(B)\,/\Delta\sigma^{\mathrm{ort}}(B)$ is a measure of the anisotropy of the quantum correction to the conductivity $\delta\sigma(B)$. A comparison of $\Delta\sigma(B)$ for both principal transport directions by plotting them on a log scale (Fig. 8.27, left) shows that its directional anisotropy (the vertical distance

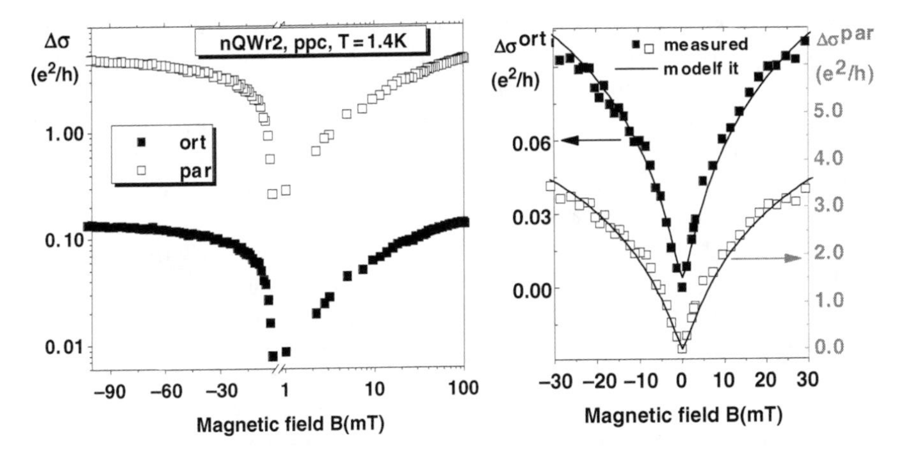

Fig. 8.27 Change of conductivity in magnetic field $\Delta\sigma(B)$ of QWr sample nQWr2 after illumination (ppc) at $T = 1.4$ K. Transport orthogonal (*solid symbols*) and parallel (*open symbols*) to the QWr is shown. *Left*: Logarithmic scale for positive B. *Right*: Lines show the fit to the model of anisotropic 2D weak localization [49]

Table 8.3 Anisotropy $\Delta\sigma^{\text{par}}(B)/\Delta\sigma^{\text{ortr}}(B)$ in magnetic field B for the QWr sample nQWr2 (compare to Fig. 8.27, left) and the QD nQD2 sample

B (mT)	nQWr2 (ppc, $A = 37.9$)	nQD2 (ppc, $A = 13.2$)
10	33.7	12.8
30	34.1	13.0
100	35.6	13.1

of the curves of both directions) remains fairly constant for the entire range of B. The anisotropy in $\Delta\sigma(B)$ is shown for several values of B and compared with the anisotropy of the total conductivity σ in Table 8.3.

It is theoretically expected that in a 2D system with directional anisotropy these values will be identical; the similarity in the experimental values is regarded as another confirmation of this theoretical prediction at the highest anisotropy to date. According to the theory of 2D weak localization, the suppression of the conductivity caused by a magnetic field is given by [113]

$$\Delta\sigma_{2D}^{\text{ort,par}}(B) = \alpha^{\text{ort,par}} \frac{e^2}{\pi h}\left[\Psi\left(\frac{1}{2} + \frac{\tau_B}{2\tau_\Phi}\right) - \Psi\left(\frac{1}{2} + \frac{\tau_B}{2\bar{\tau}}\right) + \ln\left(\frac{\tau_\Phi}{\bar{\tau}}\right)\right]. \quad (8.8)$$

In contrast to the isotropic case ($\alpha^{\text{ort,par}} = 1$), the following modifications are done to include the anisotropy [120].

$$\alpha^{\text{ort,par}} = \frac{\sigma^{\text{ort,par}}(B=0)}{\sqrt{\sigma^{\text{ort}}(B=0)\,\sigma^{\text{ort}}(B=0)}} \quad (8.9)$$

The coefficient $\alpha^{\text{ort,par}}$ scales $\Delta\sigma$ due to anisotropy, and $\bar{\tau} = \sqrt{\tau^{\text{ort}}\tau^{\text{par}}}$ is the average elastic scattering time. The other quantities used are the magnetic relaxation time τ_{B} and the diffusion constant D, $\tau_B = l_{\text{m}}^2/2D$ and $D = v_F^2\bar{\tau}/2$, with magnetic length l_{m} and Fermi velocity v_{F}. The experimental data are fit to the model of weak localization for low magnetic fields ($B < 30$ mT). The dephasing time τ_Φ was used as a fitting parameter; all other quantities ($\tau, \alpha, v_{\text{F}}$) were determined from classical transport measurements of mobility and sheet carrier concentration at $T =$ 10 K, at which the correction term is negligible due to inelastic scattering. This procedure allows a determination of the dephasing time τ_Φ and the dephasing length $l_\Phi = (D\tau_\Phi)^{1/2}$. Figure 8.27, right, shows that the model of anisotropic 2D weak localization describes the low field magneto-conductance of the sample well. The τ_Φ from both transport directions of all samples agree within less than 15%, which confirms the validity of the model for our data. A closer inspection of the formulas yields that the fundamental input for the fit reduces to the measured resistivity ρ_{xx} along with its change in magnetic field, the carrier concentration n, and the effective mass m^* of electrons at the Fermi energy. These values enter the calculation through v_{F} and τ. While the resistivity is quite reliably measured, there is some uncertainty with the carrier concentration n, and m^* is not precisely known. The theoretically calculated value of $m^* \approx 0.03\, m_{\text{e}}$ is used for the effective mass.

The real value may, however, be larger, e.g., due to band nonparabolicities. Therefore, m^* and n are considered as parameters put into the calculation. To what extent these parameters influence the results from the fit (dephasing time and length) is investigated: All involved characteristic times scale linearly with the parameter m^*. As a consequence, the ratio τ_Φ/τ, and the dephasing length l_Φ do not depend on m^*. The impact of the carrier concentration n put into the calculation as parameter was tested by varying its value. As result, a variation, e.g., by factor of 2, has almost no influence on the resulting τ_Φ for our data. Since D is independent of the parameter n the dephasing length l_Φ is quite independent of n as well. Thus, τ_Φ as obtained from the fit can be compared among different samples, and within one sample at different carrier densities, as long as they have the same effective mass. In our case, the same effective mass (which can differ from the actual effective mass) is put into all calculations. Moreover, the dephasing length l_Φ is completely independent of the exact value of the parameter m^* and only weakly depending on the parameter n put into the model.

The dephasing times and corresponding dephasing lengths are plotted as a function of temperature in Fig. 8.28.

The consequence of the data shows that typically, τ_Φ is two orders of magnitude larger than τ which is a necessary condition for weak localization, since multiple elastic-scattering events are required for return trajectories. With increasing temperature, the dephasing times decrease due to temperature-dependent inelastic scattering following roughly a $\tau_\Phi \propto T^r$ law with $r \approx= 0.8$. This value suggests electron–electron scattering with $r = -1$ to be the dominant inelastic scattering process [122]. The dephasing times are similar for all analyzed nanostructures and do not correlate to their morphology. A change of carrier concentration in nQWr2 and nQD2 significantly changes the mobility and elastic scattering times but shows

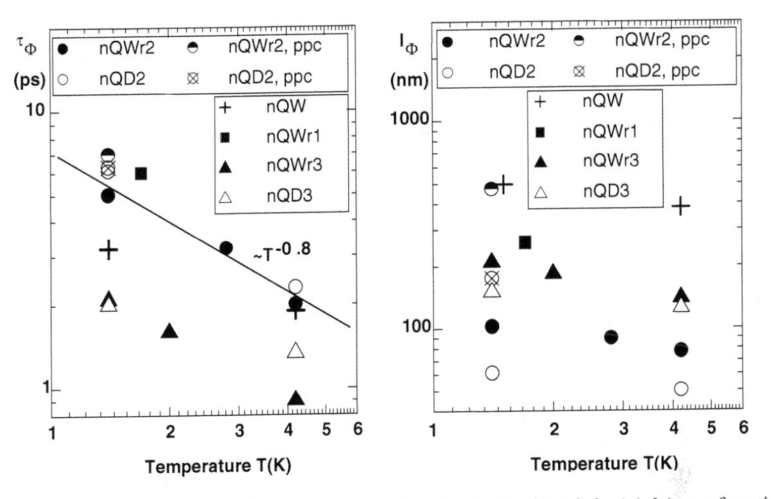

Fig. 8.28 Dephasing time τ_Φ (*left*) and corresponding dephasing length l_Φ (*right*) as a function of temperature for different nanostructures. The dephasing time and length were extracted from the analysis of the weak localization peaks [49]

only a negligible influence on the inelastic scattering time (compare full and open triangles and stars in Fig. 8.28, left). Hence, the inelastic scattering is considered merely temperature induced and independent of the nanostructure type. The dephasing lengths l_Φ (distance between subsequent inelastic scattering events) as shown in Fig. 8.28, right, are significantly larger than the wire–wire distance, and not at all related to the morphology (also compare l_Φ in nQWr2 and nQD2 for different carrier concentrations). Hence, the nanostructure–nanostructure coupling is not inelastic, which is compatible with interface roughness scattering being elastic. Despite high anisotropies in coupled nanostructures, which suggest a low degree of lateral coupling between the nanostructures, the weak localization is successfully modeled by regarding the system as a 2DEG with anisotropic conductivities. The anisotropy of weak localization (along with other quantum corrections) has the same value as the anisotropy of the total conductivity for values as large as 38; this result confirms the theory of Ref. [121] for high anisotropy in two dimensions. The independence of dephasing time on the actual morphology and coupling strength (carrier concentration) indicates that the nanostructure–nanostructure scattering is elastic. Thus, both optical and transport measurements confirm the common features of InAs/InP-coupled nanostructures: they develop the quasi-2D character both at low temperatures.

5 Application of InAs/InP Nanostructures

Peculiar band structure and carrier relaxation in the InAs/InP nanostructures support an efficient radiative recombination of electrons and holes in the InAs. In addition, the reduced dimensionality of QWrs and QDs allows to increase the temperature stability of semiconductor lasers, and to reduce their threshold currents. Thus, currently, the largest application potential of InAs nanostructures grown on InP substrates is considered to be optical devices – mainly for the telecommunication wavelength of 1.55 μm which can easily be reached in this material system. Lasers for 1.55 μm have been demonstrated based on InAs QDs and QWS [11, 123, 124], InAs QDashes [125, 126], and QWrs [127]. The usage of InAs quantum dots in infrared photodetectors has been demonstrated [128, 129]. Furthermore, the application of InAs QDs for mode lockers (as saturable absorber) [130], for optical charge storage [131], and recently as single photon source within photonic crystals [132] has been proposed. In addition, lasers with longer wavelength up to 1.9 μm, for gas sensing applications, and lasers with a broad (250 nm) gain profile have been demonstrated based on quantum dashes [133]. Recent advances on InAs/InP quantum dash (QD) materials for lasers and amplifiers, and QD device performance with particular interest in optical communication have been reviewed in Ref. [8]. The InAs/InP dashes in a barrier and dashes in well heterostructures operating at 1.5 μm were studied. These two types of QDs can provide high gain and low losses. Room temperature lasing operation on ground state of cavity length as short as 200 μm has been achieved, demonstrating the high modal gain of the active core. Buried ridge stripe (BRS)-type single-mode distributed feedback (DFB) lasers are also demonstrated in Ref. [8]. They exhibit a side-mode suppression ratio as high as 45 dB. Such DFB lasers allow the first floor-free 10-Gb/s direct modulation for back-to-back bit-error-rate and transmission over 16-km standard optical fiber. The QD Fabry–Perot lasers [8], owing to the small confinement factor and the 3D quantification of electronic energy levels, exhibit a beating linewidth as narrow as 15 kHz. Such an extremely narrow linewidth, compared to their QW or bulk counterparts, leads to the excellent phase noise and time-jitter characteristics when QD lasers are actively mode-locked. Finally, we can mention an electronic five-terminal switching device based on gate-controlled transport anisotropy proposed and investigated in Ref. [134]. By means of a gate voltage the two output voltages can be contrariwise adjusted to values between the two input voltages, yielding a possible logic application as an exchange gate. The device performance was simulated as a function of range of gate-adjustable transport anisotropy. A realization based on laterally coupled self-organized InAs QWrs with gate-controlled transport anisotropies from 5.6 to 12 was experimentally investigated. The data show that this QWr system is suitable for a realization of this device. An extended analysis that includes high source voltages in the range of the gate voltage, necessary for fast switching times, elucidates the limits and opportunities for optimization of this device.

6 Summary

In conclusion, we present a comprehensive study of the InAs/InP nanostructures, including the growth, optical characterization, and anisotropic transport in QW, QWr, and QD system. Self-assembled InAs nanostructures were grown by gas-source molecular beam epitaxy on InP(001) substrates with a focus on the influence of misorientation on the nanostructure shape. Surface steps on misoriented substrate were found to play a decisive role in the self-organized formation, shape, and alignment of InAs nanostructures. Most prominent is the formation of QDs on substrates misoriented toward [−110] and [100] under a wide range of growth conditions. On nominally oriented substrates, on the other hand, the growth conditions determine the resulting morphology: QWs, QWrs, or QDs. The decisive phases of the nanostructure formation were identified and investigated by variation of the respective growth conditions. QW, QWrs, and QD single layer and multilayer samples were grown for the investigation of their optical and transport properties. The emission and absorption properties related to interband transitions of the InAs/InP nanostructures were investigated with polarization-dependent PL and transmission spectroscopy. It was demonstrated that the hh1–e1 transition of our nanostructures covers a large wavelength range up to 2 μm, and including the technologically important 1.3 and 1.55 μm, at room temperature. Polarization-dependent PL and transmission measurements show a preferential polarization in the [−110] direction for all QWrs and QDs. The degree of polarization for our QWrs and QDs is around 10 and 20%, respectively, suggesting an anisotropic QD shape. Polarization-dependent transmission shows an additional transition that was identified as the lh1–e1 transition, and has the opposite polarization direction to the hh1–e1 transition. Temperature-dependent photoluminescence and Fourier-transformed infrared spectroscopy is used for the investigation of optically active states and concentration of states in InAs/InP(001) QWrs and parent QW system grown under deposition of 4 ML InAs on InP. Optical properties of both systems reveal much similarity in temperature behavior in spite of a qualitatively different character of the 1D and 2D DOS functions. Unusual two-branch temperature dependence of the excitonic PL band maximums with the switching from the low-energy branch to the high-energy branch is revealed in both systems and interpreted in terms of thermal activation of excitonic ground states in the array of size-distributed nanostructures. Strong modification of the absorbance line shape leading to the appearance of flat spectral parts in the absorbance spectrum is observed at room temperature both in the QW and QWr samples. In case of the QW system, such flatness is explained in terms of localized exciton decay, whereas in case of the QWr system this effect demonstrates the thermally induced change of the dimensionality: from 1D to anisotropic 2D. This change of dimensionality is also detected in the polarized absorbance measurements through disappearance or significant reduction of the polarization anisotropy in the regions of the hh1–e1 and lh1–e1 transitions in QWrs. Our simulation results show that small Gaussian inhomogeneity (~3 meV) does not qualitatively perturb the nature of 1D DOS function, although it causes the localization of the low-lying electronic states. It allows fairly good explanation of the low-temperature absorbance spectra in InAs/InP(100) QWrs. Properties of the QWrs which are dependent upon the extended nature of the wavefunction such as charge transport are expected to be altered significantly. However, the modified 1D DOS fails to explain the high-temperature absorbance data, giving grounds to introduce rather the anisotropic 2D than purely 1D DOS function. The shown optical properties suggest a high potential of InAs/InP nanostructures for optical applications ranging from lasers and photodetectors to nonlinear optical applications. The shown optical properties suggest a high potential of InAs/InP nanostructures for optical applications ranging from lasers and photodetectors to nonlinear optical applications. The in-plane transport of electrons and holes in large ensembles of InAs nanostructures, comprising QWrs, QDs, and QWs, was investigated. The transport of quantum wire-, quantum dash-, and quantum dot-containing samples is highly anisotropic with the principal axes of conductivity aligned to the <110> directions. The direction of higher mobility is [−110], which is parallel to the direction of the quantum wires. In extreme cases, the anisotropies exceed 30 for electrons and 100 for holes. These data are discussed in terms of anisotropic 2D carrier systems and in terms of a coupled 1D or 0D system. The temperature dependence and carrier-concentration dependence of mobility were analyzed with the following results: In the anisotropic 2D approximation, the dominating scattering mechanisms at low temperature are remote impurity scattering in the parallel direction for quantum wires, and interface roughness scattering in both directions for quantum dots, and in the orthogonal direction for quantum wires and quantum dashes. In p-type samples, though, the hole transport in the orthogonal direction cannot be described by scattering but is described as hopping between laterally localized states. In the coupled nanostructure framework, the transport anisotropy is demonstrated to result from directionally anisotropic tunnel coupling between adjacent nanostructures rather than from the nanostructure shape anisotropy. The weak localization contribution to conductivity in quantum wire and quantum dot samples is successfully modeled by regarding the system as a 2DEG with anisotropic conductivities. The anisotropy of weak localization has the same value as the anisotropy of the total conductivity for values as large as 38; confirming the theory for high anisotropy in two dimensions. The independence of dephasing time, extracted from the weak localization data, on the actual morphology and coupling strength (carrier concentration) indicates a strong coupling and elastic nanostructure–nanostructure scattering.

Acknowledgments We would like to appreciate many people who contributed to this work. R. Pomraenke, S. Dreßler, H. Kirmse, K.-J. Friedland, Vas. P. Kunets, V.G. Dorogan, Z.Ya. Zhuchenko, S Noda, E.A. Decuir Jr, and M.O. Manasreh are deeply thanked for their skill and participation.

References

1. Ribeiro, E., Müller, E., Heinzel, T., Auderset, H., Ensslin, K., Medeiros-Ribeiro, G., Petroff, P.M.: Phys. Rev. B **58**, 1506 (1998)
2. Shchukin, V.A., Bimberg, D.: Rev. Mod. Phys. **71**, 1125 (1999)
3. Lundstrom, T., Schoenfeld, W., Lee, H., Petroff, P.M.: Science **286**, 2312 (1999)
4. Songmuang, R., Kiravittaya, S., Schmidt, O.G.: Appl. Phys. Lett. **82**, 2892 (2003)
5. Porte, L.: J. Cryst. Growth **273**, 136 (2004)
6. Cornet, C., Schliwa, A., Even, J., Doré, F., Celebi, C., Létoublon, A., Macé, E., Paranthoën, C., Simon, A., Koenraad, P.M., Bertru, N., Bimberg, D., Loualiche, S.: Phys. Rev. B **74**, 035312 (2006)

7. Fuster, D., Alen, B., Gonzalez, L., Gonzalez, Y., Martınez-Pastor, J., Gonzalez, M., Garcıa, J.M.: Nanotechnology **18**, 035604 (2007)
8. Lelarge, F., Dagens, B., Renaudier, J., Brenot, R., Accard, A., van Dijk, F., Make, D., Le Gouezigou, O., Provost, J.-G., Poingt, F., Landreau, J., Drisse, O., Derouin, E., Rousseau, B., Pommereau, F., Duan, G.-H.: IEEE J. Selected Top. Quantum Electron. **13**, 111 (2007)
9. Chutia, S., Friesen, M., Joynt, R.: Phys. Rev. B **73**, 241304 (2006)
10. Takemoto, K., Sakuma, Y., Hirose, S., Usuki, T., Yokoyama, N.: Jpn. J. Appl. Phys. Part 2 **43**, L349 (2004)
11. Saito, H., Nishi, K., Sugou, S.: Appl. Phys. Lett. **78**, 267 (2001)
12. Temkin, H., Alavi, K., Wagner, W.R., Pearsall, T.P., Cho, A.Y.: Appl. Phys. Lett. **42**, 845 (1983)
13. Peronne, E., Pollack, T., Lampin, J.F., Fossard, F., Julien, F., Brault, J., Gendry, G., Marty, O., Alexandrou, A.: Phys. Rev. B **63**, 081307 (2001)
14. Even, J., Loualiche, S., Miska P., Platz, C.: J. Phys. Condens. Matter **15**, 8737 (2003)
15. Polimeni, A., Patane, A., Henini, M., Eaves, L., Main, P.C., Hill, G.: Physica E **7**, 452 (2000)
16. Liu, H., Steer, M.J., Badcock, T.J., Mowbray, D.J., Skolnick, M.S., Navaretti, P., Groom, K.M., Hopkinson, M., Hogg, R.A.: Appl. Phys. Lett. **86**, 143108 (2005)
17. Karachinsky, L.Y., Kettler, T., Gordeev, N.Yu., Novikov, I.I., Maximov, M.V., Shernyakov, Yu.M., Kryzhanoskaya, N.V., Zhukov, A.E., Semenova, E.S., Vasilev, A.P., Ustinov, V.M., Ledentsov, N.N., Kovsh, A.R., Shchukin, V.A., Mikhrin, S.S., Lochmann, A., Schulz, O., Reissmann, L., Bimberg, D.: Electron. Lett. **41**, 478 (2005)
18. Paranthoen, C., Bertru, N., Dehaese, O., Le Corre, A., Loualiche, S., Lambert, B., Patriarche, G.: Appl. Phys. Lett. **78**, 1751 (2001)
19. Bierwagen, O., Masselink, W.T.: Appl. Phys. Lett. **86**, 113110 (2005)
20. Notzel, R., Haverkort, J.E.M.: Adv. Funct. Mater. **16**, 327 (2006)
21. Benoit, J.M., Le Gratiet, L., Beaudoin, G., Michon, A., Saint-Girons, G., Kuszelewicz, R., Sagnes, I.: Appl. Phys. Lett. **88**, 041113 (2006)
22. Michon, A., Saint-Girons, G., Beaudoin, G., Sagnes, I., Largeau, L., Patriarche, G.: Appl. Phys. Lett. **87**, 253114 (2005)
23. Cornet, C., Levallois, C., Caroff, P., Folliot, H., Labbe, C., Even, J., Le Corre, A., Loualiche, S., Hayne, M., Moshchalkov V.V.: Appl. Phys. Lett. **87**, 233111 (2005)
24. Gutiérrez, H.R., Cotta, M.A., Bortoleto, J.R.R., de Carvalho, M.M.G.: J. Appl. Phys. **92**, 7523 (2002)
25. Çelebi, C., Ulloa, J.M., Koenraad, P.M., Simon, A., Letoublon, A., Bertru, N.: Appl. Phys. Lett. **89**, 023119 (2006)
26. Gonzalez, L., Garcıa, J.M., Garcıa, R., Briones, F., Martınez-Pastor, J., Ballesteros, C.: Appl. Phys. Lett. **76**, 1104 (2000)
27. Lei, W., Chen, Y.H., Wang, Y.L., Xu, B., Ye, X.L., Zeng, Y.P., Wang, Z.G.: J. Cryst. Growth **284**, 20 (2005)
28. Lei, W., Chen, Y.H., Wang, Y.L., Huang, X.Q., Zhao, Ch., Liu, J.Q., Xu, B., Jin, P., Zeng, Y.P., Wang, Z.G.: J. Cryst. Growth **286**, 23 (2006)
29. Garcia, J.M., Gonzalez, L., Gonzalez, M.U., Silveira, J.P., Gonzalez, Y., Briones, F.: J. Cryst. Growth **227/228**, 975 (2001)
30. Fuster, D., Gonzalez, L., Gonzalez, Y., Martinez-Pastor, J., Ben, T., Ponce, A., Molina, S.: Eur. Phys. J. B **40**, 434 (2004)
31. Fuster, D., Gonzalez, M.U., Gonzalez, L., Gonzalez, Y., Ben, T., Ponce, A., Molina, S.I.: Appl. Phys. Lett. **84**, 4723 (2004)
32. Narvaez, G.A., Zunger, A.: Phys. Rev. B **75**, 085306 (2007)
33. Li, H.X., Wu, J., Wang, Z.G., Daniels-Race, T.: Appl. Phys. Lett. **75**, 1173 (1999)
34. Ma, W., Nötzel, R., Trampert, A., Ramsteiner, M., Zhu, H., Schönherr, H.-P., Ploog, K.H.: Appl. Phys. Lett. **78**, 1297 (2001)
35. Bierwagen, O., Walther, C., Masselink, W.T., Friedland, K.-J.: Phys. Rev. B **67**, 195331 (2003)

36. Maes, J., Hayne, M., Sidor, Y., Partoens, B., Peeters, F.M., González, Y., González, L., Fuster, D., García, J.M., Moshchalkov, V.V.: Phys. Rev. B **70**, 155311 (2004)
37. Karlsson, K.F., Weman, H., Dupertuis, M.A., Leifer, K., Rudra, A., Kapon, E.: Phys. Rev. B **70**, 045302 (2004)
38. Fuster, D., Martınez-Pastor, J., Gonzalez, L., González, Y.: J. Phys. D Appl. Phys. **39**, 4940 (2006)
39. Sidor, Y., Partoens, B., Peeters, F.M., Ben, T., Ponce, A., Sales, D.L., Molina, S.I., Fuster, D., González, L., González, Y.: Phys. Rev. B **75**, 125120 (2007)
40. Mazur, Yu.I., Wang, Zh.M., Tarasov, G.G., Xiao, M., Salamo, G.J., Tomm, J.W., Talalaev, V., Kissel, H.: Appl. Phys. Lett. **86**, 063102 (2005)
41. Moison, J.M., Bensoussan, M., Houzay, F.: Phys. Rev. B **34**, 2018 (1986)
42. Tabata, A., Benyattou, T., Guillot, G., Gendry, M., Hollinger, G., Viktorovitch, P.: J. Vac. Sci. Technol. B **12**, 2299 (1994)
43. Li, C.H., Li, L., Law, D.C., Visbeck, S.B., Hicks, R.F.: Phys. Rev. B **65**, 205322 (2002).
44. Brault, J., Gendry, M., Grenet, G., Hollinger, G., Desières, Y., Benyattou, T.: Appl. Phys. Lett. **73**, 2932 (1998)
45. Yang, H., Ballet, P., Salamo, G.J.: J. Appl. Phys. **89**, 7871 (2001)
46. Yang, H., Mu, X., Zotova, I.B., Ding, Y.J., Salamo, G.J.: J. Appl. Phys. **91**, 3925 (2002)
47. Yoon, S., Moon, Y., Lee, T.-W., Yoon, E., Kim, Y.D.: Appl. Phys. Lett. **74**, 2029 (1999)
48. Anantathanasarn, S., Nötzel, R., van Veldhoven, P.J., Eijkemans, T.J., Wolter, J.H.: J. Appl. Phys. **98**, 13503 (2005)
49. Bierwagen, O.: Growth and anisotropic transport properties of self-assembled InAs nanostructures in InP, PhD Thesis, Humboldt-Universität zu Berlin-(2007)
50. Walther, C., Hoerstel, W., Niehus, H., Erxmeyer, J., Masselink, W.T.: J. Cryst. Growth **209**, 572 (2000)
51. Salem, B., Olivares, J., Guillot, G., Bremond, G., Brault, J., Monat, C., Gendry, M., Hollinger, G., Hassen, F., Maaref, H.: Appl. Phys. Lett. **79**, 4435 (2001)
52. Fréchengues, S., Bertru, N., Drouot, V., Lambert, B., Robinet, S., Loualiche, S., Lacombe, D., Ponchet, A.: Appl. Phys. Lett. **74**, 3356 (1999)
53. Grosse, F., Gyure, M.F: Phys. Status Solidi **234**, 338 (2002)
54. Alén, B., Martínez-Pastor, J., García-Cristobal, A., González, L., García, J.M.: Appl. Phys. Lett. **78**, 4025 (2001)
55. Dupuy, E., Regreny, P., Robach, Y., Gendry, M., Chauvin, N., Tranvouez, E., Bremond, G., Bru-Chevallier, C., Patricarche, G.: Appl. Phys. Lett. **89**, 123112 (2006)
56. Gendry, M., Monat, C., Brault, J., Regreny, P., Hollinger, G., Salem, B., Guillot, G., Benyattou, T., Bruchevallier, C., Bremond, G.: J. Appl. Phys. **95**, 4761(2004)
57. Poole, P.J., Williams, R.L., Lefebvre, J., Moisa, S.: J. Cryst. Growth **257**, 89 (2003)
58. Sakuma, Y., Takeguchi, M., Takemoto, K., Hirose, S., Usuki, T., Yokoyama, N.: J. Vac. Sci. Technol. B, **23**, 1741 (2005)
59. Akaishi, M., Okawa, T., Saito, Y., Shimomura, K.: IEEE J. Selected Top. Quantum Electron. **14**, 1197 (2008)
60. Saito, Y., Okawa, T., Akaishi, M., Shimomura, K.: J. Cryst. Growth **310**, 5073 (2008)
61. Miska, P., Paranthoen, C., Even, J., Dehaese, O., Folliot, H., Bertru, N., Loualiche, S., Senes, M., Marie, X.: Semicond. Sci. Technol. **17**, L63 (2002)
62. Mazur, Yu. I., Noda, S., Tarasov, G.G., Dorogan, V.G., Salamo, G.J., Bierwagen, O., Masselink, W.T., Decuir, E.A., Jr., Manasreh, M.O.: J. Appl. Phys. **103**, 054315 (2008)
63. Carlsson, N., Junno, T., Montelius, L., Pistol, M.E., Samuelson, L., Seifert, W.: J. Cryst. Growth **191**, 347 (1998)
64. Prieto, J.A., Armelles, G., Priester, C., Garcia, J.M., Gonzalez, L., Garcia, R.: Appl. Phys. Lett. **76**, 2197 (2000)
65. Born, M., Wolf, E.: Principles of Optics. Cambridge University Press, Cambridge (1999)
66. Adachi, S.: J. Appl. Phys. **53**, 5863 (1982)

67. Cornet, C., Platz, C., Caroff, P., Even, J., Labbé, C., Folliot, H., Le Corre, A., Loualiche, S.: Phys. Rev. B **72**, 035342 (2005)
68. Paki, P., Leonelli, R., Isnard, L., Masut, R.A.: J. Vac. Sci. Technol. A **18**, 956 (2000)
69. Pryor, C.E., Pistol, M.-E.: Phys. Rev. B **72**, 205311 (2005)
70. Létoublon, A., Favre-Nicolin, V., Renevier, H., Proietti, M., Monat, C., Gendry, M., Marty, O., Priester, C.: Phys. Rev. Lett. **92**, 186101 (2004)
71. Miska, P., Paranthoen, C., Even, J., Bertru, N., Le Corre, A., Dehaese, O.: J. Phys. Condens. Matter **14**, 12301 (2002)
72. Lacombe, D., Ponchet, A., Frechengues, S., Drouot, V., Bertru, N., Lambert, B., Le Corre, A.: Appl. Phys. Lett. **74**, 1680 (1999)
73. Pettersson, H., Pryor, C., Landin, L., Pistol, M.-M., Carlsson, N., Seifert, W., Samuelson, L.: Phys. Rev. B **61**, 4795 (2000)
74. Holm, M., Pistol, M.-E., Pryor, C.: J. Appl. Phys. **92**, 932 (2002)
75. Miska, P., Even, J., Platz, C., Salem, C., Benyattou, T., Bru-Chevalier, C., Guillot, G., Bremond, G., Moumanis, Kh., Julien, F.H., Marty, O., Monat, C., Gendry, M.: J. Appl. Phys. **95**, 1074 (2004)
76. Vurgaftman, I., Meyer, J.R., Ram-Mohan, L.R.: J. Appl. Phys. **89**, 5815 (2001)
77. Lin, T.Y., Fan, J.C., Fan, Chen, Y.F.: Semicond. Sci. Technol. **14**, 406 (1999)
78. Felici, M., Trotta, R., Masia, F., Polimeni, A., Miriametro, A., Capizzi, M., Klar, P.J., Stolz, W.: Phys. Rev. B **74**, 085203 (2006)
79. Gurioli, M., Vinattieri, A., Martinez-Pastor, J., Colocci, M.: Phys. Rev. B **50**, 11817 (1994)
80. Brandt, O., Tapfer, L., Cingolani, R., Ploog, K., Hohenstein, M., Phillipp, F.: Phys. Rev. B **41**, 12599 (1990)
81. Mazur, Yu.I., Dorogan, V.G., Bierwagen, O., Tarasov, G.G., Decuir, E.A., Jr, Noda, S., Ya Zhuchenko, Z., Manasreh, M.O., Masselink,.W.T.,Salamo, G.J.: Nanotechnology **20**, 065401 (2009)
82. Bastard, G.: Wave mechanics applied to semiconductor heterostructures. LesUlis: Editions de Physique (1988)
83. Pikus, G.E., Bir, G.L.: Sov. Phys. Solid State **1**, 136 (1959); ibidem **1**, 1502 (1960)
84. Mazur, Yu.I., Wang, Zh.M., Tarasov, G.G., Wen, H., Strelchuk, V., Guzun, D., Xiao, M., Salamo, G.J., Mishima, T.D., Guoda, D., Lian, M.B., Johnson.: J. Appl. Phys. **98**, 053711 (2005)
85. Daly, E.M., Glynn, T.J., Lambkin, J.D., Considine, L., Walsh, S.: Phys. Rev. B **52**, 4696 (1995)
86. Schmitt-Rink, S., Chemla, D.S., Miller, D.A.B.: Advances in Physics **38**, 89 (1989)
87. Tribe, W.R., Steer, M.J., Mowbray, D.J., Skolnick, M.S., Forshaw, A.N., Roberts, J.S., Hill, G., Pate, M.A., Whitehouse, C.R., Williams, G.M.: Appl. Phys. Lett. **70**, 993 (1997)
88. Lomascolo, M., Ciccarese, P., Cingolani, R., Rinaldi, R., Reinhart, F.K.: J. Appl. Phys. **83**, 302 (1998)
89. Rasnik, I., Rego, L.G.C., Marquezini, M.V., Triques, A.L.C., Brasil, M.J.S.P., Brum, J.A., Cotta, M.A.: Phys. Rev. B **58**, 9876 (1998)
90. Gershoni, D., Katz, M., Wegscheider, W., Pfeiffer, L.N., Logan, R.A., West, K.: Phys. Rev. B **50**, 8930 (1994)
91. Quang D.N., Tung, N.H.: Phys. Rev. B **62**, 6555 (2000)
92. Chen, Y.H., Yang, Z., Wang, Z.G., Xu, B., Liang, J.B., Qian, J.J.: Phys. Rev. B **56**, 6770 (1997)
93. Bockelmann U., Bastard, G.: Phys. Rev. B **45**, 1688 (1992)
94. Kumar, J., Kapoor, S., Gupta, S.K., Sen, P.K.: Phys. Rev. B **74**, 115326 (2006)
95. Dykman, M.I., Tarasov, G.G.: Sov. Phys. JETP **45**, 1181 (1977)
96. González, L., García, J.M., García, R., Briones, F., Martínez-Pastor, J., Ballesteros, C.: Appl. Phys. Lett. **76**, 1104 (2000)
97. Mu, X., Zotova, I.B., Ding, Y.J., Yang, H., Salamo, G.J.: Appl. Phys. Lett. **79**, 1091 (2003)
98. Nötzel, R., Eissler, D., Hohenstein, M., Ploog, K.H.: J. Appl. Phys. **74**, 431 (1993)

99. Friedland, K.-J., Schönherr, H.-P., Nötzel, R., Ploog, K.H.: Phys. Rev. Lett. **83**, 156 (1999)
100. Kim, G.-H., Ritchie, D.A., Liang, C.T., Lian, G.D., Yuan, J., Pepper, M., Brown, L.M.: Appl. Phys. Lett. **78**, 3896 (2001)
101. Wallart, X., Pinsard, B., Mollot, F.: J. Appl. Phys. **97**, 053706 (2005)
102. Zahurak, J.K., Iliadis, A.A., Rishton, S.A., Masselink, W.T.: IEEE Electron. Device Lett. **15**, 489 (1994)
103. van der Pauw, L.J.: Philips Res. Rep. **13**, 1 (1958)
104. Bierwagen, O., Pomraenke, R., Eilers, S., Masselink, W.T.: Phys. Rev. B **70**, 165307 (2004)
105. Tokura, Y.: Phys. Rev. B **58**, 7151(1998)
106. Nakamura, Y., Koshiba, S., Sakaki, H.: Appl. Phys. Lett. **69**, 4093 (1996)
107. Sakaki, H.: Jpn. J. Appl. Phys. **19**, L735 (1980)
108. Kul'bachinskii, V.A., Lunin, R.A., Kytin, V.G., Golikov, A.V., Demin, A.V., Rogozin, V.A., Zvonkov, B.N., Nekorkin, S.M., Filatov, D.O.: J. Exp. Theor. Phys. (JETP) **93**, 815 (2001)
109. Kul'bachinskii, V.A., Lunin, R.A., Rogozin, V.A., Mokerov, V.G., Kindo, K., de Visser, A.: Semiconductors **37**, 70 (2003)
110. Ye, Q., Shklovskii, B.I., Zrenner, A., Koch, F., Ploog, K.: Phys. Rev. B **41**, 8477 (1990)
111. Friedland, K.-J., Hoerike, M., Hey, R., Shlimak, I., Resnick, L.: Physica A **302**, 375 (2001)
112. Song, H.Z., Akahane, K., Lan, S., Xu, H.Z., Okada, Y., Kawabe, M.: Phys. Rev. B **64**, 085303 (2001)
113. Beenakker, C.W.J., van Houten, H.: Solid state physics, vol. 44. Academic, New York, NY (1991)
114. Ando, T.: J. Phys. Soc. Jpn. **51**, 3900 (1982)
115. Gold, A.: Phys. Rev. B **38**, 10798 (1988)
116. Sakaki, H., Noda, T., Hirakawa, K., Tanaka, M., Matsusue, T.: Appl. Phys. Lett. **51**, 1934 (1987)
117. Landau, L.: Z. Physik **64**, 629 (1930)
118. Ando, T., Fowler, A.B., Stern, F.: Rev. Mod. Phys. **54**, 437 (1982)
119. Klitzing, K., Dorda, G., Pepper, M.: Phys. Rev. Lett. **45**, 494 (1980)
120. Lee, P.A., Ramakrishnan, T.V.: Rev. Mod. Phys. **57**, 287 (1985)
121. Wölfle, P., Bhatt, R.H.: Phys. Rev. B **30**, 3542 (1984)
122. Bishop, D.J., Dynes, R.C., Lin, B.J., Tsui, D.C.: Phys. Rev. B **30**, 3539 (1984)
123. Allen, C.N., Poole, P.J., Marshall, P., Fraser, J., Raymond, S., Fafard, S.: Appl. Phys. Lett. **80**, 3629 (2002)
124. Anantathanasarn, S., Nötzel, R., van Veldhoven, P.J., van Otten, F.W.M., Barbarin, Y., Servanton, G., de Vries, T., Smalbrugge, E., Geluk, E.J., Eijkemans, T.J., Bente, E.A.J.M., Oei, Y.S., Smit, M.K., Wolter, J.H.: Appl. Phys. Lett. **89**, 073115 (2006)
125. Ukhanov, A.A., Wang, R.H., Rotter, T.J., Stintz, A., Lester, L.F., Eliseev, P.G., Malloy, K.J.: Appl. Phys. Lett. **81**, 981 (2002)
126. Schwertberger, R., Gold, D., Reithmaier, J.P., Forchel, A.: IEEE Photon. Technol. Lett. **14**, 735 (2002)
127. Suárez, F., Fuster, D., González, L., González, Y., García, J.M., Dotor, M.L.: Appl. Phys. Lett. **89**, 091123 (2006)
128. Finkman, E., Maimon, S., Immer, V., Bahir, G., Schacham, S.E., Fossard, F., Julien, F.H., Brault, J., Gendry, M.: Phys. Rev. B **63**, 045323 (2001)
129. Zhang, W., Lim, H., Taguchi, M., Tsao, S., Movaghar, B., Razeghi, M.: Appl. Phys. Lett. **86**, 191103 (2005)
130. Inoue, J., Isu, T., Akahane, K., Tsuchiya, M.: Appl. Phys. Lett. **89**, 151117 (2006)
131. Pettersson, H., Bååth, L., Carlsson, N., Seifert, W., Samuelson, L.: Phys. Rev. B **65**, 073304 (2002)
132. Dalacu, D., Frédérick, S., Bogdanov, A., Poole, P.J., Aers, G.C., Williams, R.L., McCutcheon, M.W., Young, J.F.: J. Appl. Phys. **98**, 023101 (2005)
133. Somers, A., Kaiser, W., Reithmaier, J.P., Forchel, A., Gioaninni, M., Montrosset, I.: Appl. Phys. Lett. **89**, 061107 (2006)
134. Bierwagen, O., Masselink, W.T.: Appl. Phys. Lett. **90**, 133507 (2007)

Index

Note: The letters 't' and 'f'followed by the locators represents 'tables' and 'figures'.

Z.M. Wang, A. Neogi (eds.), *Nanoscale Photonics and Optoelectronics*, Lecture Notes in Nanoscale Science and Technology 9,
DOI 10.1007/978-1-4419-7587-4,

P

Q

T

U